Smart Ceramics

Smart Ceramics

Preparation, Properties, and Applications

Edited by

Ajay Kumar Mishra

PAN STANFORD PUBLISHING

Published by

Pan Stanford Publishing Pte. Ltd.

Penthouse Level, Suntec Tower 3
8 Temasek Boulevard
Singapore 038988

Email: editorial@panstanford.com
Web: www.panstanford.com

British Library Cataloguing-in-Publication Data
A catalogue record for this book is available from the British Library.

Smart Ceramics: Preparation, Properties, and Applications

Cover image, taken by Lingyan Shi from Adrian Rodriguez-Contreras' Lab, shows a trans-cranial image of brain microvessels filled up with Texas red dextran 70 kDa and wrapped with Gcamp-6 GFAP astrocytes, using a multiphoton fluorescence microscope.

ISBN 978-981-4774-30-7 (Hardcover)
ISBN 978-1-315-16359-8 (eBook)

Contents

Preface

The term "smart ceramic materials" refers to ceramic materials fabricated from ultrafine particles. These materials have attracted the interest of researchers and scientists because of their potential to manipulate the length scale in the nanorange, leading to better and some unusual material properties. Smart ceramics have been synthesized to ensure control of particle size, surface contamination, and degree of agglomeration. The sol-gel route has been mainly utilized for the synthesis of smart ceramics because of its ability to produce a large variety of compositions and ensure homogeneous mixing of the constituent particles at low temperature.

Recent advances in nanotechnology have paved the way for the development of new smart materials. Sol-gel bioceramics play an important role in the biomedical field because of their superior biological and mechanical properties. Because of their unique physical and chemical properties, various metal oxide nanoparticles have emerged as the materials of choice in the removal of various types of pollutants from air and water. Porous nanostructured ceramics are an attractive class of materials that have found potential in various applications, ranging from simple to complexones, such as bioimaging, sensors, paints pigments, optics, and electronics, because of their surface- and size-dependent properties. Smart ceramics play a crucial role in industrial applications, particularly in the field of bone surgery, for example, the development of bone substitutes for load-bearing bone parts. This represents one of the most challenging applications, especially due to the difficulty of expressing high bioactivity and bone-like mechanical properties. Smart ceramic materials have also attracted researchers from the area of biomedical science, especially in tissue engineering, dental applications, and drug and antigen delivery using modified ceramics.

This book describes innovation in technologies through the development of functionalized ceramic materials from the perspective of energy, environment, and healthcare applications. It describes recent and expected challenges, along with potential solutions, in advanced techniques for the synthesis and characterization of nanostructured ceramics and their composites: bioceramics, bioactive ceramics, multifunctional nanoceramics, transparent ceramics, nanocoreshells, nanowires, thin films, nanotubes, and nanorods. The applications include environmental applications, healthcare applications, electrochemical sensors, high-temperature superconductors, fuel in nuclear reactors, electrical insulators, refractory material, electrical transformers, and magnetic core memory.

The book will be more beneficial to researchers, scientists, engineers, and technologists working in industry, national/international research laboratories and academia with interest in traditional and advanced smart ceramic composites. Researchers registered for their postgraduate/graduate/undergraduate degrees in the areas of smart ceramics, nanomaterials, nanoscience, and engineering will also be highly benefitted.

Ajay Kumar Mishra
University of South Africa, South Africa
2017

Chapter 1

Recent Trends in Sol-Gel-Based Nanoceramics

Pradeep Pratap Singh[a] and Ambika[b]

[a]*Department of Chemistry, Swami Shraddhanand College, University of Delhi, New Delhi 110036, Delhi, India*
[b]*Department of Chemistry, Hans Raj College, University of Delhi, New Delhi 110007, Delhi, India*

ambikasinghhrc@gmail.com

Nanoceramics refer to ceramic materials fabricated from ultrafine particles. They have attracted the interest of researchers and scientists due to potential to achieve better and some unusual material properties by manipulating length scale in the nanorange. The synthesis of nanocrystalline powders is an essential step in the processing of nanoceramics. Various methods have been reported to synthesize nanocrystalline powders to ensure appropriate control of particle size, surface contamination, and degree of agglomeration. Recently the sol-gel route has been utilized for the synthesis of nanoceramics. The sol-gel methods have the ability to produce a large variety of compositions and ensure homogeneous mixing of the constituent particles at low temperature. This

Smart Ceramics: Preparation, Properties, and Applications
Edited by Ajay Kumar Mishra

ISBN 978-981-4774-30-7 (Hardcover), 978-1-315-16359-8 (eBook)
www.panstanford.com

method can be used in ceramic processing, in manufacturing, and for producing very thin films of metal oxides for various purposes. Sol-gel-derived nanoceramics have numerous applications in different technological areas. This chapter deals with the recent trends in sol-gel-based nanoceramics and their applications.

1.1 Introduction

The field of ceramics has utilized nanoscience and nanotechnology for producing a variety of advanced materials with unique properties and performance. Ceramic nanocomposites are attracting growing interest due to their wide range of applications. "Nanoceramics" is a term used to refer to ceramic materials fabricated from particles less than 100 nm in diameter. They encompass many forms of ceramics where at least one of the dimensions is in the nanoscale range (Biest, 2013). Nanoceramics exhibit unique mechanical and surface characteristics, such as superplasticity, machinability, strength, toughness, and bioactivity due to the fine grain size, abundant grain boundaries, and controllable crystallinity (Kiani et al., 2014). However, their processing is the major challenge in research. The basic steps in nanoceramic fabrication involve synthesis of unagglomerated nanosized powders with uniform size distribution and sintering without grain growth. Various chemical methods have been adopted to synthesize nanocrystalline powders to ensure appropriate control of particle size, surface contamination, and degree of agglomeration. Recently, the sol-gel method has been utilized for the synthesis of nanoceramics (Perera, 2010; Owens et al., 2016).

Sol-gel processing is a liquid-phase processing of nanocomposite powders that utilize inorganic salts or metal-organic compounds for sol preparation. The sol is converted into a gel by hydrolysis and condensation reactions, which is then dried to eliminate the excess liquid phase. The gel shrinks and transforms to the desired phase after drying, leading to complex shapes, directly from the gel state. This process controls the homogeneity of chemical composition and lowers the processing temperatures. In addition, by controlling the gelation parameters and subsequent thermal treatments, it is possible to tailor the microstructure.

Even small quantities of dopants, such as organic dyes and rare earth elements, can be introduced in the sol and end up uniformly dispersed in the final product (Hafez, 2012). Sol-gel processing can be used in ceramic processing, in manufacturing, and for producing very thin films of metal oxides for various purposes (Palmero, 2015). Sol-gel-derived nanoceramics have diverse applications in optics, electronics, energy, space, sensors, medicine, corrosion protective coatings, engineering, and biomaterials science (Zhao et al., 2008; Simchi et al., 2011; Uche, 2013; Senthil et al., 2014).

1.2 Classification of Ceramic Nanocomposites

Ceramic nanocomposites can be broadly classified into the following categories (Basu et al., 2006):

- **Intergranular nanocomposites**: These nanocomposites possess nanosized reinforcements at the grain boundaries and triple junctions of the matrix grains, which pins in the grain boundaries from migration, which results in improved creep resistance.
- **Intragranular nanocomposites**: These consist of nanoscaled reinforcements within the matrix grains, which results in high strength and toughness at room temperature.
- **Inter-/intragranular nanocomposites**: These nanocomposites show a combination of properties of the first two classes. They have nanosized reinforcements at the grain boundaries, at the triple junctions, and within the matrix grains. The hardness, toughness, strength, fracture resistance for creep, and fatigue at high temperatures, as well as the thermal shock fracture resistance, are strongly improved for these composites (Niihara et al., 1999).
- **Nano-/nanocomposites:** In these nanocomposites both the matrix and the reinforcement particulates are in the nanosized range. They exhibit superplasticity and superior machinability due to enhanced grain boundary sliding.
- **Nano-/microcomposites:** In this type micron-size particulates are dispersed in a nanosized matrix (Basu et al., 2006).

1.3 Sol-Gel Methods and Chemistry

Sol-gel processing is one of the recent synthesis routes that are used to produce nanoceramics. All sol-gel methods involve two distinct phases, solution and gelation. A sol is a colloidal suspension of solid particles, whereas a gel is an interconnected network of solid-phase particles that form a continuous entity throughout a secondary phase that is generally liquid. The presence of a network formed by the interlocking of particles of the gelling agent gives rise to the rigidity of a gel. The nature of the particles and the type of form that is responsible for the linkages determine the structure of the network and the property of the gel. Aerogels are sol-gel-derived solid materials with porosities of about 80%–98%. The high porosity is achieved through supercritical drying of wet gel in an autoclave (Roy, 1999; Ulrich, 2004). In sol-gel technology, these phases are conserved though the chemical reactions that take place during the gel evolutions and can be manipulated in a variety of ways, such as by altering the initial precursors, time for gelation, catalysts, degree of solvation, gelation conditions, or physical processing of the gel, depending on the specific application (Fig. 1.1).

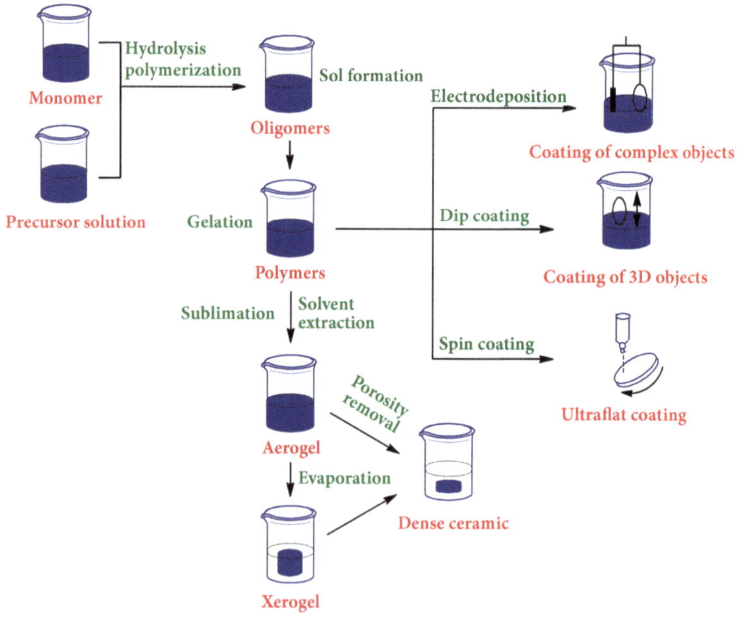

Figure 1.1 Different steps of sol-gel synthesis and its applications.

There are four methods for the production of gels: (i) flocculation of lyophilic colloids by salts or precipitating liquids, (ii) evaporation of certain colloidal solutions, (iii) chemical reactions that lead to a change in the shape of lyophilic molecules (e.g., the denaturation of albumen on heating involves some uncoiling of the protein molecules and a gel structure results), and (iv) swelling of a dry colloid (xerogel) when placed in contact with a suitable liquid (e.g., starch granules added to water).

1.3.1 Colloidal Sol-Gel Methods

Colloidal solutions can be defined as solutions containing discrete particles that do not settle but remain suspended for several years (Iler, 1979). Colloidal sol-gel methods are an active area of research for producing uniform particle sizes (Hench et al., 1990). The applicability of the colloidal methods is based on two key aspects of the process: stabilization of the colloidal particles within the sol and flocculation to form the gel. Aggregation of colloidal particles can be achieved by the removal of the solvent, altering the pH, salinity, or temperature (Colomban, 1989). In most cases, colloidal sol-gel methods contribute to stabilizing the sensitive molecules. The advantage of colloidal methods over alkoxide-based systems is that the majority of the network is already present in the sol. Therefore, certain adaptations, such as the introduction of osmoprotectants, could be applied without significantly interfering with the integrity of the inorganic capsule itself (Perullini et al., 2011).

1.3.2 Polymer-Assisted Sol-Gel Methods

These methods involve the chelation of reactive inorganic gel-forming agents within an organic polymer network, depending on the material to be produced (Omori et al., 2014). The gel-forming agents are dispersed throughout the solution, thereby preventing the precipitation of aggregates within the sol (Kakihana, 1996). However, this method requires subsequent heat treatment to remove the organic polymer following the formation of the inorganic gel. This route offers an effective means of synthesizing materials that disperse poorly in viscous solutions or that would otherwise form reaction products prior to assembly into the required form. This is especially useful for the calcium phosphates (CaPs), which have a tendency to form a diverse range of minerals when present at low concentrations in aqueous solutions (Larsen, 1986). Recently nanoscale $LiMn_2O_4$ particles using Pluronic P-123 as a stabilizing polymer and citric acid as a chelation agent have been synthesized (Yang et al., 2013). The nanoparticle exhibited a superior degree of porosity when compared to particles that were synthesized without the presence of the polymer network. Chemical properties, such as the biomimetic molar ratios of apatites, can also be achieved with polymer-assisted stabilization due to the homogeneous elemental distribution of the gel network (Omori et al., 2014). To control the spatial distribution of calcium within a silica sol-gel network synthesized via a tetraethylorthosilicate (TEOS) precursor with 3-glycidoxy-propyl-trimethoxysilane as the covalent coupling agent c-glutamic acid polymers have been utilized (Valliant et al., 2013). A homogeneous distribution of calcium in the silica network and the larger polymers also reduce the degradation rates. Inorganic networks can also be formed in situ through the polymerization of organic precursors. This method involved the formation of a 3D polyester network as a result of the reaction between ethylene glycol and citric acid (Pechini, 1967).

1.4 Applications of Sol-Gel in Nanoceramics

1.4.1 Nanosized Films and Nanostructured Coatings

Sol-gel techniques have been utilized to change the functionalities by patterning the surface of a glass or another material that provides more degrees of freedom compared with uniform thin films or coating. In the traditional sol-gel method, hydrolysis-condensation processes are followed by condensation polymerization upon film application. Sol-gel preparations are cost effective, are simple to set up, and have the ability to coat complex shapes, similar to biomimetic coatings. The major advantage of sol-gel applications over biomimetic coatings is the strength of the coating-substrate adhesion (Qu et al., 2008; Zhang et al., 2006). However, the evaporation process results in voids and channels throughout the solid gel and cannot provide adequate corrosion protection. Also, the maximum coating thickness attainable by a sol-gel is lower than 2 mm, which can be increased by the incorporation of nanoparticles in the sol without increasing the sintering temperature (Castro et al., 2004). Incorporation of nanoparticles in the hybrid sol-gel systems increases the corrosion protection properties due to lower porosity and lower cracking potential (Table 1.1) (Zheludkevich et al., 2005). Incorporation of inorganic nanoparticles can insert corrosion inhibitors, preparing inhibitor nanoreservoirs for self-repairing pretreatments with controlled release properties (Zheludkevich et al., 2006). Sol-gel films containing zirconia nanoparticles have improved barrier properties. An additional improvement to corrosion protection is observed when cerium nitrate is doped to the above film. Zirconia particles present in the sol-gel matrix act as nanoreservoirs, providing prolonged release of the cerium ions (Zheludkevich et al., 2005).

Table 1.1 Effect of different types of coatings on the surface of the material

Effect	Conventional coating	Nanocoating
Contact tension and its effect on the shape of a water droplet	 High	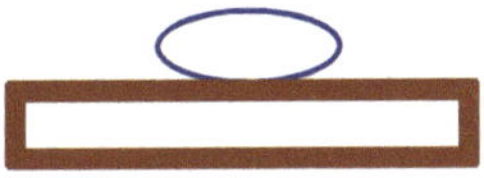 Reduced
Moisture penetration	 High	 Low
Surface roughness	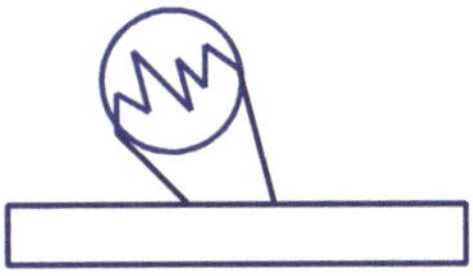 High	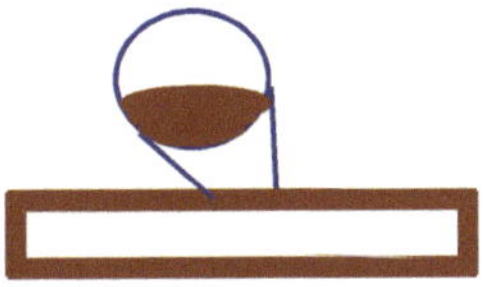 Low

The zirconium-based conversion coatings (ZrCCs) on cold-rolled steel (CRS) substrates have the maximum corrosion resistance values and uniformity (Mohammadloo et al., 2014). An inorganically cross-linked (silica) sol-gel coating has been used to generate complex patterns (Letailleur et al., 2010). Recently, an aqueous sol-gel process of forming functionalized in situ silica nanoparticles and then cross-linking the nanoparticles to form a thin film has been reported. This self-assembled nanophase particle (SNAP) promotes adhesion of overcoat layers more effectively and provides barrier-type corrosion resistance, but it does not have the ability to leach corrosion inhibitors upon coating damage and minimize corrosion of the unprotected area (Zheludkevich et al., 2005). The SNAP surface coating could replace the currently used chromate-containing surface treatment and can provide the basis for a long-lived coating for aluminum alloys (Voevodin et al., 2003, 2006). Sol-gel-derived, hexavalent, chrome-free, environment-friendly, multilayered solar selective coatings were generated on AISI SS 321 substrates using Ag-TiO_2 as the cermet layer and titania and silica as the dielectric layers with high and low refractive indices, respectively (Subasri et al., 2016). $La_2(Zr_{0.7}Ce_{0.3})_2O_7$ (LZ_7C_3) nanoceramics for potential thermal barrier coating (TBC) application have been reported via the sol-gel process. The nanosized LZ_7C_3 shows excellent thermostability with the lowest relative density and the highest sintering resistance in comparison to LZ and 8YSZ (Wang et al., 2012). SiC-based materials $YBa_2Cu_3O_{7-\delta}$ oxide with high purity and good optical properties have been reported using the sol-gel route (Mollicone et al., 2015). Single-layer (CeO_2 or MgO) and double-layer (CeO_2/MgO) ceramic thin films were formed from a precursor solution by a sol-gel process. The corrosion resistance of AZ80 magnesium alloy substrates was found to improve with ceramic coatings. The double-layer ceramic coating has a high corrosion resistance as compared to the single-layer coatings (Harada et al., 2013). Zn-Ni-Al_2O_3 nanocomposite coatings on mild steel using a novel sol-enhanced electroplating method have been reported, which produced more uniform and compact deposits, with a fine-grained microstructure when compared to Zn-Ni coatings (Ghaziof et al., 2015). Thermal evaporation and sol-gel coating techniques have been applied to fabricate a nanostructured thin ZnO film that has a rough surface of a hexagonal wurtzite structure as compared to thermally evaporated films. These films were further applied for the photodegradation of 2,4,6-trichlorophenol (TCP) in water in the presence of UV irradiation. Complete decay of TCP and its intermediates was observed after 60 min when the thermal-evaporated photocatalyst was applied. However, by operating sol-gel catalyst, the concentration of intermediates initially increased and then remained constant with irradiation time. A thermally evaporated ZnO thin film exhibited higher photocatalytic activity in comparison to sol-gel-derived films (Abdel et al., 2009). A broad range of mesoscopic objects, thin porous films, and nanocomposite coatings with the use of the sol-gel technique has been reported. Silica nanotubes, titania nanoparticles, porous titania, and zirconia coatings, as well as titania nanocomposites, were successfully synthesized by changing the type of sol-gel precursor and sol composition and applying the dip-coating deposition procedure in order to obtain thin films or coatings. An enhancement of antiwear properties was observed for 1 wt% of the nanophase content. The annealing at high temperatures is a primary factor that affects the reduction of friction and wear of titania coatings, while the presence of nanoparticles has a secondary effect (Piwonski, 2009).

The sol-gel route combined with the dip-coating technique has been widely investigated to coat metallic biomaterials for enhancement of adhesion (Koutsopoulos, 2002; Zhang et al., 2011). A sol-gel-derived hydroxyapatite (HAp) nanoscale coating to enhance osseointegration and adhesion strength has been reported (Ramakrishna et al., 2001). The sol-gel dip-coating technique is able to produce an adhesive thin coating layer without severe cracking by using a lower annealing temperature (Kim et al., 2004). Normally an extreme temperature is required for plasma spray deposition techniques, which can easily destabilize the crystal structure of HAp and can decompose it (Liu et al., 1994; Carayon et al., 2003). Recently, the bonding strengths of HAp coatings on 316L stainless steel substrates at different annealing temperatures have been compared (Liu et al., 2002). HAp coatings have been deposited onto commercially pure Ti and Ti-6Al-4V alloys (Stoch et al., 2005). HAp coated onto sand-blasted and etched acid substrates has been compared to HAp coatings (Bryington et al., 2013). In recent years, the sol-gel-modified dip-coating technique has been developed to enhance the quality of coating surfaces. The effect of poly-(ε-caprolactone) (PCL) concentration on HAp coating properties at a fixed withdrawal speed has been demonstrated (Yusoff et al., 2014). The addition of PCL to the HAp caused the formation of surface-level pores, which can promote osseointegration. A Ti_6Al_4V substrate with 30%

PCL/HAp is a good combination for coating the substrate with enhanced adhesion (Hornberger et al., 2012). The formation of the oxide layer, generally TiO_2, before the sol-gel dip-coating process improves the bonding between HAp and the Ti substrate (Wen et al., 2007). Furthermore, the addition of carbon nanotubes (CNTs) to HAp has been carried out and leads to HAp-CNT/TiO_2 double-layer coatings with excellent bonding strength (Ji et al., 2012). Also, Eu^{3+} and Tb^{3+} codoped Gd_2O_3 films have been elaborated by the sol-gel process and dip-coating technique (Murillo et al., 2009).

1.4.2 Nanoceramics as Surfaces for Self-Cleaning Function

Self-cleaning glass products are available from major glass manufacturers. There is a great interest in the design and development of surfaces that not only provide biocidal activity but are also easy to clean and even self-cleaning. Most of such coatings acquire their biocidal/self-cleaning capacity by incorporating specific nanoparticles, generally silver (Ag) and titanium oxide (TiO_2) (Morrow et al., 2003; Li et al., 2005). Nano-TiO_2 is used for developing anti-UV, antibacterial, and self-cleaning paints. These possess self-cleaning hydrophobic properties, which cause water droplets to bead off a fully cured surface, picking up dirt and other surface contaminants along the way. This self-cleaning action helps clean and maintain important surfaces and accelerate drying, leaving the surface with minimal spotting. Recently, the corona treatment technique with an inert sol-gel coating and an anatase TiO_2 layer has been utilized to activate the organic surface to allow a uniform TiO_2 sol-gel coating (Cai et al., 2006). $MTiO_2$ thin films were prepared on fused quartz via the modified reverse micellar and sol-gel methods with higher photocatalytic activity and better light-induced hydrophilicity than the simple TiO_2 thin films (Yu et al., 2002). Nanoparticles of surface-treated Al_2O_3 molecules help increase hydrophobicity and increase scratch resistance. The HfO_2 thin film was deposited on the glass substrate and exhibited good surface flatness. The coating had a smaller grain size with increasing radio frequency (RF) power. The lower oxidation state of Hf^{2+} can enhance the hydrophobicity of the HfO_2-deposited glass. The HfO_2-deposited glass exhibited hydrophobicity and had optical selection in the visible-infrared (Vis-IR) region in comparison to pure glass (Lin et al., 2013). Sol-gel processes have been preferably used to create the surface functionality, for example, in motor vehicle windscreens or exterior mirrors (Shokuhfar et al., 2012). House paints with photocatalytically active TiO_2 nanoparticles have self-cleaning properties, and they can also reduce nitrogen oxide and ozone (Mathiazhagan et al., 2011). Nano–titanium oxide is synthesized using a number of synthesis techniques, like hydrothermal, sol-gel, and sonochemical. The coating of the product on cotton fabric has been studied for different biological activities (Sherine et al., 2014). The sol-gel synthesis and structure of cordierite ($2MgO \cdot 2Al_2O_3 \cdot 5SiO_2$), tialite ($Al_2O_3 \cdot TiO_2$), and cordierite/tialite glass-ceramics as well as cordierite containing 2 wt% of Co and (Co+Cu) have been reported using different kinds of precursors, such as $Al(NO_3) \times 9H_2O$, $Si(OC_2H_5)_4$, $Ti(OC_4H_9)_4$, $Mg(NO_3)_2 \times 6H_2O$, chlorides or nitrates of corresponding metals, H_2O, C_2H_5OH, aqua regia, NH_4F, and PEG (Radev et al., 2009). Porous glass-ceramics with a skeleton of γ-$Ti(HPO_4)_2 \cdot 2H_2O$ crystals have been synthesized from glasses in the Li_2O-CaO-TiO_2-P_2O_5 system with porous TiO_2 layers of 200–500 nm thickness deposited on its surfaces. The resulting porous glass-ceramics show high activities for NH_3 adsorption as well as for photodecomposition of the adsorbed NH_3 molecules (Daiko et al., 2008).

1.4.3 Nanoceramics in Molecular Separation Membranes

Layered nanostructures are also being developed for application in filtration applications (Bhave, 1991; Burggraaf et al., 1996). These membranes are able to separate molecules and increase the exergetic efficiency of a wide range of processes in the chemical industry and thus have a large impact on energy consumption. Ceramic membranes have a graded structure with a base structure with macroporosity of variable pore diameters ranging from 2 to 50 nm. The microporous top layer with a pore diameter smaller than 2 nm is made of silica via sol-gel synthesis. Depending also on the conditions of the synthesis, including the calcination temperature and means to exclude dust particles during production, a high permeance and selectivity for gas separation could be obtained (De Vos et al., 1998; Luiten et al., 2010). Monodisperse silica sols with well-defined spherical particles ranging in size from 5 to 310 nm have been prepared through Stober's process. Both particulate and polymeric sol-gel routes have been employed for the preparation of stable silica sols, which can be used to design the predefined membrane pore structures with high thermal stability by cubic/random/close packing of monodisperse spherical particles incorporated into the polymeric network (Topuz et al., 2010). A series of anion-exchange organic-inorganic hybrid membranes was prepared through the sol-gel process from different precursors, such as the copolymer of glycidylmethacrylate and γ-methacryloxypropyltrimethoxy silane, *N*-triethoxysilylpropyl-*N,N,N*-trimethylammonium iodine (A-1100(+), monosilicon), and monophenyltriethoxysilane (EPh). Increasing the content of A-1100(+) could increase the membrane water uptake (W_R), ion exchange capacity (IEC), and membrane potential (E_m), while a decrease in the EPh content can result in an increase in W_R, E_m, and elongation at break (E_b). The membranes have relatively strong hydrophobicity and high mechanical strength (Wu et al., 2009). Alumina membrane filters in the form of thin (0.3–0.8 mm) discs of 25–30 mm diameter, suitable for microfiltration application, have been fabricated by the tape-casting technique. Further using this microfiltration membrane as a substrate, a boehmite sol coating was applied on it and an ultrafiltration membrane with a very small thickness was formed. The top thin ultrafiltration boehmite layer was prepared by a sol-gel method, with a thickness of 0.5 mm, which could be utilized for microbe separation (Das et al., 2009).

1.4.4 Nanoceramics in Abrasives

Microcrystalline corundum abrasives have grown due to high hardness, high strength, good wear resistance, and long service life, as compared to the conventional coarser single-crystalline fused-corundum abrasives. A nanocrystalline corundum abrasive has been synthesized by the two-step sintering method using the sol-gel process. The grain size is about 10 times less than the samples made by the conventional sintering technique, with a homogeneous microstructure and excellent mechanical properties and wear resistance as compared to those sol-gel-derived corundum abrasives with conventional sintering methods and fused corundum abrasives (Li et al., 2009). The corundum-based abrasives with superior tribological properties for grinding have been synthesized by sol-gel technologies. The combined effect of controlled propagation of subsurface shear cracks and plastic deformation leads to the formation of flat contact zones on which a nanocrystalline FeO debris layer adheres to the alumina (Mayer et al., 2006). Fe^{3+} ions have been used in the transformations from boehmites and pseudoboehmite xerogels via transition aluminas to corundum. The results indicated that (i) isolated Fe^{3+} ions can indicate local structural changes but have (almost) no influence on the temperature of corundum formation, (ii) the temperature of corundum formation decreases as the result of action of small α-Fe_2O_3 particles, and (iii) during thermal treatments Fe^{3+} ions are distributed between different phases or precursors of transition aluminas, corundum, Fe_2O_3, and a Fe^{3+} pool (Xera et al., 2006).

1.4.5 Nanoceramics in Engineering Materials

A nano-Al_2O_3 powder has been reported by a simple aqueous sol-gel method using inexpensive $AlCl_3 \cdot 6H_2O$ and Al powder (Shojaie-Bahaabad et al., 2008). Nanosized mullite with an average crystal size below 100 nm was synthesized by spark plasma sintering combined with a sol-gel process and the addition of glucose to the precursor materials. A suppression of grain growth was achieved by adding glucose. The sample prepared with a glucose/Al^{3+} ratio of 3:1 was the most homogeneous, leading to nearly full densified mullite with nanosized grains at a lower sintering temperature as compared to the sample prepared without glucose (Rena et al., 2015).

1.4.6 Nanoceramics in Electronics

Europium-ion doped $Y_3Al_5O_{12}$ garnet nanophosphors (YAG: Eu^{3+}) with wide-ranging size tunability have been prepared via a sol-gel pyrrolysis method using urea and polyvinyl alcohol as the fuel system with a stronger emission intensity as compared to the bulk sample. The particle size and morphology of the synthesized powders depend on the oxidizer (metallic nitrates)-to-fuel ratio. The photoluminescence properties of the prepared YAG:Eu^{3+} phosphors are dependent on the preparation conditions (Lu et al., 2004). $BaTiO_3$ nanoceramics with a homogeneous grain size have been reported that can retain ferroelectricity above room temperature (Xiao et al., 2007). $BaTi_{0.2}(Fe_{0.5}Nb_{0.5})_{0.8}O_3$ (BTFN) ceramic was prepared by a sol-gel method. A high dielectric constant with low loss was found at room temperature at a 1 kHz frequency. Also, BTFN ceramics possess a high magnetodielectric response at room temperature and a 9 kOe magnetic field (Patel et al., 2013). The precursors based on ZrO_2 and HfO_2 with a narrow size distribution of agglomerated particles (100–300 nm) have been synthesized by the citrate sol-gel method (Panova et al., 2011). Forsterite-based ceramic nanopowders have been synthesized using a sol-gel method assisted by further thermal treatment (Karbovnyk et al., 2016). Nanosized B_2O_3-doped $ZnTiO_3$ particles were successfully synthesized through the sol-gel technique. The B_2O_3 content, the high molding pressure, and the sintering temperature had great influence on microstructure and the dielectric properties of materials. As-prepared 1.0 wt% B_2O_3-doped $ZnTiO_3$ sintered at 900°C for 4 h possessed excellent dielectric properties and nanosized, uniform grains contributed to the high quality factor. The porous bulk ceramics contributed to the low dielectric constant (Wu et al., 2010). Dielectric and electrical modulus studies of sol-gel-derived fine-grain (100x) lead zirconate titanate (PZT)/$xCeO_2$ nanocomposites have been reported. The dielectric response shows superimposed local maximas in high-frequency regimes, which in low-frequency regions seems to be overshadowed by high conductivity (Goel et al., 2013). Single-phase bismuth ferrite ($BiFeO_3$, BFO) nanoparticles have been synthesized by the sol-gel technique with high magnetoelectric properties (Kumari et al., 2014). Gd_2O_3 nanopowders codoped with Eu^{3+} and Tb^{3+} were obtained via a sol-gel process using gadolinium pentanedionate as the precursor and europium and terbium nitrates as doping sources (Morales Ramirez et al., 2010). The growth mechanism of sol-gel-derived DNA on crystalline ZnO powder, and the influence of pH value of the sol on the crystallite size, morphology, and structure of ZnO powder, is reported. The maximum-sized nanocrystallite (14 nm) of ZnO powder could be obtained at pH 9. The bandgap increases with a decrease in the size of the ZnO nanocrystallites. The growth mechanism of ZnO particles from $Zn(OAc)_2 \cdot 2H_2O$ precursor by the sol-gel process has been reported in terms of solvation, hydrolysis, and polymerization. The synthesized ZnO powders can be utilized as the electrode material for dye-sensitized solar cells (Rani et al., 2008).

1.4.7 Nanoceramics in Biomaterials

CaPs have attracted a great interest because of their applications as biomaterials such as tricalcium phosphate (TCP), HAp, and calcium tetraphosphate ($Ca_4P_2O_9$), respectively (Dorozhkin, 2010). Bioresorbable and bioactive materials, including β-tricalcium phosphate (β-TCP) and biphasic calcium phosphate (BCP) and their combinations with resorbable polymers, which were used as scaffolds, have been widely investigated (Ramay et al., 2004; Chen et al., 2009). It resulted in great interest in the convenient synthesis of the apatite powders. HAp, TCP, and BCP nanocrystalline powders have been prepared by a simple sol-gel approach. Three different kinds of nanostructured CaP powders can be achieved by changing the ratio of calcic and phosphorous sources (Chen et al., 2011). Nanocrystalline powders of HAp using a simple sol-gel approach have been prepared using different starting materials, such as $Ca(NO_3)_2 \cdot 4H_2O$ and P_2O_5, phosphoric pentoxide (P_2O_5), and calcium nitrate tetrahydrate ($Ca(NO_3)_2 \cdot 4H_2O$) (Kim et al., 2004). The HAp prepared using latter starting materials has the added advantage of a simpler and shorter time of synthesis without pH control and a long hydrolysis time (Feng et al., 2005). HAp nanohexagonal rods of 70–90 nm diameter and 400–500 nm length are synthesized using a sol-gel route with calcium nitrate and potassium dihydrogenphosphate (Padmanabhan et al., 2009). The $Ca_{(10-x)}B_x(PO_4)_6(OH)_2$ nanopowder ceramics were prepared by a sol-gel method with good UV shielding, insulating, and ferroelectric properties. The above ceramics can be used in drug delivery systems and for composite bioceramic applications (Hammad, 2016). The sol-gel processing route is suitable for the fabrication of calcium hydroxyapatite ($Ca_{10}(PO_4)_6(OH)_2$, CHAP) thin films on a Si substrate by the spin-coating technique (Petruleviciene et al., 2015). Porous nanocrystalline HAp has been synthesized by the sol-gel route using equimolar solutions of $Ca(NO_3)_2 \cdot 4H_2O$ and $(NH_4)_2HPO_4$ dissolved in ethanol. It is important for biomedical applications as they could enhance the adhesion between the natural and synthesized bone apatite (Anee Kuriakose et al., 2004). Pure calcium pyrophosphate ($Ca_2P_2O_7$) and Mn-doped $Ca_2P_2O_7$-based ceramics were synthesized by sol-gel method. The average crystallite size and crystallinity are affected by the addition of a lower amount of Mn, whereas the content of Mn in higher amounts did not cause a remarkable effect on these parameters. The ceramics exhibit a paramagnetic behavior at room temperature, and magnetic susceptibility increases with the addition of Mn. The microstructure of the ceramics is affected by the content of Mn (Kaygili, 2015). Nanostructured HAp in the presence of polyethylene glycol synthesized by the sol-gel technique is reported (Kumar et al., 2010). Single-crystal strontium hydroxyapatite (SrHAP) nanorods have been synthesized by a water-based low-temperature sol-gel process using strontium nitrate and diammonium hydrogen phosphate. The aging time plays a significant role in regulating the morphology of the nanoparticles. A SrHAP sample aged at 60°C for 48 h consisted of monodispersed nanorods with lengths of 120–180 nm and 30 nm in diameter (Zhang et al., 2014). Strontium-doped hydroxyapatite (Sr-doped HAp) nanopowder has been synthesized using a sol-gel method. The substitution of Sr atoms for Ca atoms stabilizes the HAp phase, leading to the inhibition of the appearance of the β-TCP phase upon high-temperature calcination (Mardziah et al., 2009). Both undoped HAp and three Ce-substituted HAp samples with variable amounts of Ce were synthesized by a sol-gel method. In all samples, the amount of HAp exceeded 92%, while the amount of admixture β-TCP was always below 8% and no changes were observed by the addition of Ce. The microstructure and elemental composition of all the samples were found to be affected by the addition of Ce (Kaygili et al., 2014). Diopside has been prepared by a modified sol-gel method owing to the higher surface area. Compared to conventional micron-size diopside, nanosized diopside could mimic more efficiently the nanocrystal structure and function of natural bone apatite (Ghorbaniana et al., 2012). The introduction of multiwalled carbon nanotubes (MWCNTs) into the HAp matrix and dip coating of a nanocomposite on a titanium alloy (Ti_6Al_4V) plate have been conducted to improve the performance of the HAp-coated implant via a sol-gel method. The evaluation of the mechanical properties of HAp and HAp/MWCNT composite coatings with different weight percentages of MWCNTs showed that the addition of low concentrations of MWCNTs (0.5 and 1 wt%) had an improved effect on the mechanical properties of nanocomposite coatings and is biocompatible (Abrishamchian et al., 2013). Effective production of both biocompatible ceramic-mineral composites and apatites with predefined stoichiometry has been achieved with polymer-assisted sol-gel methods (Kong et al., 2005; Vargas-Becerril et al., 2014). Nanostructured forsterite bulks were prepared by two-step sintering of a sol-gel-derived forsterite nanopowder that is a bioactive ceramic with good biocompatibility that can be used as a bone tissue engineering material (Kharaziha et al., 2010).

1.4.8 Hybrid Nanoceramics in Drug and Protein Delivery

Nanoporous hybrid ceramics are of high interest in biotechnological and medical applications as delivery and carrier systems for small and large biomolecules, as they provide controlled surface chemistry as well as an intrinsic porosity on a nanometer scale, which results in high surface areas (Garcia-Bennett, 2011). Large biomolecules often are very sensitive to environmental parameters such as pH, temperature, and ionic strength. They can be stabilized by encapsulation in silica gel or silica-gel-based hybrid materials with organic polymers (Lillis et al., 2000; Harris et al., 2012). Silica gel forms a protective cage around the biomolecules and prevents their aggregation or unfolding. The silica gel has several advantages in comparison to organic polymers, such as improved mechanical strength, chemical stability, biological inertness, low-temperature synthesis, and no swelling in aqueous solutions (Livage et al., 2001). The biological activity of biomolecules should not be affected during the sol-gel process—this is the most important factor for successful encapsulation. The synthesis reactions (especially the condensation) are too harsh for most proteins and enzymes as the alcohol generated during hydrolysis of TEOS can lead to denaturation (Ferrer et al., 2002). Recently, a method has been developed for the encapsulation of sensitive proteins in which addition of extra alcohol is not required, and after the condensation reaction, a buffer is added to increase the pH prior to encapsulation (Ellerby et al., 1992). The encapsulation of enzymes such as glucose-6-phosphatase and horseradish peroxidase can also be carried out without any alcohol on the base of sodium silicate (Bhatia et al., 2000).

Silica sol-gel can be utilized as a delivery or carrier system, but its geometric character has to be adapted according to the biological agent. Density, pore size distribution, and nanostructure as release-determining parameters can be controlled by changes in pH, water and alcohol content, and catalyst selection (Brinker et al., 1990; Tebbe et al., 2007). The release of heparin from an organically modified porous silica matrix poly(methacryloxypropyl)-poly(silsesquioxane) (PMAPS) has been reported with retained biological activity that can be utilized for coating the drug-eluting coronary stents. The matrices were obtained from methacryloxypropyltrimethoxysilane (MAS) via a sol-gel process under acidic conditions, followed by photochemical polymerization and cross-linking of the organic matrix. For controlled release of heparin the polymerization degree of the organic matrix can be modulated in the range of 0%–71%. Higher drug loads and decreased polymerization degree lead to a faster release profile of heparin. The heparin concentration has no influence on the degradation rate of the matrix. Silica-based sol-gel ceramics have been utilized for the encapsulation of enzymes as biological catalysts with increased stability in comparison to the free enzyme that could be attributed to multiple-point covalent bonding on the carrier matrix. The adsorption and biological activity of immobilized enzymes is strongly influenced by the pH and electrostatic interactions (Tipton et al., 1979; Yanagisawa et al., 1990; Malmsten et al., 1998). The protein folding can be avoided by building an enzyme-polyelectrolyte complex, whereby the critical charged sites are protected (Chen et al., 1998). In relation to the pH in these nanoscaled pores, it is important to consider that the pH could be lowered by the overlapping electrical double layer, induced by the negatively charged silica gel surface and hence the excess of protons (Bhatia et al., 2000).

1.4.9 Miscellaneous

Synthesis of nano-α-alumina particles by a sol-gel method using aqueous solutions of aluminum isopropoxide and aluminum nitrate has been reported. The nature of surfactant, stabilizing agents, and different stirring times affects the size and shape of particles formed and also the degree of aggregation (Mirjalili et al., 2010). Barium hexaferrite nanostructures have been synthesized via the sol-gel autocombustion method based on the reaction between $Ba(NO_3)_2$ and $Fe(NO_3)_3 \cdot 9H_2O$ in water, in the presence of carbohydrate sugar as fuel and reducing and capping agent. The Ba^{2+} molar ratio to Fe^{3+} influences the phase and morphology of products. When the molar ratio of Fe^{3+} to Ba^{2+} reaches 12, uniform nanoparticles are produced when maltose is used for reduction, resulting in a hard magnetic material as the product (Mandizadeh et al., 2015). Nanoceramic $Ba_{0.5}Sr_{0.5}(Co_{0.8}Fe_{0.2})_{1-x}Zr_xO_{3-\delta}$ (BSCF) powders have been synthesized by a sol-gel process using nitrate-based powdered chemicals to obtain low-potential cathode materials suitable for cathodes (Sharma et al., 2015). Also, nanoceramic powders of proton-conducting $BaZr_{0.8}Y_{0.2}O_{3-\delta}$ (BZY_{20}) have been prepared at a low crystallization temperature (below 130°C) by sol-gel synthesis using an all-alkoxide route. Due to a very low crystallization temperature, hydroxyl defects in the lattice crystal are incorporated in situ during the synthesis process (Cervera et al., 2007). Metal-doped alumina nanopowders where metal could be calcium or rare earths have been prepared by a new sol-gel route using polyhydroxoaluminum (PHA) and $CaCl_2$ or $ReCl_3$ solutions under α-alumina seeding that can be utilized for fabricating dense, fine-grained alumina ceramics (Odaka et al., 2008, 2009). Among the rare-earth dopants studied, Lutetium yields the most suitable nanopowders for low-temperature densification. Peptization and doping of lanthana on colloidal sol-gel titania have been derived from titanyl sulphate. Doped titania also shows an excellent photoactivity as evaluated by the extent of decomposition of methylene blue when exposed to UV light (Baijua et al., 2005). Al-substituted $LiTi_2(PO_4)_3$ (LATP) has been widely investigated and utilized for solid electrolyte in Li-ion or Li-air batteries. Major challenges such as unsatisfactory phase purity, low sintering activity, and high production costs are faced during the fabrication of LATP. $Li_{1.5}Al_{0.5}Ti_{1.5}(PO_4)_3$ (LATP05) has been conveniently prepared on a large scale by the novel sol-gel method, with high phase purity, active densification behavior, and high conductivity (Ma et al., 2016). Binary ytterbium-aluminate glass microspheres with eutectic composition were prepared by flame synthesis in methane-oxygen flame. The starting powder was prepared by a sol-gel method in order to obtain homogeneous glass particles with a narrow interval of particle size distribution. Small portions of YbAG, α-Al_2O_3, and a nonidentified phase were present in glass microspheres after flame synthesis (Prnova et al., 2013). Holmium-doped Bi_2O_3 nanoceramic powders derived from a sol-gel method have been reported with stable, high oxygen-ion conductivity. The translation motion of the charge carrier, oxygen vacancies, and space charge polarization are responsible for the change in activation energy as a function of temperature (Tascıoglua et al., 2012).

1.5 Future Aspects

Nanoceramics are increasingly playing an important role in medicine. Nanoceramic-based coatings on metallic implants in dentistry, orthopedics, and drug delivery could be a new area of research. Chemical modification in synthetic methodologies can affect the particle size, shape, or surface properties of the nanoceramics, which could influence their biological properties and activities; therefore, this is an important area for further research. Surface area and pore diameter are critical factors for drug adsorption and release in drug delivery systems. Functionalization of nanoceramics can also be exploited in the development of sensing devices to allow not only therapy but also simultaneous diagnosis or theranostics. Particles, including inorganic compounds, may be produced with a precise shape and uniform size by a sol-gel technique that may serve as a basis for nanoenabled vaccines and therapeutics. New nanoceramic-based nanocarriers should be developed with long circulation time, low reticuloendothelial system sequestration, ability to extravasate into tumor tissue, and cancer-cell-specific internalization. Nonviral gene delivery systems are also an important area, but the attention paid to it is much lower than the viral vector counterpart. Gas separation by nanoceramic-based molecular membranes could play a useful role in energy conversion processes and can be exploited for further development. Nanoceramics could also offer improvements in the future to the world construction systems.

1.6 Conclusions

Sol-gel chemistry offers a flexible approach for obtaining a diverse range of materials. A variety of nanoceramics can be synthesized by sol-gel methods, which opens up opportunities for new technological applications, such as in optics, electronics, medicines, and sensors. The major advantages of the sol-gel synthesis process are the tunable structures that allow control over morphology, porosity, and size, with high purity of the starting materials and low processing temperatures. Low-temperature synthesis also allows the inclusion of materials, including drugs, growth factors, and proteins, that would otherwise be destroyed by high temperatures. The drugs can be loaded during the sol-gel preparation process and released in a controlled manner locally to treat different diseases, such as infections, osteoporosis, and cancer. Nanoceramic coatings have been used in corrosion protection of metals, which can further be utilized in different industries, such as marine, building, and defense. Nanoceramics have been used for the fabrication of advanced electronic nanodevices, which could open new ways in the development of nations. Thus, sol-gel-based nanoceramics have a huge potential with regard to economic development, infrastructure, and human health.

References

Abdel A, Sawsan A, Mahmoud A, Ahmed Aboul-Gheit K. (2009). Sol-gel and thermally evaporated nanostructured thin ZnO films for photocatalytic degradation of trichlorophenol. *Nanoscale Res. Lett.*, **4**: 627–634.

Abrishamchian A, Hooshmand T, Mohammadi M, Najafi F. (2013). Preparation and characterization of multi-walled carbon nanotube/hydroxyapatite nanocomposite film dip coated on Ti–6Al–4V by sol-gel method for biomedical applications: an *in vitro* study. *Mater. Sci. Eng. C*, **33**: 2002–2010.

Anee Kuriakose T, Narayana Kalkuraa S, Palanichamy M, Arivuoli D, Dierks K, Bocelli G, Betzel C. (2004). Synthesis of stoichiometric nano crystalline hydroxyapatite by ethanol-based sol-gel technique at low temperature. *J. Cryst. Growth*, **263**: 517–523.

Baijua KV, Sibua CP, Rajesha K, Pillaia PK, Mukundana P, Warriera KGK, Wunderlich W. (2005). An aqueous sol-gel route to synthesize nanosized lanthanadoped titania having an increased anatase phase stability for photocatalytic application. *Mater. Chem. Phys.*, **90**: 123–127.

Basu ABM. (2006). Bulk nanoceramic composites for structural applications: a review. *Proc. Indian Natl. Sci. Acad.*, **72**: 97–111.

Basu B, Venkateswaran T, Kim DY. (2006). Microstructure and properties of spark plasma-sintered ZrO_2-ZrB_2 nanoceramic composites. *J. Am. Ceram. Soc.*, **89**: 2405–2412.

Bhatia RB, Brinker CJ, Gupta AK, Singh AK. (2000). Aqueous sol-gel process for protein encapsulation. *Chem. Mater.*, **12**: 2434–2441.

Bhave RR. (1991). *Inorganic Membranes: Synthesis, Characterization and Applications*. van Nostrand Reinhold, New York.

Biest OV. (2013). Nanoceramics: issues and opportunities. *Int. J. Appl. Ceram. Technol.*, **10**: 565–576.

Brinker CJ, Scherer GW. (1990). *Sol-Gel Science: The Physics and Chemistry of Sol-Gel Processing*. Gulf Professional, Houston, TX.

Bryington MS, Hayashi M, Kozai Y, Vandeweghe S, Andersson M, Wennerberg A, Jimbo R. (2013). The influence of nano hydroxyapatite coating on osseointegration after extended healing periods. *Dent. Mater.*, **29**: 514–520.

Burggraaf AJ, Cot L. (1996). *Fundamentals of Inorganic Membrane Science and Technology*. Elsevier Science and Technology Series No. 4. Elsevier, Amsterdam.

Cai R, Van GM, Aw PK, Itoh K. (2006). Solar-driven self cleaning coating for a painted surface. *C. R. Chim.*, **9**: 829–835.

Carayon M, Lacout J. (2003). Study of the Ca/P atomic ratio of the amorphous phase in plasma-sprayed hydroxyapatite coatings. *J. Solid State Chem.*, **172**: 339–350.

Castro Y, Ferrari B, Moreno R, Duran A. (2004). Coatings produced by electrophoretic deposition from nano-particulate silica sol-gel suspensions. *Surf. Coat. Technol.*, **182**: 199–203.

Cervera RB, Oyama Y, Yamaguchi S. (2007). Low temperature synthesis of nanocrystalline proton conducting $BaZr_{0.8}Y_{0.2}O_{3-\delta}$ by sol-gel method. *Solid State Ionics*, **178**: 569–574.

Chen J, Wang Y, Chen X, Ren L, Lai C, He W, Zhang Q. (2011). A simple sol-gel technique for synthesis of nanostructured hydroxyapatite, tricalcium phosphate and biphasic powders. *Mater. Lett.*, **65**: 1923–1926.

Chen JD, Wang Y, Chen X. (2009). In situ fabrication of nano-hydroxyapatite in a macroporous chitosan scaffold for tissue engineering. *J. Biomater. Sci. Polym. Ed.*, **20**: 1555–1565.

Chen Q, Kenausis GL, Heller A. (1998). Stability of oxidases immobilized in silica gels. *J. Am. Chem. Soc.*, **120**: 4582–4585.

Colomban P. (1989). Gel technology in ceramics, glass-ceramics and ceramic composites. *Ceram. Int.*, **15**: 23–50.

Daiko Y, Yajima H, Kasuga T. (2008). Preparation of porous titanium phosphate glass-ceramics for NH_3 gas adsorption with self-cleaning ability. *J. Eur. Ceram. Soc.*, **28**: 267–270.

Das N, Maiti HS. (2009). Ceramic membrane by tape casting and sol gel coating for microfiltration and ultrafiltration application. *J. Phys. Chem. Solids*, **70**: 1395–1400.

De Vos RM, Verwey H. (1998). High-selectivity, high-flux silica membranes for gas separation. *Science*, **279**: 1710–1711.

Dorozhkin S. (2010). Nanosized and nanocrystalline calciumortho-phosphates. *Acta Biomater.*, **6**: 715–734.

Ellerby L, Nishida C, Nishida F, Yamanaka SA, Dunn B, Valentine JS, Zink JL. (1992). Encapsulation of proteins in transparent porous silicate-glasses prepared by the sol-gel method. *Science*, **255**: 1113–1115.

Feng W, Mu-sena L, Yu-penga L, Yong-xin Q. (2005). A simple sol-gel technique for preparing hydroxyapatite nanopowders. *Mater. Lett.*, **59**: 916–919.

Ferrer ML, del Monte F, Levy D. (2002) A novel and simple alcohol-free sol-gel route for encapsulation of labile proteins. *Chem. Mater.*, **14**: 3619.

Garcia-Bennett AE. (2011). Synthesis, toxicology and potential of ordered mesoporous materials in nanomedicine. *Nanomedicine*, **6**: 867–877.

Ghaziof S, Kilmartin PA, Gao W. (2015). Electrochemical studies of sol-enhanced Zn-Ni-Al_2O_3 composite and Zn-Ni alloy coatings. *J. Electroanal. Chem.*, **755**: 63–70.

Ghorbaniana L, Emadi R, Razavib M, Shin H, Teimouri A. (2012). Synthesis and characterization of novel nanodiopsidebioceramic powder. *J. Nanostruct.*, **2**: 357–361.

Goel P, Biradar AM. (2013). A typical dielectric behavior in sol-gel derived fine grain PZT/CeO_2 nanocomposites. *J. Phys. Chem. Solids*, **74**: 854–861.

Hafez HS, El-fadaly E. (2012). Synthesis, characterization and color performance of novel Co^{2+}-doped alumina/titania nanoceramic pigments. *Spectrochim. Acta Part A*, **95**: 8–14.

Hammad MS. (2016). Nanostructure hydroxyapatite based ceramics by sol gel method. *J. Alloys Compd.*, **661**: 251–256.

Harada Y, Kumai S. (2013). Effect of ceramics coating using sol–gel processing on corrosion resistance and age hardening of AZ80 magnesium alloy substrate. *Surf. Coat. Technol.*, **228**: 59–67.

Harris JM, Lopez GP, Reichert WM. (2012). Silica-dispersed glucose oxidase for glucose sensing: in vitro testing in serum and blood and the effect of condensation pH. *Sens. Actuators B*, **174**: 373–379.

Hench LL, West JK. (1990). The sol-gel process. *Chem Rev.*, **90**: 33–72.

Hornberger H, Virtanen S, Boccaccini A. (2012). Biomedical coatings on magnesium alloys: a review. *Acta Biomater.*, **8**: 2442–2455.

Iler RK. (1979). *The Chemistry of Silica: Solubility, Polymerization, Colloid and Surface Properties, and Biochemistry*. Wiley, New York, Chichester.

Ji X, Lou W, Wang Q, Ma J, Xu H, Bai Q, Liu C, Liu J. (2012). Sol-gel derived hydroxyapatite-carbon nanotube/titania coatings on titanium substrates. *Int. J. Mol. Sci.*, **13**: 5242–5253.

Kakihana M. (1996). Invited review "sol-gel" preparation of high temperature superconducting oxides. *J. Sol-Gel. Sci. Technol.*, **6**: 7–55.

Karbovnyk I, Borshchyshyn I, Vakhula Y, Lutsyuk I, Klymc H, Bolesta I. (2016). Impedance characterization of Cr^{3+}, Y^{3+} and Zr^{4+} activated forsterite nanoceramics synthesized by sol-gel method. *Ceram. Int.*, **42**: 8501–8504.

Kaygili O, Dorozhkin SV, Keser S. (2014). Synthesis and characterization of Ce-substituted hydroxyapatite by sol-gel method. *Mater. Sci. Eng. C*, **42**: 78–82.

Kaygili O. (2015). Synthesis and characterization of paramagnetic Mn doped $Ca_2P_2O_7$ ceramics by sol-gel method. *J. Ceram. Proc. Res.*, **16**: 54–58.

Kharaziha, M, Fathi MH. (2010). Improvement of mechanical properties and biocompatibility of forsterite bioceramic addressed to bone tissue engineering materials. *J. Mech. Behav. Biomed. Mater.*, **3**: 530–537.

Kiani A, Rahmani M, Manickam S, Tan B. (2014). Nanoceramics: synthesis, characterization, and applications. *J. Nanomater.*, **2014**: 528348.

Kim HW, Kong YM, Bae CJ, Noh YJ, Kim HE. (2004). Sol-gel derived fluor-hydroxyapatite biocoatings on zirconia substrate. *Biomaterials*, **25**: 2919–2926.

Kim I, Kumt PN. (2004). Sol-gel synthesis and characterization of nanostructured hydroxyapatite powder. *Mater. Sci. Eng. B*, **111**: 232–236.

Kong YM, Bae CJ, Lee SH, Kim HW, Kim HE. (2005). Improvement in biocompatibility of ZrO_2-Al_2O_3 nano-composite by addition of HA. *Biomaterials*, **26**: 509–517.

Koutsopoulos S. (2002). Synthesis and characterization of hydroxyapatite crystals: a review study on the analytical methods. *J. Biomed. Mater. Res.*, **62**: 600–612.

Kumar AR, Kalainathan S. (2010). Sol-gel synthesis of nanostructured hydroxyapatite powder in presence of polyethyleneglycol. *Physica B*, **405**: 2799–2802.

Kumari B, Mandal PR, Nath TK. (2014). Magnetic, magnetocapacitance and dielectric properties of $BiFeO_3$ nanoceramics. *Adv. Mater. Lett.*, **5**: 84–88.

Larsen MJ. (1986). An investigation of the theoretical background for the stability of the calcium-phosphate salts and their mutual conversion in aqueous-solutions. *Arch. Oral. Biol.*, **31**: 757–761.

Letailleur A, Teisseire J, Chemin N, Barthel E, Sondergard E. (2010). Chemorheology of sol-gel silica for the patterning of high aspect ratio structures by nanoimprint. *Chem. Mater.*, **22**: 3143–3151.

Li R, Chen L. (2005). A paint containing nano titanium oxide and nano silver, and its preparation method. Chinese Patent, CN 10027622.

Li Z, Li Z, Zhang A, Zhu Y. (2009). Synthesis and two-step sintering behavior of sol-gel derived nanocrystalline corundum abrasives. *J. Eur. Ceram. Soc.*, **29**: 1337–1345.

Lillis B, Grogan C, Berney H, Lane WA. (2000). Investigation into immobilisation of lactate oxidase to improve stability. *Sens. Actuators B*, **68**: 109–114.

Lin SS, Liao CS. (2013). The hydrophobicity and optical properties of the HfO_2-deposited glass. *Ceram. Int.*, **39**: 353–358.

Liu D, Chou H, Wu J. (1994). Plasma-sprayed hydroxyapatite coating: effect of different calcium phosphate ceramics. *J. Mater. Sci. Mater. Med.*, **5**: 147–153.

Liu DM, Yang Q, Troczynski T. (2002). Sol-gel hydroxyapatite coatings on stainless steel substrates. *Biomaterials*, **23**: 691–698.

Livage J, Coradin T, Roux C. (2001) Encapsulation of biomolecules in silica gels. *J. Phys.: Condens. Matter*, **13**: R673–R691.

Lu C, Hsu W, Dhanaraj J, Jagannathan R. (2004). Sol-gel pyrolysis and photoluminescent characteristics of europium-ion doped yttrium aluminum garnet nanophosphors. *J. Eur. Ceram. Soc.*, **24**: 3723–3729.

Luiten MWJ, Benes NE, Huiskes C, Kruidhof H, Nijmeijer A. (2010). Robust method for micro-porous silica membrane fabrication. *J. Membr. Sci.*, **348**: 1–5.

Ma Q, Xu Q, Tsai C, Tietz F, Guillon O. (2016). A novel sol-gel method for large-scale production of nanopowders: preparation of $Li_{1.5}Al_{0.5}Ti_{1.5}(PO_4)_3$ as an example. *J. Am. Ceram. Soc.*, **99**: 410–414.

Malmsten M, Burns N, Veide A. (1998). Electrostatic and hydrophobic effects of oligopeptide insertions on protein adsorption, *J. Colloid Interface Sci.*, **204**: 104–111.

Mandizadeh S, Soofivand F, Salavati-Niasari M. (2015). Sol-gel auto combustion synthesis of $BaFe_{12}O_{19}$ nanoceramics by using carbohydrate sugars as a novel reducing agent. *Adv. Powder Technol.*, **26**: 1348–1354.

Mardziah CM, Sopyan I, Ramesh S. (2009). Strontium-Doped hydroxyapatite nanopowder *via* sol-gel method: effect of strontium concentration and calcination temperature on phase behavior. *Trends Biomater. Artif. Organs*, **23**: 105–113.

Mathiazhagan A, Joseph R. (2011). Nanotechnology: a new prospective in organic coating: review. *Int. J. Chem. Eng. Appl.*, **2**: 225–237.

Mayer J, Engelhorn R, Bot R, Weirich T, Herwartz C, Klocke F. (2006). Wear characteristics of second-phase-reinforced sol-gel corundum abrasives. *Acta Mater.*, **54**: 3605–3615.

Mirjalili F, Hasmaliza M, Abdullah LC. (2010). Size-controlled synthesis of nano α-alumina particles through the sol-gel method. *Ceram. Int.*, **36**: 1253–1257.

Mohammadloo AA, Sarabia R, Mohammad Hosseinia M, Saraylooa H, Sameieb R, Salimiba. (2014). A comprehensive study of the green hexafluorozirconic acid-based conversion coating. *Prog. Org. Coat.*, **77**: 322–330.

Mohd Yusoff MF, Abdul Kadir MR, Iqbal N, Hassan MA, Hussain R. (2014). Dip coating of poly(ε-caprolactone)/hydroxyapatite composite coating on Ti_6Al_4V for enhanced corrosion protection. *Surf. Coat. Technol.* **245**: 102–107.

Mollicone J, Lenormanda P, Ansart F, Rousseau B. (2015). Functionalization of SiC-based materials by a selective $YBa_2Cu_3O_{7-\delta}$ coating *via* sol gel route in order to optimize their optical properties. *Thin Solid Films*, **596**: 18–23.

Morales Ramirez AJ, Garcia Murillo A, Carrillo Romoa FJ, Garcýa Hernandez M, Jaramillo Vigueras D, Chaderyron G, Boyer D. (2010). Properties of Gd_2O_3: Eu^{3+}, Tb^{3+} nanopowders obtained by sol-gel process. *Mater. Res. Bull.*, **45**: 40–45.

Morrow WH, McLean, LJ. (2003). Self-cleaning UV reflective coating, its applying methods, and UV irradiating device prepared therefrom. US Patent, US 2003059549.

Murillo AG, Ramírez, AJM, Romo FJC, Hernandez MG, Dominguez MAC. (2009). Synthesis, structural and optical studies of sol-gel Gd_2O_3: Eu^{3+}, Tb^{3+} films. *Mater. Lett.*, **63**: 1631–1634.

Niihara K, Suzuki Y. (1999). Strong monolithic and composite $MoSi_2$ materials by nanostructure design. *Mater. Sci. Eng. A*, **261**: 6–15.

Odaka A, Yamaguchi T, Fujita T, Taruta S, Kitajima K. (2008). Densification of Ca-doped alumina nanopowders prepared by a new sol-gel route with seeding. *J. Eur. Ceram. Soc.*, **28**: 2479–2485.

Odaka A, Yamaguchi T, Fujita T, Taruta S, Kitajima K. (2009). Densification of rare-earth (Lu, Gd, Nd)-doped alumina nanopowders obtained by a sol-gel route under seeding. *Powd. Techn.*, **193**: 26–31.

Omori Y, Okada M, Takeda S, Matsumoto N. (2014). Fabrication of dispersible calcium phosphate nanocrystals via a modified Pechini method under non-stoichiometric conditions. *Mater Sci Eng C: Mater Biol Appl.*, **42**: 562–568.

Owens GJ, Singh RK, Foroutan F, Alqaysi M, Han C, Mahapatra C, Kim H, Knowles JC. (2016). Sol-gel based materials for biomedical applications. *Prog. Mater. Sci.*, **77**: 1–79.

Padmanabhan SK, Balakrishnan A, Chub M, Lee YJ, Kim TN, Cho S. (2009). Sol-gel synthesis and characterization of hydroxyapatite nanorods, *Particuology*, **7**: 466–470.

Palmero P. (2015). Structural ceramic nanocomposites: a review of properties and powders' synthesis methods. *Nanomat.*, **5**: 656–696.

Panova TI, Morozova LV, Drozdova IA, Shilova OA. (2011). Sol-gel synthesis of solid solutions based on zirconium and hafnium dioxides. *Glass Phys. Chem.*, **37**: 505–511.

Patel PK and Yadav KL. (2013). Giant dielectric permittivity and room temperature magnetodielectric study of $BaTi_{0.2}(Fe_{0.5}Nb_{0.5})_{0.8}O_3$ nanoceramic. *Mater. Res. Bull.*, **48**: 1435–1438.

Pechini MP. (1967). Method of preparing lead and alkaline earth titanates and niobates and coating method using the same to form a capacitor. In Co SE (ed.). US Patent, US 3330697.

Perera Y. (2010). Preparation of nano-ceramics *via* aqueous sol-gel method modified with surfactants: an overview. *Mater. Sci. Forum*, **644**: 79–84.

Perullini M, Amoura M, Roux C, Coradin T, Livage J, Japas ML, Jobbagy M, Bilmes SA. (2011). Improving silica matrices for encapsulation of *Escherichia coli* using osmoprotectors. *J. Mater. Chem.*, **21**: 4546–4552.

Piwonski I. (2009). Mesoscopic objects, porous layers and nanocomposites-possibilities of sol-gel chemistry. *Appl. Surf. Sci.*, **256S**, S86–S91.

Prnova A, Bodisova K, Skratek M, Klement R, Migat M, Veteska P, Galusek D, Bruneel E, Driessche IV. (2013). Preliminary study of thermal properties of Al_2O_3-YB_2O_3 glass microspheres. *Measurement: Proceedings of the 9th International Conference, Smolenice, Slovakia*, pp. 327–330.

Qu H, Wei M. (2008). Improvement of bonding strength between biomimetic apatite coating and substrate. *J. Biomed. Mater. Res. Part B*, **84**: 436–443.

Radev L, Samuneva B, Mihailova I, Pavlova L, Kashchieva E. (2009). Sol-gel synthesis and structure of cordierite/tialite glass ceramics. *Process. Appl. Ceram.*, **3**: 125–130.

Ramakrishna S, Mayer J, Wintermantel E, Leong KW. (2001). Biomedical applications of polymer-composite materials: a review. *Compos. Sci. Technol.*, **61**: 1189–1224.

Ramay HRR, Zhang M. (2004). Biphasic calcium phosphate nano-composite porous scaffolds for load-bearing bone tissue engineering. *Biomaterials*, **25**: 5171–5180.

Rani S, Suri P, Shishodia PK, Mehra RM. (2008). Synthesis of nanocrystalline ZnO powder via sol-gel route for dye-sensitized solar cells. *Sol. Energy Mater. Sol. Cells*, **92**: 1639–1645.

Rena L, Fua Z, Wanga Y, Zhangb F, Zhanga J, Wanga W, Wang H. (2015). Effect of the addition of glucose on the fabrication of nanosized mullite froma sol–gel by spark plasma sintering. *J. Eur. Ceram. Soc.*, **35**: 1915–1921.

Roy R. (1999). In *Proceedings of World Congress on Superconductivity*. World Scientific, Singapore, p. 29.

Senthil T, Anandhan S. (2014). Structure-property relationship of sol-gel electrospun ZnO nanofibers developed for ammonia gas sensing. *J. Colloid Interface Sci.*, **432**: 285–296.

Sharma DRK, Rao PVB, Reddy MN, Srikanth D. (2015). Synthesis and characterization of $Ba_{0.5}Sr_{0.5}$ $(CO_{0.8}Fe_{0.2})_{1-x}Zr_xO_{3-\delta}$ (BSCF) nanoceramic cathode powders by sol-gel process for solid oxide fuel cell. International Conference on Nano Science & Engineering Applications (ICONSEA-2014). *Mater. Today: Proc.*, **2**: 4442–4450.

Shojaie-Bahaabad M, Nassaj TE. (2008). Economical synthesis of nano alumina powder using an aqueous sol-gel method. *Mater. Lett.*, **62**: 3364–3366.

Shokuhfar A, Alzamani M, Eghdam E, Karimi M, Mastali S. (2012). SiO_2-TiO_2 nanostructure films on windshields prepared by sol-gel dip-coating technique for self-cleaning and photocatalytic applications. *Nanosci. Nanotechnol.*, **2**: 16–21.

Simchi A, Tamjid E, Pishbin F, Boccaccini AR. (2011). Recent progress in inorganic and composite coatings with bactericidal capability for orthopaedic applications. *Nanomed. Nanotechnol. Biol. Med.*, **7**: 22–39.

Stoch A, Jastrzebski W, Długon E, Lejda W, Trybalska B., Stoch GJ, Adamczyk A. (2005). Sol-gel derived hydroxyapatite coatings on titanium and its alloy Ti_6Al_4V. *J. Mol. Struct.*, **744–747**: 633–640.

Subasri R, Soma Raju KRC, Reddy DS, Hebalkar, NY, Padmanabham G. (2016). Sol-gel derived solar selective coatings on SS 321 substrates for solar thermal applications. *Thin Solid Films*, **598**: 46–53.

Tascýoglua I, Arýb M, Usluc I, Kocyigitc S, Dagdemirb Y, Orumlub VC, Altýndal S. (2012). Temperature dependent conductivity and structural properties of sol-gel prepared holmium doped Bi_2O_3 nanoceramic powder. *Ceram. Int.*, **38**: 6455–6460.

Tebbe D, Thull R, Gbureck U. (2007). Correlation between heparin release and polymerization degree of organically modified silica xerogels from 3-methacryloxy propyl polysilsesquioxane. *Acta Biomater.*, **3**: 829–837.

Tipton KF, Dixon HBF. (1979). Effects of pH on enzymes. In *Methods in Enzymology*, Purich DL (ed.). Academic Press, New York, pp. 183–234.

Topuz B, Iftcioglu MC. (2010). Preparation of particulate/polymeric sol-gel derived microporous silica membranes and determination of their gas permeation properties. *J. Membr. Sci.*, **350**: 42–52.

Uche DV. (2013). Sol-gel technique: a veritable tool for crystal growth. *Adv. Appl. Sci. Res.*, **4**: 506–510.

Ulrich DR. (1984). Flame synthesis of fine particles. *Chem. Eng. News*, **62**: 22–29.

Valliant EM, Romer F, Wang D, McPhail DS, Smith ME, Hanna JV, Jones JR. (2013). Bioactivity in silica/poly(gamma-glutamic acid) sol-gel hybrids through calcium chelation. *Acta Biomater.*, **9**: 7662–7671.

Vargas-Becerril N, Tellez-Jurado L, Reyes-Gasga J, Alvarez-Perez MA, Rodriguez-Lorenzo LM. (2014). Synthesis of nanosized carbonated apatite by a modified Pechini method: hydroxyapatite nucleation from a polymeric matrix. *J. Sol-Gel Sci. Technol.*, **72**: 571–580.

Voevodin N, Balbyshev VN, Khobaib M, Donley MS. (2003). Nanostructured coatings approach for corrosion protection. *Prog. Org. Coat.*, **47**: 416–423.

Voevodin NN, Kurdziel JW, Mantz R. (2006). Corrosion protection for aerospace aluminum alloys by modified self-assembled nanophase particle (MSNAP) sol-gel. *Surf. Coat. Technol.*, **201**: 1080–1084.

Wang C, Wang Y, Fan X, Binglin Zou WH, Cao X. (2012). Preparation and thermophysical properties of $La_2(Zr_{0.7}Ce_{0.3})_2O_7$ ceramic *via* sol-gel process. *Surf. Coat. Technol.*, **212**: 88–93.

Wen CE, Xu W, Hu WY, Hodgson PD. (2007). Hydroxyapatite/titania sol-gel coatings on titanium-zirconium alloy for biomedical applications. *Acta Biomater.*, **3**: 403–410.

Wu SP, Luo JH, Cao SX. (2010). Microwave dielectric properties of B_2O_3-doped $ZnTiO_3$ ceramics made with sol-gel technique. *J. Alloys Compd.*, **502**: 147–152.

Wu Y, Wu C, Xu T, Fu Y. (2009). Novel anion-exchange organic-inorganic hybrid membranes prepared through sol-gel reaction of multi-alkoxy precursors. *J. Membr. Sci.*, **329**: 236–245.

Xera RS, Nofzb M, Feista M, Scholz G. (2006). Fe^{3+} assisted formation of α-Al_2O_3, starting from sol-gel precursors. *J. Solid State Chem.*, **179**: 652–664.

Xiao CJ, Zhang WW, Chi ZH, Li FY, Feng SM, Jin CQ, Wang XH, Li LT, Chen RZ. (2007). Ferroelectric $BaTiO_3$ nanoceramics prepared by a three-step high-pressure sintering method. *Phys. Status Solidi A*, 1–7.

Yanagisawa T, Shimizu T, Kuroda K, Kato C. (1990). The preparation of alkyltrimethylammonium-kanemite complexes and their conversion to microporous materials. *Bull. Chem. Soc. Jpn.*, **63**: 988–992.

Yang Z, Jiang Y, Xu HH, Huang YH. (2013). High-performance porous nanoscaled $LiMn_2O_4$ prepared by polymer-assisted sol-gel method. *Electrochim. Acta*, **106**: 63–68.

Yu JC, Yu J, Ho W, Zhao J. (2002). Light-induced super-hydrophilicity and photocatalytic activity of mesoporous TiO_2 thin films. *J. Photochem. Photobiol. A*, **148**: 331–339.

Zhang JX, Guan RF, Zhang XP. (2011). Synthesis and characterization of sol-gel hydroxyapatite coatings deposited on porous NiTi alloys. *J. Alloys Compd.*, **509**: 4643–4648.

Zhang S, Xianting Z, Yongsheng W, Kui C, Wenjian W. (2006). Adhesion strength of sol-gel derived fluoridated hydroxyapatite coatings. *Surf. Coat. Technol.*, **200**: 6350–6354.

Zhang W, YaminChai N, Xu X, Wang Y. (2014). Synthesis of nanosize single-crystal strontium hydroxyapatite via a simple sol-gel method. *Ceram. Int.*, **40**: 16061–16064.

Zhao Y, Sugiyama S, Miller T, Miao X. (2008). Nanoceramics for blood-borne virus removal. *Expert Rev. Med. Devices*, **5**: 395–405.

Zheludkevich ML, Miranda Salvado IM, Ferreira MGS. (2005). Sol-gel coatings for corrosion protection of metals. *J. Mater. Chem.*, **15**: 5099–5111.

Zheludkevich ML, Serra R, Montemor MF, Miranda Salvado IM, Ferreira MGS. (2006). Corrosion protective properties of nanostructured sol-gel hybrid coatings to AA2024-T3. *Surf. Coat. Technol.*, **200**: 3084–3094.

Zheludkevich ML, Serra R, Montemor MF, Yasakau KA, Miranda Salvado IM, Ferreira MGS. (2005). Nanostructured sol-gel coatings doped with cerium nitrate as pre-treatments for AA2024-T3 corrosion protection performance. *Electrochim. Acta*, **51**: 208–217.

Chapter 2

Ceramic Materials: General Introduction, Properties, and Fabrication Methods

Deepak Pathania, Rishu Katwal, and Pankaj Thakur

School of Chemistry, Shoolini University, Solan 173212, Himachal Pradesh, India

dpathania74@gmail.com

This comprehensive review is focused on the recent progress in the properties of and synthetic approaches to ceramic materials, which can be used to understand effectively the fabrication methods. Ceramic materials are inorganic, nonmetallic materials (SiC, Si_3N_4, UO_2, TiC, ZnO, MgB_2, ZnO, ZrO_2, Al_2O_3, etc.). The knowledge of designing and fabrication techniques of ceramics is the need of the hour. Ceramics offer many advanced properties as compared to other materials, for example, they are hard, refractory, wear-resistant, and brittle, and they can be used as electrical insulators; thermal insulators; and nonmagnetic, oxidation-resistant, thermal, and chemically stable insulators. The properties of ceramics have made them famous for some much more astonishing applications. It has the potential to revolutionize

Smart Ceramics: Preparation, Properties, and Applications
Edited by Ajay Kumar Mishra

ISBN 978-981-4774-30-7 (Hardcover), 978-1-315-16359-8 (eBook)
www.panstanford.com

a series of medical and engineering tool procedures so that they are portable, cheap, safe, and easy to administer. Ceramic materials are used in different industries, including medicine, mining, refineries, the aerospace industry, food and chemical industries, electronics, packaging science, transmission electricity, and lightwave communication. It is hard to think of an area of modern life that has not been touched by ceramics. Various fabrication methods have been discussed in detail for a better understanding of ceramic materials.

2.1 Introduction

The word "ceramics" comes from the Greek word *keramikos*, meaning "pottery." Its origin is a Sanskrit term meaning "to burn." Mainly traditional ceramic products were made from clay (or clay mixed with other materials). Tableware and decorative ceramics are still made this way. Clay is an imperative ceramic and is the backbone of the customary ceramic industry. In modern ceramic engineering, ceramics is the art and science of building objects from inorganic, nonmetallic materials by the action of heat. There is a long history of ceramic art in almost all advanced cultures, and often evidence is left from missing cultures. Cultures mainly distinguished for ceramics are the Chinese, Cretan, Greek, Persian, Mayan, Japanese, and Korean cultures, as well as the modern Western cultures. Until the sixteenth century, in Europe small quantities of luxurious Chinese porcelain were imported. From the sixteenth century onward attempts were made to reproduce it in Europe, including soft-paste and the Medici porcelain made in Florence (Kelly et al., 1996).

Ceramic materials were used more than 10,000 years ago, during the Stone Age (Anusavice, 2003). In 1723, Pierre Fauchard illustrated the enameling of metal denture bases (Kelly, 1996). In 1789, De Chemant, a French dentist, introduced the first porcelain denture tooth. In Paris, in 1808, Fonzi introduced *tetro-metallic incorruptibles*, which are ceramic teeth with embedded platinum pins. Dr. Charles Land introduced the first ceramic crown in 1903 (Anusavice, 2003). The historical point of view till 1950 was that the most important category of ceramics was traditional clays, which were made into pottery, bricks, tiles, etc. The products made from ceramics play a significant role in the historical consideration of the technology and culture of the people who lived many years ago, and household items help to understand the culture well as the historical developments of a particular civilization (Singh, n.d.).

2.2 Ceramics in the Past

Figure 2.1 Porcelain imports of the East India Company from China to Europe (1600–1800): >3,000,000 pieces, annual dividends up to 750% pa on the invested money. From "Ceramic Materials," Filser and Gauckler, ETH-Zurich, lecture notes 2007.

During this century, worldwide studies on ceramics (Fig. 2.1), polymers, and metals have resulted in the establishment of materials science as a scientific discipline. A feature of these studies is the interdisciplinary nature and modern chemistry is making an increasingly significant contribution to the research, development, and manufacture of ceramic materials. Ceramics are related with mixed binding: a combination of covalent, ionic, and metallic. The majority of ceramics are compounds of metals of metalloids and nonmetals. Diamond and graphite are classified as ceramics. The properties of ceramic materials are ordered by the types of atoms present, bonding between the atoms, and the way the atoms are crammed together. Most ceramics are made up of two or more elements. For example, aluminum atoms and oxygen atoms combine to form an alumina (Al_2O_3) compound. The properties of a specific ceramic depend not just on the materials from which it is made but also on its crystalline structure. The structure that plays a vital role in the concluding property of a material is called its microstructure (Licciulli, 2003). Ceramic materials are used for a large number of applications due to a variety of properties. For example, ceramic materials are:

- Hard
- Brittle
- Wear resistant
- Refractory
- Thermal insulators
- Oxidation resistant
- Nonmagnetic
- Electrical insulators
- Chemically stable
- Prone to thermal shock

Ceramics are inelastic at room temperature, that is, they fail suddenly, when subjected to tension, with little or no plastic deformation prior to fracture. On the other hand, metals are ductile, and they own this tremendously useful property due to imperfections called dislocations within their crystal lattices (Mason, n.d.). Hardness is an important property where we focus on the use of wear-resistant materials. So, when we want to have wear resistance, we tend to increase the hardness. So, hardness is suitable for ceramics. If high hardness is a prerequisite ceramics have an edge. Ductility, which is another valuable property, is low for ceramics and high for metals.

Ceramics are excellent insulators and poor conductors of electricity. There are two ways that ceramics can be made electrically conductive: (i) oxygen vacancies, leading to ionic conductivity at effectively high temperatures, and (ii) point defects. Good examples of conductive transition metal oxide ceramics are the copper-based superconductors. Ceramics are used in many composites, often as the reinforcement. In a few composites, the ceramic is substituted as the matrix. Ceramic composites are those where there is a continuous ceramic matrix with isolated inclusions or, at least, a bicontinuous skeletal microstructure with one ceramic phase.

2.3 Classification

- Glasses: Based on SiO_2 (Li_2O-Al_2O_3-$n$$SiO_2$, MgO-Al_2O_3-$n$$SiO_2$, ZnO-Al_2O_3-$n$$SiO_2$, SrO-BaO-Al_2O_3-SiO_2)
- Traditional ceramics (clay products):
  - o Porous ceramics (bricks, pottery, china)
  - o Compact ceramics (porcelain, earthware)
- Advanced ceramics: refractory ceramics (SiC, BeO, Al_2O_3, ZrO_2, MgO):
  - o Electro-optics: $LiNbO_3$
  - o Piezoelectrics and ferroelectrics: $BaTiO_3$, $SrTiO_3$
  - o Abrasive ceramics: Nitrides and carbides SiC, Si_3N_4
  - o Biomaterials: Hydroxyapatite
  - o Superconductive ceramics ($YBa_2Cu_3O_7$)
  - o Molecular membranes (produced from inorganic materials such as titania, alumina, zirconia oxides, silicon carbide, or some glassy materials)

2.4 Synthesis Method of Ceramic Materials

Various methods of synthesis of ceramic materials are discussed in this section.

2.4.1 The Sol-Gel Method

Colloid science is necessary for the booming application of chemistry to ceramic synthesis. The synthesis engages the nucleation and growth of the particles' size via colloidal dispersions and other synthesis methods like the a sol-gel process. A sol is a stable dispersion of colloidal particles or polymers in a solvent, and these particles may be crystalline or amorphous. A gel consists of a continuous 3D network that encircles a liquid phase. The system is built from the agglomeration of colloidal particles in a colloidal gel. Usually, the sol particles may interact with hydrogen bonds or van der Waals forces. The gelation process may be reversible if other interactions are involved. The sol-gel process has been widely used for the synthesis of metal alkoxide powders. For example, $Y_3Fe_5O_{12}$ was synthesized by hydrolysis of a mixture of $Y(OC_4H_9)_3$ and $Fe(OC_2H_5)_3$ in C_2H_5OH in xylene (Taketomi et al., 1994). The particle size was calculated to be 9 nm. Van De Graaf et al. (1980) also used metal alkoxides as precursors for the fabrication of ZrO_2-Y_2O_3 ceramics. Researchers synthesized zirconia ceramics using precursors zirconium carboxylates, and zirconium carboxylates precursors were prepared by the reaction of zirconium oxychloride with the sodium or ammonium carboxylic acid salts of the acids (Kulmeteva et al., 2009; Madhusudhana et al., 2014). Sirota et al. (2013) prepared ZrO_2 ceramics from nitrate zirconium, ammonia, yttrium acetate, and surfactants of oxyethylene nonyl phenol, NH_4OH. The results revealed zirconia with a tetragonal crystallite size ranging from 3 to 15 nm. Purwasasmita et al. (2013) used zirconium (IV) oxychloride as a precursor for preparation of zirconia powder. In the reaction calcium nitrate was used as a stabilizer and sodium alginate and CaO as a chelating agent by a sol-gel method. The results revealed that the zirconia that was formed had a tetragonal structure. The particle sizes of the zirconia were ranging from 27 nm to 189 nm. Aluminum nitride has been synthesized by carbothermal reduction followed by nitridation of alumina gel at a temperature range of 1200°C–1350°C (Chaudhuri et al., 2013). Buckley and Greenblatt (1994) prepared tetraethylorthosilicate by a sol-gel method that involved hydrolysis and a condensation reaction. The intermediates generated silanols

(Si-OH). Condensation may take place for two silanols or a silanol and an ethoxy to form bridging oxygen or siloxane (Si-O-Si), and water or ethanol is eliminated. The mechanism of sol-gel polymerization of tetraethylorthosilicate is as follows (Rahman and Padavettan, 2012):

Acid-Catalyzed Hydrolysis

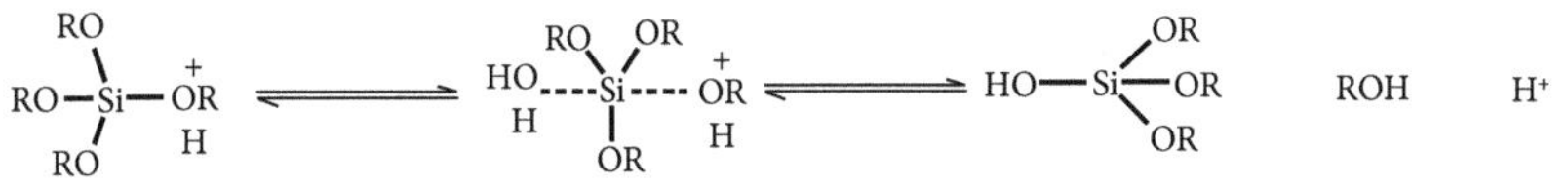

Acid-Catalyzed Condensation

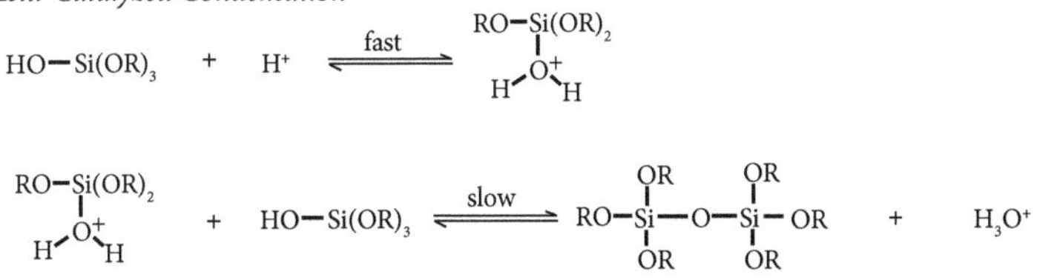

Base-Catalyzed Hydrolysis

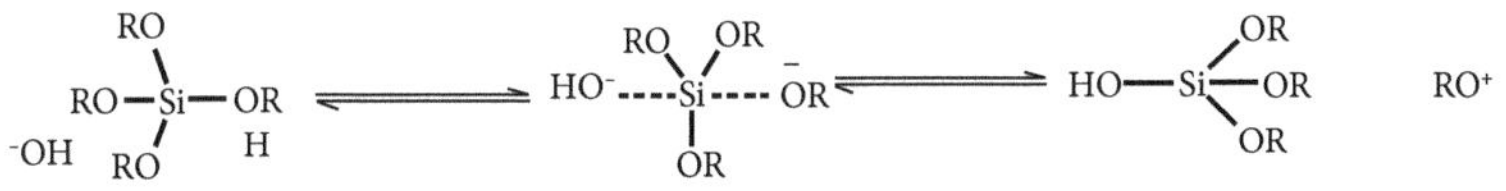

Base-Catalyzed Condensation

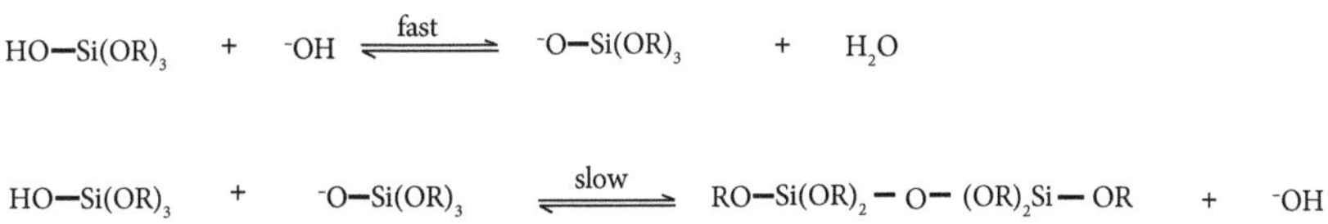

The sol-gel process is extensively used to fabricate silica and homogeneous products at mild conditions. The route engages the hydrolysis and condensation of metal alkoxides ($Si(OR)_4$), such as tetraethylorthosilicate, or inorganic salts, such as sodium silicate, in the presence of a mineral acid or base as the catalyst (Rahman and Padavettan, 2012). Various glasses, like SiO_2, SiO_2-Al_2O_3, SiO_2-P_2O_5, SiO_2-P_2O_3, B_2O_2-P_2O_3-Na_2O, and SiO_2-Ai_2O_3-B_2O_3-K_2O-Na_2O, have been developed by the sol-gel method

(Zelinski et al., 1984). Hayashi and Saita (1985) have reported the synthesis of glasses in the system $CaO\text{-}SiO_2$ from the alkoxide method. Vijayalakshmi et al. (2009) synthesized porous silica gels using tetraethylorthosilicate and polyacrylic acid by a sol-gel method. $Ce_{0.9}Gd_{0.1}O_{1.95}$ ceramics were prepared by a mixture of CeO_2 and Gd_2O_3 in acetone (Huang et al., 1998). The results revealed spherical powders of $Ce_{0.9}Gd_{0.1}O_{1.95}$ with an average diameter of 250 nm were obtained at 700°C and a cubic unit cell $a = 5.422 \pm 0.034$ Å.

2.4.2 The Electrochemical Method

The electrochemical method offers important and unique possibilities in the development of ceramic materials. The electrochemical method deals with the reactions, proceeding at the expense of electrical energy. In the electrochemical deposition process with the help of an electric field, the ions move through a solution. The electrochemical cell consists of a cathode and an anode, and the voltage is measured between the two electrodes. The electrochemical deposition involves the different electron transfer steps between the electrode and electroactive species present in the solution. Electrochemical deposition of ceramic oxides can be carried out under oxidizing conditions as well as reducing conditions from the alkaline solutions. Two mechanisms, depending on the species in the solution, may occur, as follows (Therese and Kamath, 2000):

$$H^+ + e^- \rightarrow H_{ads} \tag{2.1}$$

$$2H^+ + 2e^- \rightarrow H_2' \tag{2.2}$$

$$2H_2O + 2e^- \rightarrow H_2 + 2OH' \tag{2.3}$$

$$O_2 + 2H_2O + 4e^- \rightarrow 4OH' \tag{2.4}$$

If a nitrate salt is present in the solution then hydronium ions can be consumed or hydroxide ions produced by

$$NO_3^- + 2H^+ + 2e^- \rightarrow NO_2^- + H_2O \tag{2.5}$$

$$NO_3^- + 10H^+ + 8e^- \rightarrow NH_4^+ + 3H_2O \quad (2.6)$$

$$NO_3^- + H_2O + 2e^- \rightarrow NO_2^- + 2OH^- \quad (2.7)$$

$$NO_3^- + 7H_2O + 8e^- \rightarrow NH_4^+ + 10OH^- \quad (2.8)$$

Electrodeposition of lanthanum oxide (La_2O_3) in aqueous solutions has been reported by various researchers (Bocchetta et al., 2007; Gonzalez-Rovira et al., 2008; Liu et al., 2010; Yao et al., 2011). Bocchetta et al. first demonstrated that at a cathodic current of 1 mA/cm^2 on an Al substrate using galvanostatic deposition, it was feasible to obtain $La(OH)_3$ from a solution of lanthanum nitrate. Yao et al. used a copper substrate instead of an Al substrate and used lanthanum nitrate and ammonia nitrate to obtain $La(OH)_3$ nanorods. Liu et al. (2010) used F-doped SnO_2 substrates from a bath containing $La(NO_3)_3$ and dimethyl sulphoxide (DMSO) to formed La_2O_3 from $La(OH)_3$ nanospindles and nanorods. X-ray diffraction (XRD) results revealed a hexagonal structure of La_2O_3, and it also showed how much the DMSO percentage affected the nanostructure. Nanospindles produced at a higher concentration and a lower concentration produced nanorods. Zhitomirsky et al. (1998–2001) used the aqueous and mixed alcohol-water solutions of cerium chloride or nitrate in cathodic electrodeposition of CeO_2 and Gd-doped ceria films. Zhitomirsky (1998) also proposed two- or four-electron reduction-deposition of $CeO_2 \cdot nH_2O$ or $Ce(OH)_3OOH$ (Zhitomirsky, 1999; Zhitomirsky and Petric, 1999, 2001). Lu et al. (2008) electrodeposited TbO_{2-x} coatings onto copper substrates from a nitrate solution. Scanning electron microscopy (SEM) images of TbO_{2-x} confirmed flower-like structures in which the leaf-shaped nanosheets were around 1 μm in width and 100 nm in thickness.

Wade et al. (1992) electrodeposited aluminum nitride ceramic in an electrolyte solution containing NH_3 and NH_4Br at an Al electrode and produced a mixture of $Al(NH_3)_6Br_3$ and $[Al(NH_2)(NH)]_n$. It has been noticed that calcination above 800°C results in the sublimation of $Al(NH_3)_6Br_3$ and transfer of the polymer ceramic precursor in $[Al(NH_2)(NH)]_n$. Metal nitride ceramic precursors (AlN and NbN) have been prepared by anodic electrophoretically deposition in liquid NH_3 electrolyte solutions on *n*Si substrates on applying a voltage higher than 10 V/cm between electrodes. Later, Wade and Crooks (1996) electrophoretically deposited AlN coatings onto Pt substrates. They used poly(acrylic acid)-coated AlN powders emulsified in ethanol. Anode electrophoretic deposition resulted in a $[CH_2CH(COOH)]_n/AlN$ coating, which was sintered at 1850°C in N_2 to yield an AlN coating 1100°C.

The cathodic electrodeposition process has been used to produce Au particles on a titanium dioxide substrate, Ag/AgCl reference electrode, and a Pt counterelectrode in a sodium acetate solution. SEM results showed that with deposition time both particle size and density increased. It has been noticed that a deposition time below 1 s generated particle sizes below 10 nm while a 1 min. deposition time resulted in particle sizes above 150 nm (Baeck et al., 2005).

Switzer (1986) deposited Tl_2O_3 films by anodic oxidation of an alkaline Tl(I) acetate solution at current densities of 1–10 mA/cm^2 on a titanium electrode. Sakai et al. synthesized a new compound $Pb_8Tl_5O_{24}$ with a cubic fluorite structure by anodic oxidation of Tl(I) to Tl_2O_3 and Pb(II) to PbO_2 on a platinum plate potentiostatically.

By using both galvanostatic and potentiostat techniques V_2O_5 has been deposited on different substrates, such as stainless steel, glassy carbon, platinum, nickel, graphite, and SnO_2 glass, by anodic oxidation of a saturated aqueous NH_4VO_3 solution (Andrukaitis et al., 1988).

The inert cathodic electrochemical deposition of zirconia and yttria-doped zirconia films has been achieved by using $ZrOCl_2$ and YCl_3 salts in an ethyl alcohol-water solvent (Zhitomirsky and Petric, 2002). Recently, Sharma et al. (2015) synthesized copper sulfide-selenide ($CuS_{0:2}Se_{0:8}$) thin films on a fluorine-doped tin oxide (FTO)-coated glass substrate in a $CuSO_4$, SeO_2, and $Na_2S_2O_3$ solution and stainless steel substrates using the electrodeposition technique. SEM showed nanoplatelet structures of a thickness of 140 to 150 nm. The thickness of the nanoplatelets is in the range of 140 to 150 nm. The nanoplatelets in the film are randomly spread, and they collide with each other. The average thickness of an electrodeposited $CuS_{0:2}Se_{0:8}$ thin film was 975 nm.

Cho and his research group (1995, 1996a, 1996b, 1997a, 1997b, 1997c, 1997d, 1997e) electrodeposited $CaWO_4$, $SrWO_4$, and $Ca(1_x)Sr_x$ WO_4 on tungsten substrates and $CaMoO_4$, $SrMoO_4$, and $BaMoO_4$ films on molybdenum foils. The development of barium molybdate films in N_2 gas at a current density of 1 mA/cm^2 on a metallic molybdenum substrate in a barium hydroxide aqueous solution has been studied by Xia and Fuenzalida (2003). Nickel-zinc ferrites $Ni_{0.5}Zn_{0.5}Fe_2O_4$ thin films have been synthesized via electrodeposition from an aqueous solution containing $NiSO_4$, $ZnSO_4$, and $(NH_4)_2Fe(SO_4)_2$ at constant currents ranging from 0.25 to 1 A. The ferrites were annealed at different temperatures from 800°C to 1000°C. A SEM micrograph showed an octahedral-like structure. XRD results showed the size of the films increased with an increasing annealing temperature. Hence, it increased from 27 nm at 800°C to 32 nm at 850°C and then to 39 nm at 1000°C (Saba et al., 2012).

2.4.3 The Combustion Method

This technique was discovered in the former Soviet Union by Merzhanov et al. (1971). The growth of this method led to the emergence of a new scientific route that incorporates the aspects of ignition of materials science. Combustion synthesis is an efficient method to produce a broad variety of superior materials that contain products of ceramics, intermetallics, and composites. This method is founded on the gelling and combustion of an aqueous solution containing metal salts and some organic fuel. The combustion process involves an exothermal redox reaction. The reaction promoted the disintegration of the overinflated precursor and produced a large volume of gasses and after calcination yielded a nanocrystalline powder. The combustion technique was controlled by the size of the container and the mass of the mixture. Studies carried out by Kingsley and Patil (1988) confirmed that the mass/volume ratio is critical for the occurrence of combustion synthesis, as the containers of 300 mL containing compositions less than 5 g did not undergo the ignition process. The researchers improve the existing synthesis methods and develop new resourceful technologies by increasing requirements for the quality and operation properties of nitride ceramics (Merzhanov and Borovinskaya, 1972, 1975, 1990).

Tyagi et al. (2007) have used a combustion process to prepare Gd_2O_3 and $YCrO_3$ using $Y(NO_3)_3$, $Gd(NO_3)_3$, $Fe(NO_3)_{39}H_2O$, and glycine as starting materials. The average crystallite size of nanocrystalline $YCrO_3$ was found to be 36 nm. The transmission electron microscopy (TEM) microstructure also revealed the presence of an onion-like structure made of concentric rings. The crystallite sizes of $GdFeO_3$ were in the range 40 to 65 nm.

Calcium phosphate has been prepared by combustion synthesis that contained $Ca(NO_3)_2 \cdot 4H_2O$ and $(NH_4)_2HPO_4$ salts. The particles were observed with an average particle size of 0.45 mm (Tas et al., 2000). γ-Alumina powders were synthesized by a solution combustion reaction between aluminum nitrate as the oxidizer and urea and glycine fuel as a reducing agent. γ-Alumina has an extremely porous microstructure with a particle size of 5 nm (Baburao et al., 2013). Layek and Verma (2012) synthesized $BiFeO_3$ particles using bismuth nitrate, ferric nitrate, and citric acid by a combustion method without using any solvent. These particles are rhombohedral with a distorted perovskite structure with an average particle size of 47 nm.

2.4.4 Coprecipitation

In this process, the required metal cations are coprecipitated as hydroxides, oxalates, carbonates, formates, or citrates from a standard medium (Zhang et al., 2006). These precipitates are subsequently calcined at appropriate temperatures to powder. Coprecipitation consists of the dissolution of the compound salt precursor in the aqueous media and subsequent precipitation of material from the solution by varying the pH. The coprecipitation involves the formation of multicomponent ceramic oxides using hydrous oxides or oxalates through the intermediate precipitates so that an intimate mixture of components is formed during precipitation and chemical homogeneity is maintained on calcination.

Balmer et al. (1995) used the coprecipitation route to develop ZrO_2/Al_2O_3 ceramics using aluminum nitrate hydrate and zirconium acetate. The author has observed tetragonal zirconia and γ-Al_2O_3 phases at temperatures below 1100°C and grain sizes ranging between 40 and 100 nm. At 1200°C large and thin plate-like Al_2O_3 grains containing nanometer zirconia inside the zirconia matrix are observed. Recently, Han et al. (2015) reported Al_2O_3/ZrO_2 powders synthesized from coprecipitation using NH_4HCO_3, NH_4OH, and $(NH_3)_2CO_3$ as precipitant agents. Al_2O_3/ZrO_2 showed spheres with diameters of 47 nm. Geethalakshmi et al. (2012) used zirconium oxychloride octahydrate ($ZrOCl_2 \cdot 8H_2O$) and sodium hydroxide (NaOH) as starting materials for the synthesis of zirconium dioxide. The results indicated that on calcination of ZrO_2 at different temperatures (700, 1000, and 1200°C/h) the average grain size was estimated to be 25 nm, 42 nm, and 46 nm, respectively. $LiNbO_3$ ceramic was prepared in an aqueous mixture of ammonium carbonate along with ammonium hydroxide. $LiNbO_3$ powders were sintered at 600°C, 700°C, 800°C, 900°C, and 1000°C. The particle size of $LInBo_3$ ceramic sintered at 800°C was shown to be 68 nm (Prakash et al., 2012). The coprecipitation process had been used for the synthesis of $SrBi_2Nb_2O_9$ in the mixture of niobium (V) oxide, bismuth (III) nitrate, and strontium chloride. The average grain size was found to be 4 μm (Gaikwad et al., 2005). Joseph et al. (2006) synthesized polycrystalline $Zn_{0.95}Cr_{0.05}O$ with ZnO and Cr_2O_3 powders. The crystallite sizes obtained on annealing at different temperatures were 24 nm, 33 nm, 38 nm, 40 nm, and 43 nm at 400°C, 500°C, 600°C, 700°C, and 900°C, respectively. Tabrizi and Nassaj (2015) prepared an Al_2O_3-CeO_2 composite by using $AlCl_3 \cdot 6H_2O$, Al powder, $CeCl_3 \cdot 7H_2O$, and NH_4OH. The results exhibited transition phases of alumina to α-Al_2O_3 take place at 1200°C. TEM images showed that the powders were composed of nanoparticles in the range of 30–70 nm. SEM results of sintered Al_2O_3-CeO_2 showed a submicronic/micronic microstructure.

$BaFe_{0.5}Nb_{0.5}O_3$ powders were prepared using barium chloride, ferric nitrate, niobium chloride, and citric acid anhydrous. It has been observed that the pure perovskite phase was obtained by the calcination at 850°C. SEM noted an average grain size of the ceramic of about 1.53 μm (Charoenthai and Traiphol, 2011). Barbier et al. (2009) reported the synthesis of $CaCu_3Ti_4O_{12}$ ceramics via the coprecipitation method. Metal chlorides ($CaCl_2$, $TiCl_3$, and $CuCl_2$) and oxalic acid were used as the precipitation agent. The morphology exhibits an average grain size of about 30 nm. The density of the films is 4.8 g cm^{-3}.

Yttrium oxide nanopowder has been prepared using yttrium nitrate and ammonium hydroxide and ammonium hydrogen carbonate. TEM results revealed the spherical shape of yttria particles and particle sizes were about 40–80 nm (Khajelakzay et al., 2015). $BaFe_{12}O_{19}$ was synthesized by coprecipitation in a solution of $NaOH/Na_2CO_3$, $FeCl_3$, and $BaCl_2$. Particles with a hexagonal morphology and particle sizes between 1 and 2 nm were observed.

MgO-ZrO_2 and Y_2O_3-MgO-ZrO_2 ceramic powders were synthesized by the method of coprecipitation. Magnesium chloride, yttrium chloride, and zirconium oxychloride were used as starting materials. The presence of clusters of rounded morphology with fine particles was observed. The average size of clusters was 20 to 30 nm, and the concrete surface was in the range of 60 to 70 $m^2\ g^{-1}$ (Yamagata et al., 2014).

The coprecipitation reaction of $Zr(OCl)_2$ and YCl_3 in a $NaNO_3$-KNO_3 eutectic has been carried out at 450°C to produce Y_2O_3-ZrO_2 ceramic. The specific surface areas were calculated from 110 $m^2\ g^{-1}$, and particles of sizes 6–10 nm have been observed (Descemond et al., 1993).

2.4.5 The Spray Pyrolysis Method

The spray pyrolysis method is based on the ultrasonic generation of micrometric-size aerosol droplets, and the evaporation, precipitation, drying, and decomposition occur in a dispersed phase decomposition at intermediate temperatures (400°C–800°C). The spray pyrolysis process makes it possible to control particle properties such as size, morphology, and chemical composition. The particles are generated through the thermally induced processes of nuclei formation, collision, and coalescence after confirming the processes of solvent evaporation/drying and solute precipitation and decomposition (Koch et al., 1990; Castaneda et al., 2003; Lopez Ibanez et al., 2004). Yang et al. (2015) synthesized $SrTiO_3$, $SrTi_{0.5}Zr_{0.5}O_3$, and $SrZrO_3$ ceramic composites by the spray pyrolysis method using strontium nitrate and zirconium nitrate solution. The microstructure shown in Fig. 2.2 was obtained after sintering at 1250°C.

Figure 2.2 SEM micrographs of prepared (a) $SrTiO_3$, (b) $SrTi_{0.5}Zr_{0.5}O_3$, and (c) $SrZrO_3$ ceramics at 1250°C.

Cerium (IV) ammonium nitrate, cerium (III) nitrate hydrate, and cerium (III) acetate hydrate were used to prepared ceria powders CeAN, CeA, and CeN. TEM image results revealed that particles obtained from CeAN are all solid spherical. This may be because ammonium nitrate acts as a surfactant to form micelles and because of the good solubility of CeAN in an aqueous solution (Shiha et al., 2012).

Spray pyrolysis was used to produce ZrO_2 and CeO_2 particles using zirconium hydroxyl acetate and cerium acetate. ZrO_2 particles are spherical in shape, and CeO_2 particles are bowl-like in shape with an average diameter of 100 nm (Chen et al., 2006). Tok et al. (2006) reported the prepration of Al_2O_3 particles with a size range of 5 to 30 nm. Anhydrous $AlCl_3$ particles were used to prepare the Al_2O_3 particles. The particle size ranged from 10 to 30 nm. Martin et al. (2010) used aluminum nitrate nonahydrate for the preparation of Al_2O_3 particles. TEM images showed particles of 480 nm diameter. The presence of an 8–11 nm thick crust was observed on the particle surface. Kim et al. (1998) prepared TiO_2 and $BaTiO_3$ powder by ultrasonic spray pyrolysis from a metal alkoxide solution in the temperature range of 400°C–900°C. The particles size was about 0.02 nm. Jokanovic et al. (1996) prepared alumina from $AlCl_3$ and $Al(NO_3)$. The particle size was about 0.3 nm. Ceramic particles of titania-toughened alumina (TTA) and zirconia-titania-toughened alumina (ZTTA) have been perpared at various temperatures using solutions of zirconium, titanium, and aluminum. TTA particles were spherical shaped with the average particle size 2 nm. ZTTA particles were of a diameter smaller than 1 nm, and a surface area of 3.434 m^2/g was determined (Shim and Lee, 2002). Zirconia oxide (ZrO_2) nanofibers were synthesized using a zirconia substrate at 900°C for 1 h in N_2 gas and zirconia oxide calcined at various temperatures (400°C, 600°C, 800°C, and 1000°C). The size of the particles obtained at 400°C was 19 nm and increased to 20 nm at 600°C, 35 nm at 800°C, and 85 nm at 1000°C, respectively (Zhang et al., 2014).

2.5 Ceramic Properties

The ceramic materials are the most advanced in scientific knowledge as well as in commercial applications. Ceramics have prime physical properties, including electrical, thermal, magnetic, and mechanical.

2.5.1 Magnetic Properties

Magnetic materials are made from ceramic magnetic materials, which are widely used in various applications of industrial and medical equipment. Ceramics are tremendously high, showing great rigidity under compression and bending. The strength of ceramic materials is determined by the bend strength, the amount of pressure required to deform a material. The magnetic properties, including coercivity (H_c), magnetic saturation (M_S), and magnetic remnance (M_r), all are important magnetic characteristics of the material. M_r is the remaining magnetization of a ferromagnetic material when the applied magnetic field is removed. The material could generate maximum magnetization M_S. The composition of the magnetic material and the processing, such as the nature of atoms, electron structure, and crystal structure, are dependent on both M_r and M_s. H_c is the intensity of the external field required to remove all the magnetization of the material of the external field. The magnetic properties of ceramic materials are classified by their response to an external magnetic field as diamagnetic, paramagnetic, and ferromagnetic.

Mazumdar and Hossain (2012) prepared Y-type hexagonal ferrites of $Ba_2Ni_{2-x}Zn_xFe_{12}O_{22}$ to study their magnetic properties. Magnetization has been measured using a superconducting quantum interference device magnetometer. Magnetization is a function of the applied magnetic field *M–H* for polycrystalline $Ba_2Ni_{2-x}Zn_xFe_{12}O_{22}$ at room temperature (300 K). The results revealed that magnetization of $Ba_2Ni_{2-x}Zn_xFe_{12}O_{22}$ increases linearly when the applied magnetic field is increased up to 0.1 T and conquers its saturation value for fields higher than 1.5 T. It is clear from the *M–H* curve that polycrystalline $Ba_2Ni_{2-x}Zn_x Fe_{12}O_{22}$ compositions are in the ferrimagnetic state at room temperature.

$SrFe_{12}O_{19}$ + $12SrB_{1.5}Bi_{0.5}O_4$ ceramics have been tested for magnetic properties by Zaitsev et al. (2004). It has been observed that the coercive field of the samples increases from 188 to 504 kA/m as the heat treatment temperature is raised from 600°C to 850°C. The coercive fields of the $SrFe_{12}O_{19}$ + $12SrB_{1.5}Bi_{0.5}O_4$ ceramic annealed at different temperatures and their mass magnetizations in a field of 720 kA/m are listed in Table 2.1. The magnetization was seen to increase with annealing temperature. It has also been investigated that the anneals at 850°C for 4 and 8 h had no effect on the mass magnetization of the samples but reduced their coercive fields to 476 and 440 kA/m, respectively.

Magnetic properties of $(Bi_{1-x}La_xFeO_3)1\text{–}y(PbTiO_3)y$ were first reported by Singh et al. (2008) in solid solutions. The authors observed symmetric magnetic hysteresis loops, with magnetization increasing with La^{3+} doping. Mishra et al. (2011) reported nonsaturated magnetic hysteresis loops for $(Bi_{1-x}La_xFeO_3)_{0.5}$ $(PbTiO_3)_{0.5}$ solid solutions with $x = 0, 0.5$ at low temperatures. The results inferred the magnetization measured in field-cooling and zero-field-cooling conditions, which point toward a spin-glass behavior of the solid solution with $x = 0.2$.

Table 2.1 Coercive field H_c and mass magnetization M at 720 kA/m in glass samples of nominal composition $SrFe_{12}O_{19}$ + $12SrB_{1.5}Bi_{0.5}O_4$ annealed at different temperatures

t_{ann} (°C)	M (A m^2/kg)	H_c (kA/m)
600	5.6	188
630	6.1	268
700	7.0	392
750	6.3	452
800	7.9	472
850	9.8	504

Markiewicz et al. (2013) reported magnetic measurements of $(Bi_{1-x}La_xFeO_3)_{0.5}(PbTiO_3)_{0.5}$ ceramics using a vibrating sample magnetometer probe mounted on a quantum design physical property measurement system. Magnetization measurements were performed for $(BiFeO_3)_{0.5}(PbTiO_3)_{0.5}$ and $(Bi_{1-x}La_xFeO_3)_{0.5}$ $(PbTiO_3)_{0.5}$, with x = 0.2 at constant temperatures of 300, 200, 100, and 4 K. The $M(H)$ curves demonstrate linear dependence in a broad range of magnetic fields and indicate an antiferromagnetic ordering.

Ferrites are primarily used for permanent magnets, are of low cost, and offer good magnetic properties. Barium ferrites ($BaFe_{12}O_{19}$) have been tested for magnetic properties. It has been observed that for samples calcined for 1.5 h, the value of intrinsic coercivity decreases from 440 kA/m (5.5 kOe) calcined at 800°C to 162 kA/m (2.04 kOe) at 1100°C. This result shows that particle growth occurs with a consequent reduction of the intrinsic coercivity with the increase of the calcination temperature (Janasi et al., 2002). Mallik et al. (2012) studied magnetic properties of cobalt ferrite. The magnetic coercivity was found to be 926 Oe and the magnetic saturation value about 42 emu/g. It has been observed that with the increase in the calcination temperature the magnetic coercivity value decreases and increases in the magnetic saturation value.

Nehal (2015) characterized magnetic properties of various ferrite ceramics such as Zn ferrite ($ZnFe_2O_4$), hematite (Fe_2O_3), and Ba-hexaferrite ($BaFe_{12}O_{19}$). An increase in saturation magnetization from 66.5 emu/g for heat-treated Zn ferrite and hematite and 19 emu/g for Ba-hexaferrite ($BaFe_{12}O_{19}$) and hematite was measured with a vibrating scanning magnetometer. The results confirm that a recycle rolling mill scales into hard and soft magnetic glass ceramics with the magnetic phase content reaching 92.4 wt%. Wu et al. (2012) obtained $BiFeO_3$ with a saturation magnetization of ~2.914 emu/g, which showed that the crystallized structure can also increase the ferromagnetism of $BiFeO_3$. As represented in Fig. 2.3a, the magnetization versus magnetic field (*M–H*) curves were considered at room temperature using the vibrating sample magnetometer (VSM). Loops of all samples exhibited apparent hysteresis, and the coercive field was S1 = 10 Oe, S2 = 140 Oe, S3 = 5 Oe, and S4 = 18 Oe. An unyielding saturation magnetization of 2.914 emu/g was obtained in subcrystallized $BiFeO_3$, which was much larger, as shown in Figs. 2.3a and 2.3c. In all the samples, zero-field-cooled and field-cooled magnetization curves exhibited spin-glass freezing behavior.

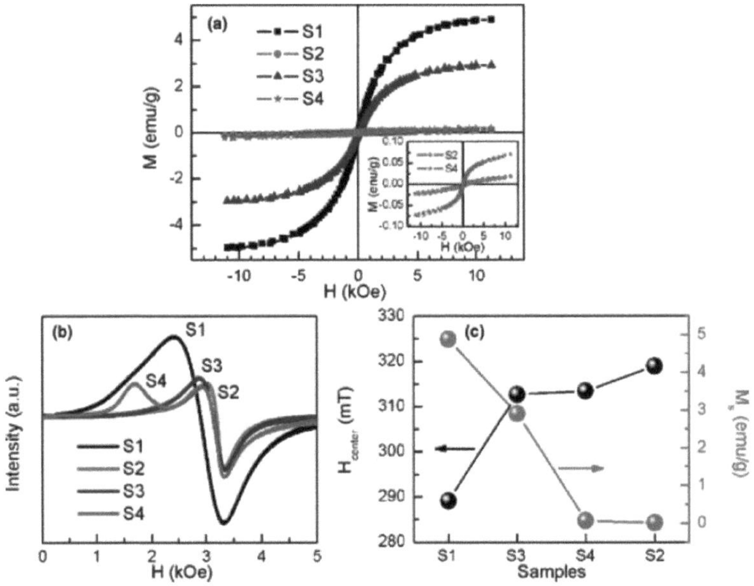

Figure 2.3 (a) The *M–H* curves of S1, S2, S3, and S4 measured at RT. (Inset) The magnified view of S2 and S4. (b) ESR spectrums of as-prepared samples (S1, S2, S3, and S4). (c) The calculated resonance field and saturation magnetization (M_S) for samples S1–S4.

Recently, Li et al. (2015) measured magnetic properties of β-Co_2P nanorods performed in the temperature range of 5 to 300 K under 500 Oe on a commercial superconducting quantum interference device. It has been observed that for β-Co_2P, the value is significantly lower than that expected for uncoupling cobalt ions. It has been observed that between the adjacent cobalt ions through P ions there is a strongly coupled antiferromagnetic interaction. There is a field dependence of the magnetization of β-Co_2P nanorods at 5 K. The *M–H* curve indicates that this material comprises a combination of weak magnetic and antiferromagnetic constituents called the canted antiferromagnetic state Bahgat et al. (2014). Strontium hexaferrite is the popular materials for permanent magnets. The authors measured magnetic values at 300 and 4.2 K, which replicate excellent magnetic behavior. The calculated energy product values for the formed samples ranged from 0.3 to 2.4.

Emad et al. (2008) reported the preparation of $Mn_{0.8}Zn_{0.2}$ Fe_2O_4. It was calcined at 1050°C, 1100°C, 1150°C, and 1200°C and measured for magnetic properties at room temperature in a maximum field of 15 kOe. It has been concluded that the rising in the annealing temperature increases the M_s properties of the compacts from 44.89 emu/g at 1050°C to a maximum value of about 58.19 emu/g at 1150°C. At low temperature (1050°C) a small M_s value was observed due to the increase in the presence of a magnetic species of Fe_2O_3. An unexpected decrease in the M_s (49.7 emu/g) at 1200°C was observed as the annealing temperature increases, which was attributed to the nucleation and grain growth by a reduction in density.

Magnetic properties of $NiFe_2O_4$ powders were reported by Georgea et al. (2006). It has been observed that the specific saturation magnetization of the $NiFe_2O_4$ decreased as the grain size decreased. As the annealing process increased the grain size, the coercivity was found to rise to a maximum value and then decrease.

Layek and Verma (2012) that observed $BiFeO_3$ is an antiferromagnet below 370°C. The increase in the magnetization leads to a turn down in the particle size. The coercive field value was calculated to be 180 Oe from the hysteresis loop. It has been observed that at room temperature magnetic hysteresis shows weak ferromagnetism though the magnetization does not saturate up to 1.75 T applied field.

Akhtar et al. (2014) showed that magnetic properties of the aluminum-substituted yttrium iron garnet (YIG) ferrites depend on particle size, superexchange interaction, and magnetic dilution. It has been observed that above Curie temperature (T_c) a ferromagnetic material becomes paramagnetic. YIG produces heat due to both Neel and Brown relaxations when exposed to an external magnetic field. Magnetic possessions of aluminum-substituted YIG ($Y_3Fe_{5-x}Al_xO_{12}$) were reported by Ortega et al. (2014). Authors observed that due to the enhancement of the surface spin effects, the saturation magnetization increased as the Fe/Al ratio was raised. Particle efficiency increases, ranging from 91.4% to 95.9%, as the Fe/Al ratio decreases.

Khanahmadzadeh et al. (2012) indicate a superparamagnetic behavior of nickel titanate ($NiTiO_3$)/polyimide. It has been observed that superparamagnetic particles do not show hysteresis in the *M–H* curves. Hence, H_C and M_R are near zero. The saturation magnetization is 0.2 emu/g. Magnetic properties of Mn-Zn ferrite were tested for saturation magnetization by Pissurlekar (2015). The saturation magnetization values are in the range 22.47–41.63 emu/g and hysteresis loss is found to be small for all the samples. T_c is in the range 473–558 K.

Ghosh and Dasgupta (2010) used nickel ferrite for magnetic measurements. The VSM graph of in Fig. 2.4a shows nanosized nickel ferrite (30 nm) calcined at 400°C carried out in a field of 1 T at 300 K. The graph shows a saturation magnetization (M_s) of 31.437 emu·g^{-1} for the first hysteresis loop. Magnetometric measurements (Fig. 2.4b) were carried out in a field of 6 T. It demonstrates a saturation magnetization of 43.36 emu·g^{-1}. The bulk $NiFe_2O_4$ exhibits a saturation magnetization of 50 emu·g^{-1}.

Mn-Zn ferrites were studied by Chao et al. (2015), and they obtained a saturation magnetization of 55.02 emu/g, a coercivity of 8.20 G, and a remanent magnetization of 1.71 emu/g. With the increase of pH coercivity decreased. The result was a saturation magnetization of 55.02 emu/g and a residual magnetization of 1.71 emu/g when pH was 7.5. The hysteresis curve of the samples indicate that the samples exhibit good ferromagnetic behavior.

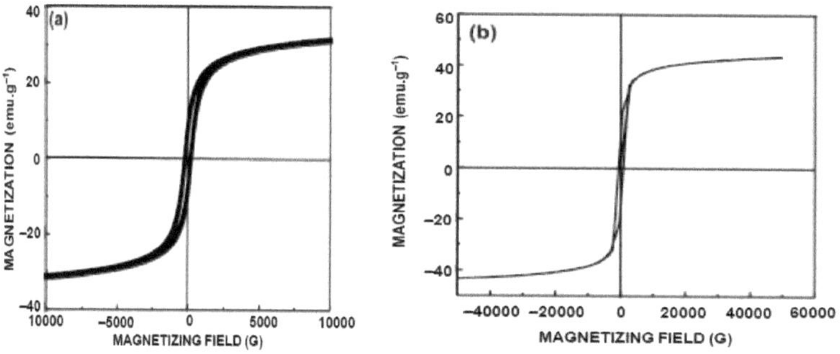

Figure 2.4 (a) VSM of $NiFe_2O_4$ at 1 T and (b) SQUID magnetometric measurement of the same sample at 6 T. Measurements were done at 300 K.

2.5.2 Thermal Properties

Ceramic materials are essential in everyday life because of their flexible structural features. Other than these properties, they do play a significant role because of their physical properties. The thermal properties of ceramic materials are diverse and have uses in different applications. Thermal conductivity transports heat energy from a high-temperature region to a low-temperature region. Thermal conduction is engendered by the struggle of electrons and the transfer of lattice vibrations. Metals with little electrical resistance and crystals in which lattice vibrations are transferred easily display high thermal conductivity. Thermal conductivity values for some ceramic materials are displayed in Table 2.2. The thermal conductivities range from approximately 2 to 50 W/m-K at room temperature. Other amorphous ceramics and glass have lower conductivities than crystalline ceramics. The phonon scattering is much more efficient when the atomic structure is highly tangled and irregular (Kingery et al., 1976).

Aggarwal et al. (2012) conducted thermal conductivity of $SrTiO_3$, $BaTiO_3$, and $KTaO_3$ ferroelectric perovskites. The results are evident that thermal conductivity decreases with the increase of temperature in $SrTiO_3$, $BaTiO_3$, and $KTaO_3$. This strange increase in the value will reduce phonon scattering by half, and hence thermal conductivity will be anomalously large near T_c. The effect of an increasing electric field is to increase the total thermal conductivity. The rise in thermal conductivity is noticeable in the vicinity of T_c for all three cases. In the vicinity of T_c, the thermal conductivity increases anomalously in agreement with the maximum in conductivity versus the temperature curve for $SrTiO_3$ near 32 K (T_c). Sun et al. (2013) studied the stability and thermal conductivity of ytterbia- and yttria-codoped zirconia. Results revealed that codoped samples showed much lower thermal conductivities than those single-doped samples. Among them, Yb_2O_{3+3} Y_2O_3 codoped ZrO_2 showed the lowest thermal conductivity, which was more than 20% lower than that of Y_2O_3-ZrO_2. One of the possible reasons is that defect clusters reduce the heat conductivity indeed, as shown in Fig. 2.5. For the single-doped samples, the thermal conductivity of 5Yb is slightly lower than that of 4.55Y. The codoped samples show much lower thermal conductivities than the single-doped sample.

Table 2.2 Thermal conductivity values for some ceramic materials

	c_p	α_l	k	L
Material	**(J/kg-K)**	**[(°C)$^{-1}$ × 10^{-6}]**	**(W/m-K)**	**[Ω-W/(K)^2 × 1^{-8}]**
Alumina	775	7.6	39	—
Magnesia	940	13.5	37.7	—
Spinel	790	7.6	15.0	—
Fused silica	740	0.4	1.4	—
Soda-lime glass	840	9.0	1.7	—
Borosilicate (Pyrex)	850	3.3	1.4	—

Zheng et al. (2012) synthesized Y_6UO_{12} to evaluate the thermal conductivity (k) at room temperature to 1173 K using the laser flash method. It has been observed that the k of Y_6UO_{12} reduced with increasing temperature. The room temperature k value of Y_6UO_{12} was 4.90 Wm. With increasing temperature in the whole temperature range the k of Y_6UO_{12}, Y_6WO_{12}, and Yb_6WO_{12} decreased, which indicates that the phonon contribution to the thermal conduction was predominant. Since Y_6UO_{12}, Y_6WO_{12}, and Yb_6WO_{12} treated were as insulators, the measured k was practically equal to the lattice thermal conductivity.

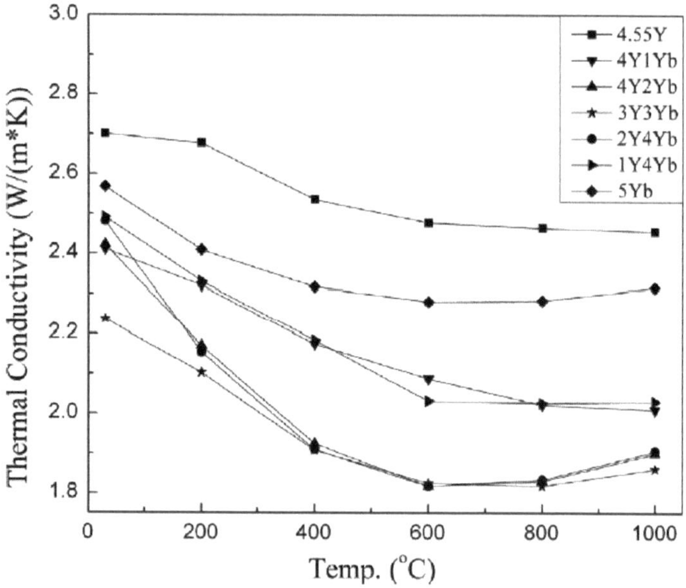

Figure 2.5 Thermal conductivities of Yb_2O_3-Y_2O_3-ZrO_2 at various temperatures.

$Sm_2(Zr_{0.6}Ce_{0.4})_2O_7$ ceramic thermal conductivity was measured by Zhang et al. (2009). The results revealed that the synthesized ceramic had a pure fluorite structure. It has been investigated that the average thermal conductivity of $Sm_2(Zr_{0.6}Ce_{0.4})_2O_7$ was lower than that of yttria-stabilized zirconia. It was seen that with increasing temperature the thermal conductivity of $Sm_2(Zr_{0.6}Ce_{0.4})_2O_7$ and $Sm_2Zr_2O_7$ decreases gradually from 1.96 to 1.52 W/m-K. $Sm_2(Zr_{0.6}Ce_{0.4})_2O_7$ showed a higher thermal conductivity as compared to $Sm_2Zr_2O_7$, which can be credited to the crystal structure transformation from pyrochlore to fluorite.

2.5.3 Electrical Properties

Electrical properties allow an electrical charge to move from atom to atom in a particular material. Ceramics are insulating materials in general, but some assortments exhibit electrical conductivity according to changes in temperature. Materials are categorized on the basis of their electrical properties as conductors, semiconductors, and insulators. New to this group are superconductors. The electrical conductivity of a material is defined in terms of ease of charge flow through it. The charge that flows comprises electrons, ions, charged holes, and their combinations. Electrical properties might include capacitance, conductivity, resistivity, dielectric constant, permittivity, piezoelectric constant, Seebeck coefficient, permeability, etc. The effect of the electric field on the nature of the *PE* curve, saturation, and remanent polarization on strontium barium niobate was studied by Kulkarni et al. (2008). The ferroelectric hysteresis loop (*PE*) was recorded with a frequency of 100 Hz and an applied pulse voltage at the maximum electric field of 15 kV/cm. It has been observed that the *PE* loop shows a low value of polarization. It may be due to the occurrence of abnormal grain growth and linked microcracks, which result in the loss of polarization due to leakage of the material. A leakage test is carried out by applying a constant DC bias voltage across the sample. This shows the highest leakage current of 10^{-5} A to 10^{-7} A.

Dielectric and ferroelectric properties of strontium barium niobate have been reported by Patro et al. (2007). The studies revealed the broad dielectric peak indicates the second-order phase transformation, that is, characteristics of ferroelectrics. Ferroelectric hysteresis loops at an electric field of 15 kV/cm at different frequencies were observed. The maximum saturation polarization (P_s) observed for CPAH1350 was 5.6 C/cm^2. Similarly, the P_s of CPAH1250 and CPAH1300 were 2.93 C/cm^2 and 4.03 C/cm^2 for 1 Hz frequency, respectively. The leakage current varied from 10^{-6} to 10^{-11} A for synthesized strontium barium niobate. The hysteresis in the curve shows the charge storage nature of ferroelectrics. Leakage current tests are done on the samples at the applied electric field of 15 kV/cm at the soaking time and the measurement time of 1000 ms. A leakage current of the order of 10^{-6} A is observed. Dielectric properties of $Ba_5GdTi_3V_7O_{30}$ were investigated at different temperatures (33°C–500°C) in a wide frequency range (102–106 Hz). Ferroelectric-paraelectric phase transition was confirmed by the appearance of hysteresis loops. At an applied electric field of 5.6 kV/cm the polarization was found to be 2Pr = 0.0539 μC/cm^2, respectively. An accurate hysteresis loop was not obtained because of the lossy characteristics of the material (Mohanty et al., 2012). Polycrystalline $Ba_5MTi_3A_7O_{30}$ (M = Ce, Pr, Nd, Sm, Gd, Dy, and Bi; A = Nb and Ta) ceramics were tested for dielectric properties in the frequency range 100 Hz–1 MHz from room temperature to 250°C. At 1 MHz, the loss tangent (tan δ) varies from 10^{-3} to 10^{-2} and the dielectric constant (ε_r) varies from 164 to 887. The study indicates a ferroelectric-paraelectric diffuse phase transition in $Ba_5MTi_3Nb_7O_{30}$ (M = Ce, Pr, Nd, Sm, Gd, Dy, and Bi) ceramics above room temperature. But $Ba_5MTi_3Ta_7O_{30}$ (M = Ce, Pr, Nd, Sm, Gd, Dy, and Bi) ceramics show a linear variation of ε_r at room temperature. The measured dielectric properties show that the above materials are feasible candidates for the production of discrete multilayer capacitors in microelectronic machinery (Bijumon et al., 2004). Chaudhari and Bichile (2013) studied the electrical properties of $PbTiO_3$, which indicate that at high temperature the material exhibits conduction due to bulk and grain boundary effects. The result shows that the peaks shift toward the high-frequency side, indicating a relaxation process in the system. The peaks' broadening suggests the presence of a temperature-dependent relaxation process in the system.

$Ba_4SrRTi_3V_7O_{30}$ (R = Sm and Dy) ceramics has been recorded by an impedance analyzer over a wide frequency range (103–106 Hz). Detailed studies of the dielectric properties suggest that they have endured ferroelectric-paraelectric phase transition well above room temperature (i.e., 432°C and 355°C at frequency 100 kHz). These ceramics exhibit high dielectric constants and low dielectric losses at a higher temperature. The occurrence of ferroelectricity in the materials is also supported by a substantial contribution of electron-phonon anharmonicity (Sahoo et al., 2008). The polycrystalline Ba_5SmTi_3 V_7O_{30} material's electrical properties were studied by using a complex impedance spectroscopy. The results revealed that with a rise in temperature the relative dielectric constant of the material increases. The bulk resistance of the material decreases with increasing temperature, exhibiting a typical negative temperature coefficient of resistance behavior. The mixed-type (ionic-polaronic and space charge) activation energy has been observed in the temperature variation of conductivity. The polarization study confirmed the existence of ferroelectricity in the compound (Sahoo et al., 2011).

$Ba_3Sr_2LaTi_3V_7O_{30}$ (BSLTV) tungsten-bronze (TB) has been investigated by Sahoo et al. (2014). The temperature dependence of Nyquist plots of BSLTV was measured at various temperatures (300°C–5000°C). At a high temperature the single semicircular arcs detected in the Nyquist plot of this compound show that the electrical response is maybe due to the grain effect. It is noted that the AC conductivity of the material increases with a rise in temperature. The estimated value of activation energy of the compound was found to be 0.0418 eV, 0.2653 eV, 0.0419 eV, and 0.2895 eV at 10 and 100 kHz in the para- and ferroelectric regions. The polycrystalline ceramic of $Sr_5EuCr_3Nb_7O_{30}$ tested for the dielectric constant (ε) and tangent loss (tan δ) measured as a function of temperature (300–700 K). The measurement of the conductivity suggests that ceramics have a negative temperature coefficient of resistivity. It was observed that the value of ε decreases monotonically with increasing f, with ε = 115 at f = 0.1 kHz and ε = 55 at f = 100 kHz. It was observed that tan δ decreases from a value of 0.51798 at f = 0.1 kHz to 0.02027 at f = 102 kHz, similar to the ε value (Raju et al., 2003).

2.5.4 Mechanical Properties

Mechanical properties are essential in structural and building materials. Numerous properties used to explain the strength of materials, such as elasticity/plasticity, tensile strength, compressive strength, fracture toughness and ductility, and hardness. Mechanical properties are determined through a series of standardized mechanical tests and arise as a result of the physical properties built into each material. Mechanical properties of titanium aluminum carbide (Ti_3AlC_2) ceramics have been tested by Zhou and Wang (2003). Fracture toughness was measured by the single-edge notched-beam method, and flexural strength was measured by using the three-point bending method with a span length of 30 mm and a crosshead speed of 0.5 mm/min. The density was 4.24 ± 0.02 g/cm^3, the flexural strength was 552 ± 30 Mpa, and the fracture toughness (K_{IC}) was 9.1 ± 0.3 MPa·m$^{1/2}$. It has been observed that high fracture toughness gives high damage tolerance proprieties to Ti_3AlC_2. Ganguly et al.

(2004) prepared Ti_3GeC_2, $Ti_3Si_{0.5}Ge_{0.5}C_2$, and $Ti_3Si_{0.75}Ge_{0.25}C_2$ to study the mechanical properties. The hardness values of both solid solution compositions (2.5 ± 0.2 GPa) were in between those of Ti_3SiC_2 (3.0 ± 0.3 GPa) and Ti_3GeC_2 (2.2 ± 0.5 GPa). From the results it has been concluded that no solid solution strengthening occurs in this system and all samples were smash tolerant and thermal shock resistant. A 300 N Vickers indentation in a 1.5 mm thick, four-point bend bar decreases its strength by anywhere from 25 to 35%. It has been observed that compressive strengths decrease with increasing Ge content. With increasing applied load, the Vickers microhardness values initially decrease but then asymptote at higher loads, as shown in Fig. 2.6. The asymptotic values (2.5 ± 0.2 GPa) for the two solid solution compositions are statistically indistinguishable. The grain sizes and hardness values explored fall in between those of Ti_3SiC_2 (3 ± 0.25 GPa) and Ti_3GeC_2 (2.2 ± 0.5 GPa), which indicates that no solid solution hardening effect is operative in this system.

MgO-ZrO_2 and Y_2O_3-MgO-ZrO_2 ceramic was tested for mechanical properties using the Vickers indentation technique by Yamagata et al. (2014). It has been observed that the addition of Y_2O_3 leads to an increase in hardness and a decrease in fracture toughness of ceramics. Vickers hardness is greater than 10 GPa and fracture toughness is 6.63 and 3.01 for Mg-PSZ and Y-Mg-PSZ, respectively. Zhang et al. (2004) studied the hardness and toughness of carbon nanotube alumina (CNT-Al_2O_3) composites. The maximum density of 3.14 g/cm^3, fracture toughness of 4.7 MPam$^{1/2}$, and Vickers hardness of 9.98 GPa were obtained for a 7.39 wt.% CNT-Al_2O_3 composite. An 8.4% increase in hardness and a 21.1% increase in toughness is observed due to the addition of CNTs over that of the pure Al_2O_3. Yousefi et al. (2009) studied the mechanical properties of glass-ceramic composites in the SiO_2-CaO-MgO-(Na_2O) system. A 10% improvement in the mechanical properties was observed with the addition of aluminum titanate to the glass-ceramic matrix up to the fracture toughness of TiB_2 ± AlN ± SiC (TAS) composites. The higher fracture toughness (6.35. 0.74 MPa·m$^{1/2}$ and 6.49.0.73 MPa·m$^{1/2}$) was obtained in AlN and SiC in the TAS composites. A thermal expansion of 8.1, 4.0, and 4.03 × 10^{-60} C^{-1} has been observed in TiB_2, SiC, and AlN, respectively (Yue et al., 1999).

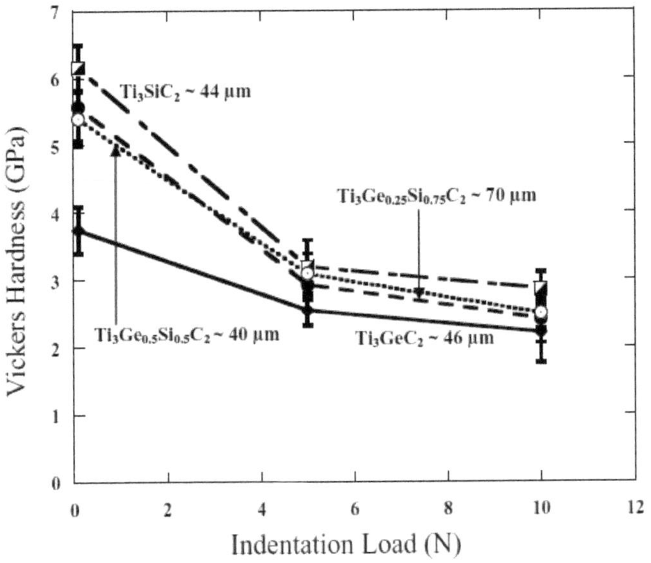

Figure 2.6 Vickers hardness versus indentation load for Ti_3GeC_2, $Ti_3Si_{0.5}Ge_{0.5}C_2$, Ti_3SiC_2, and CG $Ti_3Si_{0.75}Ge_{0.25}C_2$ samples.

References

Aggarwal GK, Kumar A, Naithani UC. (2012). Field dependent thermal conductivity of $SrTiO_3$, $BaTiO_3$ and $KTaO_3$ ferroelectric perovskites. *Inter. J. Eng. Res. Dev.*, **4**(1): 61–67.

Akhtar MN, Khan MA, Ahmad M, Murtaza G, Raza R, Shaukat SF, Asif MH, Nasir N, Abbas G, Nazir MS, Raza MR. (2014). $Y_3Fe_5O_{12}$ nanoparticulate garnet ferrites: comprehensive study on the synthesis and characterization fabricated by various routes. *J. Magn. Magn. Mater.*, **368**: 393–400.

Andriamasinoro D, Kieffer R, Kiennemann A, Poix P. (1993). Preparation of stabilized copper-rare earth oxide catalysts for the synthesis of methanol from syngas. *Appl. Catal.*, **106**: 201–212.

Andrukaitis E, Jacobs, PWM, Lorimer, JW. (1988). Vanadium pentoxide electrodes. I. Formation of vanadium pentoxide from anodic electrodeposits, *Solid State Ionics*, **27**(1–2): 19–30.

Baeck SH, Jaramillo TF, Shwarsctein AK, Farland E. (2005). Automated electrochemical synthesis and characterization of TiO_2 supported Au nanoparticle electrocatalysts. *Meas. Sci. Technol.*, **16**: 54–59.

Bahgat M, Awan FM, Hanafy HA, Alzeghaibi ON. (2014). Synthesis of hard magnetic material from secondary resources. World academy of science, engineering and technology. *Int. J. Chem. Mol. Nucl. Mater. Metall. Eng.*, **8**(9): 936–941.

Balmer ML, Lange FF, Jayaram V, Levi CG. (1995). Development of nano-nomposite microstructures in ZrO_2-Al_2O_3 via the solution precursor method. *J. Am. Ceram. Soc.*, **78**: 1489–1495.

Barbier B., Combettes C, Guillemet-Fritsch S, Chartier T, Rossignol F, Rumeaud A, Lebey T, Dutarde E. (2009). $CaCu_3Ti_4O_{12}$ ceramics from co-precipitation method: dielectric properties of pellets and thick films. *J. Eur. Ceram. Soc.*, **29**(4): 731–735.

Bijumon PV, Kohli V, Parkash O, Varma MR, Sebastian MT. (2004). Dielectric properties of $Ba_5MTi_3A_7O_{30}$ [M = Ce, Pr, Nd, Sm, Gd, Dy and Bi; A = Nb, Ta] ceramics. *Mater. Sci. Eng. B*, **113**(1): 13–18.

Bind JM, Dupin T, Schafer J, Titeux M. (1987). Industrial synthesis of coprecipitated $BaTiO_3$ powders. *J. Metals*, **54**: 60–61.

Bocchetta P, Santamaria M, Di Quarto F. (2007). Template electrosynthesis of $La(OH)_3$ and $Nd(OH)_3$ nanowires using porous anodic alumina membranes. *Electrochem. Commun.*, **9**: 683–688.

Buckley AM, Greenblatt M. (1994). Sol-gel preparation of silica gels. *J. Chem. Educ.*, **71**(7): 599.

Castaneda L, Alonso JC, Ortiz A, Andrade E, Saniger JM, Banuelos JG. (2003). Spray pyrolysis deposition and characterization of titanium oxide thin films. *Mater. Chem. Phys.*, **77**(3): 938–944.

Chang TX, Fuenzalida VM. (2003). Room temperature electrochemical growth of polycrystalline $BaMoO_4$ films. *J. Eur. Ceram. Soc.*, **23**, 519–525.

Chao X, Longjun X, Yongjun Y, Xiangyang L, Shuyun W. (2015). Influence of pH on properties of Mn–Zn ferrites synthesized from low-grade manganese ore. *Chin. J. Geochem.*, **34**(2): 219–223.

Charoenthai N, Traiphol R. (2011). Progress in the synthesis of $Ba(Fe_{0.5}Nb_{0.5})O_3$ ceramics: a versatile co-precipitation method. *J. Ceram. Proc. Res.*, **12**(2): 191–194.

Chaudhari VA, Bichile GK. (2013). Synthesis, structural, and electrical properties of pure $PbTiO_3$ ferroelectric ceramics. *Smart Mater. Res.*, **2013**: 147524(1–9).

Chaudhuri MG, Basu J, Das GC, Mukherjee S, Mitra MK. (2013). A novel method of synthesis of nanostructured aluminum nitride through sol-gel route by in situ generation of nitrogen. *J. Am. Ceram. Soc.*, **96**(2): 385–390.

Chen CY, Tseng TK, Tsai SC, Lin CK, Lin HM. (2006). Effect of precursor characteristics on zirconia and ceria particle morphology in spray pyrolysis. *Ceram. Int.*, **34**(2): 409–416.

Cho WS, Yashima M, Kakihana M, Kudo A, Sakata T, Yoshimura M. (1995). Room temperature preparation of the highly-crystallized luminescent $CaWO_4$ film by a electrochemical method. *Appl. Phys. Lett.*, **66**(9): 1027–1029.

Cho WS, Yashima M, Kakihana M, Kudo A, Sakata T, Yoshimura M. (1996). Room temperature preparation of the crystallized luminescent $Sr_{1-x}Ca_xWO_4$ solid-solution film by an electrochemical method. *Appl. Phys. Lett.*, **68**(1): 137–139.

Cho WS, Yashima M, Kakihana M, Kudo A, Sakata T, Yoshimura M. (1997b). Active electrochemical dissolution of molybdenum and application for room-temperature synthesis of crystallized luminescent calcium molybdate film. *J. Am. Ceram. Soc.*, **80**(3): 765–769.

Cho WS, Yashima M, Kakihana M, Kudo A, Sakata T, Yoshimura M. (1997e). Active electrochemical dissolution of molybdenum and application for room-temperature synthesis of crystallized luminescent calcium molybdate film. *J. Am. Ceram. Soc.*, **80**: 765–769.

Cho WS, Yoshimura M. (1996b). Room-temperature synthesis of crystallized luminescent $SrMoO_4$ film by active electrochemical dissolution of molybdenum. *Jpn. J. Appl. Phys., Part 2*, **35**(11B): L1521–L1523.

Cho WS, Yoshimura M. (1997a). Structure evolution of highly crystallized $BaWO_4$ film prepared by an electrochemical method at room temperature. *J. Am. Ceram. Soc.*, **80**(9): 2199–2204.

Cho WS, Yoshimura M. (1997c). Structural evolution of crystallized $SrWO_4$ film synthesized by a solution reaction assisted by electrochemical dissolution of tungsten at room-temperature. *Eur. J. Solid State Inorg. Chem.*, **34**(9): 895–904.

Cho WS, Yoshimura M. (1997d). Preparation of highly crystallized $BaMoO_4$ film using a solution reaction assisted by electrochemical dissolution of molybdenum. *Solid State Ionics*, **100**(1–2): 143–147.

Descemond M, Brodhag C, Thevenot F, Durand B, Jebrouni M, Roubin M. (1993). Characteristics and sintering behaviour of 3 mol% Y_2O_3-ZrO_2 powders synthesized by reaction in molten salts. *J. Mater. Sci.*, **28**: 2283–2288.

Emad MME, Mahmoud MH, Abdel-Hady A, El-Geassy. (2008). In-situ synthesis of magnetic Mn-Zn ferrite ceramic object by solid state reaction. *J. Aust. Ceram. Soc.*, **44**(1): 57–62.

Filser F, Gauckler LJ. (2007). *Material Science I, Ceramic Materials.* ETH-Zürich, Departement Materials.

Gaikwad SP, Dhage SR, Potdar HS, Violet Samuel, Ravi V. (2005). Co-precipitation method for the preparation of nanocrystalline ferroelectric $SrBi_2Nb_2O_9$ ceramics. *J. Electroceram.*, **14**(1): 83–87.

Ganguly A, Zhen T, Barsoum MW. (2004). Synthesis and mechanical properties of Ti_3GeC_2 and $Ti_3(Si_xGe_{1-x})C_2$ ($x = 0.5, 0.75$) solid solutions. *J. Alloys Compd.*, **376**: 287–295.

Geethalakshmi K, Prabhakaran T, Hemalatha J. (2012). Dielectric studies on nano zirconium dioxide synthesized through co-precipitation process. *Int. J. Chem. Mol. Nucl. Mater. Metall. Eng.*, **6**(4).

Georgea M, Johna AM, Naira SS, Joyb PA, Anantharaman MR. (2006). Finite size effects on the structural and magnetic properties of sol–gel synthesized $NiFe_2O_4$ powders. *J. Magn. Magn. Mater.*, **302**: 190–195.

Ghosh S, Dasgupta S. (2010). Synthesis, characterization and magnetic properties of nickel ferrite by soft chemical method. *Trans. Indian Ceram. Soc.*, 25–28.

Gonzalez-Rovira L, Sanchez-Amaya JM, Lopez-Haro M, Hungria AB, Boukha Z, Bernal S, Botana FJ. (2008) Formation and characterization of nanotubes of $La(OH)_3$ obtained using porous alumina membranes. *Nanotechnology*, **19**: 495305-1–495305-9.

Han X, Liang Z, Feng L, Wang W, Chen J, Xue C, Zhao H. (2015). Co-precipitated synthesis of Al_2O_3–ZrO_2 composite ceramic nanopowders by precipitant and drying method regulation: a systematic study. *Ceram. Int.*, **41**: 505–513.

Hassanzadeh Tabrizi SA, Taheri Nassaj E. (2015). Sintering of Al_2O_3-Ceo_2 nanopowder prepared by co-precipitation method. *Proceedings of the World Congress on Mechanical, Chemical, and Material Engineering (MCM 2015),* Paper No. 318-1. Barcelona, Spain.

Hayashi H, Kanoh Ma, Quan CJ, Inaba H, Wang S, Dokiya M, Tagawa H. (2000). Thermal expansion of Gd-doped ceria and reduced ceria. *Solid State Ionics*, **132**: 227–233.

Hayashi T, Saito H. (1980). Preparation of CaO-SiO_2 glasses by the gel method. *J. Mater. Sci.*, **15**(8): 1971–1977.

Huang K, Feng M, Goodenough JB. (1998). Synthesis and electrical properties of dense $Ce_{0.9}Gd_{0.1}O_{1.95}$ ceramics. *J. Am. Ceram. Soc.*, **81**(2): 357–362.

Janasi SR, Emura M, Landgraf FJG, Rodrigues D. (2002). The effects of synthesis variables on the magnetic properties of coprecipitated barium ferrite powders. *J. Magn. Magn. Mater.*, **238**: 168–172.

Jokanovic V, Janac kovic D, Spasicc AM, Uskokovicc D. (1996). Synthesis and formation mechanism of ultrafine spherical Al_2O_3 powders by ultrasonic spray pyrolysis. *Mater. Trans.*, **37**(4): 627–635.

Joseph DP, Naveenkumar S, Sivakumar N, Venkateswaran C. (2006). Synthesis of $Zn_{0.95}Cr_{0.05}O$ DMS by co-precipitation and ceramic methods: structural and magnetization studies. *Mater. Chem. Phys.*, **97**: 188–192.

Kelly JR, Nishimura I, Campbell SD. (1996). Ceramics in dentistry: historical roots and current perspectives. *J. Prosthet. Dent.*, **75**(1): 18–32.

Khajelakzay M, Razavi RS, Barekat M, Naderi M. (2015). Synthesis of yttria nanopowders by two precipitation methods and investigation of synthesis conditions. *Int. J. Appl. Ceram. Technol.*, 1–10.

Khanahmadzadeh S, Simani S, Enhessari M, Khezri B. (2012). $NiTiO_3$/polyimide nanocomposite: synthesis and magnetic properties. *2012 4th International Conference on Chemical, Biological and Environmental Engineering IPCBEE*, Vol. 43. IACSIT Press, Singapore.

Kim DJ, Kim H. (1989). Fine ceramic particles of zirconia toughened alumina (ZTA), titania toughened alumina (TTA). *Korean Ceram. Soc.*, **26**(5): 691.

Kingery WD, Bowen HK, Uhlmann DR. (1976). Chapters 12 and 16. In *Introduction to Ceramics*, 2nd ed. John Wiley & Sons, New York.

Kingsley JJ, Patil KC. (1988). A novel combustion process for the synthesis of fine particle α-alumina and related oxide materials. *Mater. Lett.*, **6**: 427–432.

Koch W, Friedlander SK. (1990). Particle growth by coalescence and agglomeration. *J. Aerosol. Sci.*, **21**(1): S73–S76.

Korner R, Ricken M, Noelting J, Riess I. (1989). Phase transformations in reduced ceria: determination by thermal expansion measurements. *J. Solid State Chem.*, **78**: 136–147.

Kulkarni AR, Patro PK, Gupta SM, Harendranath CS. (2008). Electrical properties of ferroelectric $Sr_{0.5}Ba_{0.5}Nb_2O_6$ ceramic materials synthesized by different techniques. *IEEE International Symposium on Applications of Ferroelectrics.*

Kulmeteva VB, Porozova SE, Krasnyi BL, Tarasovskii VP, Krasnyi AB. (2009). Preparation of zirconia ceramics from powder synthesized by a sol-gel method. *Refract. Ind. Ceram.*, **50**(6): 438–440.

Layek S, Verma HC. (2012). Magnetic and dielectric properties of multiferroic $BiFeO_3$ nanoparticles synthesized by a novel citrate combustion method. *Adv. Mater. Lett.*, **3**(6): 533–538.

Li T, Liao JP, Wang YF. (2015). Solvothermal synthesis and magnetic properties of β-Co_2P nanorods. *Mater. Sci. Poland*, **33**(2): 312–316.

Licciulli A. (2003). *Introduction to Ceramics*, Ceramic Science and Technology course.

Liu Z, Zheng D, Su Y. (2010). Facile and efficient electrochemical synthesis of lanthanum hydroxidenanospindles and nanorods. *Electrochem. Solid-State Lett.*, **13**: E15–E18.

Lopez Ibanez R, Ramos Barrado JR, Martin F, Brucker F, Leinen D. (2004). Oxide barrier coatings on steel strip by spray pyrolysis. *Surf. Coat. Technol.*, **188/189**: 675–683.

Lu X, Li G, Yu X, Tong Y. (2008). Electrochemical synthesis and characterization of TbO_{2-x} flowerlike nanostructures. *Electrochem. Solid-State Lett.*, **11**: K85–K88.

Madhusudhana R, Sangamesha MA, Gopal Krishne R, Krishnamurthy L, Shekar GL. (2014). Synthesis and characterization of zirconia (ZrO_2) by simple sol-gel route. *Int. J. Adv. Res.*, **2**(4): 433–436.

Mallik RK. (2012). *Synthesis and Magnetic Properties of Cobalt Ferrite with Different Morphology*. BTech thesis, National Institute of Technology, Rourkela.

Mao G, Zhang H, Lia H. (2012). Selective synthesis of morphology and species controlled $La_2O_3:Eu^{3+}$ and $La_2O_2CO_3:Eu^{3+}$ phosphors by hydrothermal method. *J. Electrochem. Soc.*, **159**: J48–J53.

Markiewicza E, Andrzejewskia B, Hilczera B, Balcerzakb M, Pietraszkoc A, Jurczykb M. (2015). Dielectric and magnetic properties of $(Bi_{1-x}La_xFeO_3)_{0.5}(PbTiO_3)_{0.5}$ ceramics prepared from mechanically synthesized powders. *J. Electroceram.*, **35**(1): 33–44.

Martin MI, Gomez LS, Milosevic O, Rabanal ME. (2010). Nanostructured alumina particles synthesized by the spray pyrolysis method: microstructural and morphological analysis. *Ceram. Int.*, **36**(2): 767–772.

Mason TO, Ceramic composition and properties, Ceramics. http://www.britannica.com/topic/composition-and-properties-103137s.

Mazumdar SC, Hossain AKM. (2012). Synthesis and magnetic properties of $Ba_2Ni_{2-x}Zn_xFe_{12}O_{22}$ hexaferrites. *World J. Condens. Matter Phys.*, **2**: 181–187.

Merzhanov A, Borovinskaya I. (1972). Self-propagating high-temperature synthesis of inorganic compounds. *Dokl. Akad. Nauk SSSR*, **204**(2): 366–369.

Merzhanov A, Borovinskaya I. (1975). A new class of combustion processes. *Combust. Sci. Technol.*, **10**(5–6): 195–200.

Merzhanov A, Mukasyan A. (2007). *Solid Flame Combustion*, 2nd ed. Torus Press, Moscow, p. 336.

Merzhanov A. (1990). In *Combustion and Plasma Synthesis of High-Temperature Materials*. Munir ZA, Holt JB (eds.). Wiley-VCH, New York, pp. 1–53.

Merzhanov A. (2000). *Solid Flame Combustion*. Institute of Structural Macrokinetics, Russian Academy of Sciences, Chernogolovka, p. 238.

Mimani T, Ghosh S. (2000). Combustion synthesis of cobalt pigments: blue and pink. *Curr. Sci.*, **78**(7): 892–896.

Mishra KK, Satya AT, Bharathi A, Sivasubramanian V, Murthy VRK, Arora AK. (2011). Vibrational, magnetic, and dielectric behavior of La-substituted $BiFeO_3$-$PbTiO_3$. *J. Appl. Phys.*, **110**(123529): 1–11.

Mohanty BB, Sahoo PS, Sahoo MPK, Choudhary RNP. (2012). Synthesis and electrical analysis of $Ba_5GdTi_3V_7O_{30}$ ceramics. *Mater. Sci. Appl.*, **3**: 173–178.

Muttay EP, Tsai T, Barnett SA. (1999). A direct-methane fuel cell with a ceria-based anode. *Nature*, **400**: 649–651.

Nehal AE. (2015). Preparation and characterization of magnetic glass ceramics derived from iron oxides bearing rolling mill scales wastes, *Nano Res. App.*, **1**(1): 1–6.

Ortega Pedro PS, Ramirez MA, Foschini CR, Garcia FG, Cilense G, Simoes AZ. (2014). Synthesis, structure and magnetic properties of $Y_3Fe_{5-x}Al_xO_{12}$ garnets prepared by the soft chemical method, *Process. Appl. Ceram.*, **8**(4): 211–218.

Patro PK, Kulkarni AR, Gupta SM, Harendranath CS. (2007). Effect of homogeniety on microstructure, dielectric, and ferroelectric properties of strontium barium niobate. *Defence Sci. J.*, **57**(1): 79–87.

Anusavice K. (2003). *Phillips' Science of Dental Materials*, 11th ed. Saunders.

Pissurlekar VJ. (2015). Synthesis and magnetic properties of Mn-Zn ferrite obtained by decomposition of precursor by sunlight. *Int. J. Sci. Res.*, **4**(11): 660–664.

Prakash BJ, Buddudu S. (2012). Synthesis and analysis of $LiNbO_3$ ceramic powders by co-precipitation method. *Indian J. Pure Appl. Phys.*, **5**: 320–324.

Purwasasmita, BS, Larasati LD, Septawendar R, Nugraha AB, Muhammad RA. (2013). Synthesis and characterization of zirconia crystal using base hot water treatment (BHWT) method. *J. Aust. Ceram. Soc.*, **49**(2): 89–94.

Rahman IA, Padavettan V. (2012). Synthesis of silica nanoparticles by sol-gel: size-dependent properties, surface modification, and applications in silica-polymer nanocomposites: a review. *J. Nanomater.*, **2012**: 132424(1–15).

Raju MRR, Choudhary RNP, Ram S. (2003). Dielectric and electrical properties of $Sr_5EuCr_3Nb_7O_{30}$ nanoceramics prepared using a novel chemical route. *Phys. Status Solidi*, **239**(2): 480–489.

Roy S, Mitra M, Basu D. (2005). Characterization of mechanical properties of alumina based hip joint prostheses. *Trends Biomater. Artif. Organs*, **18**(2): 166–173.

Saba A, Elsayed IE, Moharam M, Rashad MM. (2012). Electrochemical synthesis of nanocrystalline $Ni_{0.5}Zn_{0.5}Fe_2O_4$ thin film from aqueous sulfate bath international scholarly research network, *ISRN Nanotechnol.*, **2012**: 532168(1–8).

Sahoo PS, Mohanty BB, Panda J, Choudhary RNP. (2014). Synthesis and analysis of electrical properties of lead free $Ba_3Sr_2LaTi_3V_7O_{30}$ ceramics. *Research Inventy: Inter. J. Eng. Sci.*, **4**(12): 38–42.

Sahoo PS, Panigrahi A, Patri SK, Choudhary RNP. (2008). Structural, dielectric, electrical and piezoelectric properties of $Ba_4SrRTi_3V_7O_{30}$ (R=Sm, Dy) ceramics. *Cent. Eur. J. Phys.*, **6**(4): 843–848.

Sahoo PS, Panigrahi A, Patri SK, Choudhary RNP. (2010). Structural and electrical properties of $Ba_5SmTi_3V_7O_{30}$ ceramics. *J. Mater. Sci. Mater. Electron.*, **21**: 160–167.

Sakai M, Sekine T, Yamazaki Y. (1983). A new electrodeposited oxide film containing lead and thallium with the cubic fluorite-type structure. *J. Electrochem. Soc.*, **130**: 1631–1633.

Sharma AK, Yewale MA, Chavan GA, Kambale DB, Potdar S. (2015). Electrochemical synthesis of nanoplatelets-like $CuS_{0.2}Se_{0.8}$ thin film for photoluminescence applications. *Mater. Sci. Poland*, **33**(2): 397–401.

Sherikar BN, Umarji AM. (2013). Synthesis of γ-alumina by solution combustion method using mixed fuel approach (urea+glycine fuel). IC-RICE Conference Issue. *IJRET: International Journal of Research in Engineering and Technology*, eISSN: 2319-1163, pISSN: 2321-7308.

Shiha SJ, Wua YY, Chenb CY, Yua CY. (2012). Controlled morphological structure of ceria nanoparticles prepared by spray pyrolysis. *Procedia Eng.*, **36**: 186–194.

Shim IS, Lee CS. (2002). Synthesis and characterization of Al_2O_3/ZrO_2, Al_2O_3/TiO_2 and $Al_2O_3/ZrO_2/TiO_2$ ceramic composite particles prepared by ultrasonic spray pyrolysis. *Bull. Korean Chem. Soc.*, **23**(8): 1127–1134.

Singh A, Gupta A, Chatterjee R. (2008). Enhanced magnetoelectric coefficient in the modified $BiFeO_3$–$PbTiO_3$ system with large La substitution. *Appl. Phys. Lett.*, **93**: 022902(1–3).

Singh I, *Ceramics: Properties and processing*. Department of Mechanical and Industrial Engineering, Indian Institute of Technology, Roorkee, Module 3, Lecture 1.

Sirota V, Ivanisenko V, Pavlenko I. (2013). Synthesis of nanosized stabilized zirconium dioxide under hydrothermal conditions, *NANOCON 2013*, October 16–18, 2013, Brno, Czech Republic, EU.

Stecura S, Campbell WJ. (1961). Thermal expansion and phase inversion of rare-earth oxides. *Bur. Mines Rep. Invest.*, **5847**: 1–47.

Sun J, Qiu XP, Zhu WT. (2005). H_2 from steam reforming of ethanol at low temperature over Ni/Y_2O_3, Ni/La_2O_3 and Ni/Al_2O_3 catalysts for fuel-cell application. *Int. J. Hydrogen Energy*, **30**: 437–445.

Sun L, GuO H, Peng H, Gong S, Xu H. (2013). Phase stability and thermal conductivity of ytterbia and yttria co-doped zirconia. *Prog. Nat. Sci.: Mater. Int.*, **23**(4): 440–445.

Switzer JAJ. (1986). Made transparent rectifying contacts with n-silicon, and ohmic contacts with p-silicon using electrodeposited thallic oxide (US Patent 4,706,104). Measured barrier height, dark saturation current, and diode quality factor. *Electrochem. Soc.*, **133**: 722–729.

Taketomi S, Kawasaki K, Ozaki Y, Yuasa S, Otani Y, Miyajima H. (1994). Characteristics of yttrium iron garnet ultrafine particles prepared by the alkoxide method. *J. Am. Ceram. Soc.*, **77**(7): 1787–1792.

Tas AC. (2000). Combustion synthesis of calcium phosphate bioceramic powders. *J. Eur. Ceram Soc.*, **20**: 2389–2394.

Therese GHA, Kamath PV. (2000). Electrochemical synthesis of metal oxides and hydroxides. *Chem. Mater.*, **12**: 1195–1204.

Tok AIY, Boey FYC, Zhao XL. (2006). Novel synthesis of Al_2O_3 nanoparticles by flame spray pyrolysis. *J. Mater. Process. Technol.*, **178**: 270–273.

Touloukian YS, Kirby RK, Taylor RE, Lee TYR, Touloukian YS, Ho CY. (1977). Thermal expansion nonmetallic solids. In *Handbook of Thermophysical Properties of Matter*. Springer, New York, p. 1810.

Tyagi AK. (2007). Combustion synthesis: a soft-chemical route for functional nano-ceramics. *Bhabha Atomic Research Centre,* **285**: 39–48.

Van De Graaf MACG, Termaat JHH, Burggraaf AJ. (1985). Microstructure and sintering kinetics of highly reactive ZrO_2–Y_2O_3 ceramics, *J. Mater. Sci.*, **20**(5): 1407–1418.

Vijayalakshmi U, Balamurugan A, Rajeswari S. (2005). Synthesis and characterization of porous silica gels for biomedical applications. *Trends Biomater. Artif. Organs*, **18**(2): 101–105.

Wade T, Crooks RM. (1996). Electrophoretic deposition of metal nitride ceramic precursors. *Chem. Mater.*, **8**: 832–835.

Wade T, Park J, Garza G, Ross CB, Smith DM, Crooks RM. (1992). Electrochemical synthesis of aluminum nitride in liquid ammonia electrolyte solutions. *MRS Proc.*, 271.

Wu L, Dong C, Chen H, Yao J, Jiang C, Xue D. (2012). Hydrothermal synthesis and magnetic properties of bismuth ferrites nanocrystals with various morphology. *J. Am. Ceram. Soc.*, **95**(12): 3922–3927.

Yamagata C, Mello-Castanho SRH, Paschoal JOA. (2014). Synthesis and mechanical properties of stabilized zirconia ceramics: MgO-ZrO_2 and Y_2O_3-MgO-ZrO_2. *21ºCBECIMAT-Congresso Brasileiro de Engenharia e Ciência dos Materiais Cuiabá*, MT, Brasil.

Yang Y, Luo S, Dong F, Ding Yi, Li X. (2015). Synthesis of high-phase purity $SrTi_{1-x}Zr_xO_3$ ceramics by sol-spray pyrolysis method. *Mater. Manuf. Processes*, **30**: 585–590.

Yao CZ, Wei BH, Ma HX, Gong QJ, Jing KW, Sun H, Meng LX. (2011). Facile fabrication of $La(OH)_3$ nanorod arrays and their application in wastewater treatment. *Mater. Lett.*, **65**: 490–492.

Yousefi M, Alizadeh P, Eftekhari Yekta B, Molaie F, Ghafoorian N, Montazerian M. (2009). Synthesis and characterization of diopside glass–ceramic matrix composite reinforced with aluminum titanate. *Ceram. Int.*, **35**: 1447–1452.

Yue XM, Zhang GJ, Wang YM. (1999). Reaction synthesis and mechanical properties of TiB_2 ± AlN ± SiC composites. *J. Eur. Ceram. Soc.*, **19**: 293–298.

Zaitsev DD, Kazin PE, Tretyakov Y D, Maksimov YV, Suzdalev IP, Jansen M. (2004). Synthesis and magnetic properties of $SrFe_{12}O_{19}$ + $12SrB_{1.5}Bi_{0.5}O_4$ glass-ceramics. *Inorg. Mater.*, **40**: 1111–1115.

Zelinski BJJ, Uhlmann DR. (1984). Gel technology in ceramics. *J. Phys. Chem. Solids*, **45**(10): 1069–1090.

Zhang HS, Sun K, Xu Q, Wang F, Liu L. (2009). Preparation and thermal conductivity of $Sm_2(Zr_{0.6}Ce_{0.4})_2O_7$ ceramic. *J. Mater. Eng. Perform.*, **18**: 1140–1143.

Zhang J, Li W, Tanji T. (2014). Synthesis of zirconia oxide (ZrO_2) nanofibers on zirconnia substrates by ultrasonic spray pyrolysis. *Mater. Sci. Appl.*, **5**: 193–198.

Zhang T, Kumari Latha, Du GH, Li WZ, Wang QW, Balani K, Agarwal A. (2009). Mechanical properties of carbon nanotube-alumina nanocomposites synthesized by chemical vapor deposition and spark plasma sintering. *Composites Part A*, **40**: 86–93.

Zheng Y, Kurosaki K, Ohishi Y, Muta H, Yamanaka S. (2012). Synthesis and thermal conductivity of Y_6UO_{12}. *J. Nucl. Sci. Technol.*, **49**(5): 526–530.

Zhitomirsky I, Petric A. (1999). Electrolytic and electrophoretic deposition of CeO_2 films. *Mater. Lett.*, **40**: 263–268.

Zhitomirsky I, Petric A. (2000). Electrochemical deposition of yttrium oxide. *J. Mater. Chem.*, **10**: 1215–1218.

Zhitomirsky I, Petric A. (2001). Electrochemical deposition of ceria and doped ceria films. *Ceram. Int.*, **27**: 149–155.

Zhitomirsky I. (1998). Cathodic electrosynthesis of titanium and ruthenium oxides. *Mater. Lett.*, **33**: 305–310.

Zhou A, Wang C. (2003). Synthesis and mechanical properties of Ti_3AlC_2 by spark plasma sintering. *J. Mater. Sci.*, **38**: 3111–3115.

Chapter 3

Ceramic-Based Nanomaterials for Multifunctional Application

Sangeeta Adhikari,[a] Ajay Kumar Mishra,[b] and Debasish Sarkar[a]

[a]*Department of Ceramic Engineering, National Institute of Technology, Rourkela, India*
[b]*Nanotechnology and Water Sustainability Research Unit, College of Engineering, Science and Technology, University of South Africa, Johannesburg, South Africa*

adhikari.sangeeta8@gmail.com

Innovation in technologies through the development of functionalized ceramic nanoparticles from the perspectives of energy, environment, and health is believed to hold promise. The unusual properties of nanoparticles associated with spatial confinement of electrons, phonons, and electric fields around the particles, including a large surface-to-volume ratio, allow directional/dimensional (0D, 1D, 2D, 3D) growth of nanoparticles, explaining the structure-function relationship. The tuning of nanoscale assembly and nanoparticle growth will be explained in order to impart applicability in specific energy, environment, and clinical needs of humans.

Smart Ceramics: Preparation, Properties, and Applications
Edited by Ajay Kumar Mishra

ISBN 978-981-4774-30-7 (Hardcover), 978-1-315-16359-8 (eBook)
www.panstanford.com

3.1 Introduction

One of the major directions of the development of modern ceramic materials in industrially developed countries is their incessant improvement in the production of new ceramic materials for engineering, opto- and microelectronics, defense, aerospace, chemical and radio engineering, and others. Capital investment in the development of new advanced materials is continuously increasing. Today, the technological world using ceramic nanomaterials advocates nanotechnology for reduced use of energy and resources (Kulik et al., 1999). Theoretically it has been proved that nanomaterials have more potential over bulk conventional materials because of their reduced sizes. A new working definition to the word "nanomaterials" has been given by the US Environmental Protection Agency (EPA) in April 2010 as "an ingredient that contains particles that have been intentionally produced to have at least one dimension measurable between approximately 1 and 100 nanometers" for the facilitation of its use in products commercially. As the world is looking forward to the miniaturization, lightweight and less energy-consuming materials are explored more. The global market for ceramic nanotechnology did not match the hype initially, but several permeable products in consumer areas have grown the market. Thus, the goal is to develop materials that can contribute to the progress of environment harmony and ecologically clean, sustainable, and chemical technologies with high energy efficiency using the abundant energy available (Leonov et al., 2009).

3.2 Opportunities and Future Perspectives

The hunt for new materials began ever since the base resources for production of metals and their alloys became limited. There has been a continuous improvement in the technology for the development of modern materials in lieu of the upcoming industrial branches of ceramic engineering. Recent exploration reveals that the manufacturing of ceramic nanomaterials has an unexpected footprint over the ecological system that relates to the specialized production environments, high water and energy demands with low yield of product, and high generation of waste. These nanomaterials are liable to be used in smaller quantities than substances used conventionally such that the overall life cycle assessment of the products would contribute to an accurate impression of the total energy and environmental impacts (Chang et al., 2010). The possibility of successful commercialization of nanomaterials is only when production and application development proceed in parallel with each other.

3.2.1 Overview of Ceramic-Based Nanomaterials

In the state of art and science, ceramics are solid components composed of inorganic nonmetallic materials. The types of bonds that are usually associated with ceramics are a combination of covalent, ionic, and sometimes metallic. They do not consist of discrete molecules; however, they have interconnected atoms arranged in arrays. The majority of ceramic materials are compounds of metals, nonmetals, or metalloids. Earlier, the key element of materials science and engineering was the interrelationship between structure and properties, which classified ceramics as traditional and advanced. The higher-volume items such as whitewares, pottery, bricks, and tiles were the traditional ceramics based on clay and silica. But the advanced ceramics, like piezoelectric ceramics, ceramics for dynamic random access memories (DRAMs), and laser host materials, are considered advanced ceramics because of superior properties.

Today, the ceramic nanomaterials in various fields are known for their technical applicability in all human aspects and are thus named "technical" or "advanced" ceramics. These are further categorized into "functional" and "structural" ceramics. Functional ceramics mainly contribute to advanced electronics, optics, magnetics, and others, whereas structural ceramics are mostly used in mechanical application or load-bearing applications such as components like bioimplants, valves, rollers, and catalyst carriers. Functional ceramics include electromagnetically sensitive, surface-active, temperature- and pressure-sensitive, photosensitive, and similar materials. Structural ceramics include materials suitable for service in machine and mechanism components under extreme conditions, which are often created by the simultaneous effect of high temperature, mechanical loading, and corrosive and abrasive media (Wachtmann, et al. 1983; Yin et al., 2010).

Most of the frequently used ceramics are oxides, nitrides, and carbides that are explicitly used in areas like structural ceramics, electronic ceramics, bioceramics, coatings and films, composites and nanoceramics. A more accurate term for ceramic-based nanomaterials is probably "ceramics with high service characteristics." This name is applicable to all ceramic nanomaterials whose properties are specially optimized, and some undesirable properties are specially restricted. Some of the renowned semiconductor oxide and nonoxide materials also fall under the category of smart ceramics, as shown in Fig. 3.1.

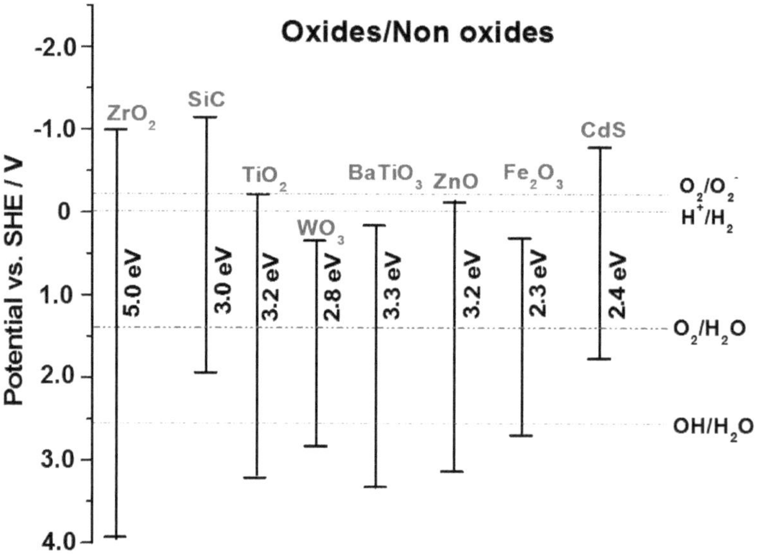

Figure 3.1 The bandgap of a few renowned smart ceramic oxides/nonoxides.

3.3 What Makes Ceramic-Based Nanomaterials Persuading?

In general, structural ceramic materials are known for their specific properties, such as brittleness in room temperature (not in elevated temperatures), poor electrical and thermal conduction (except for materials like ReO_3, an oxide ceramic, and the high-temperature superconductor $YBa_2Cu_3O_7$, a mixed oxide that has conductivity similar to Cu_2), high compressive strength, insensitivity toward harsh chemical and thermal environments, and transparency due to a large E_g, that is, bandgap (e.g., sapphire watch covers, optical fibers, and precious gemstones). The conventional ceramic materials behave indifferently compared to metals and polymers. However, the possibility of their atypical behavior can be observed upon nanostructuring. Enhancement in their structural performance for functional applications was achieved through nanostructuring that does not exist in the bulk form. In comparison to the bulk material, nanostructuring provides the following fascinating features (Zheng et al., 2011):

- An increased surface-to volume ratio, which provides more surface area for both chemical and physical interactions
- Significantly altered surface energies that allow tuning and engineering of the material's properties, as atomic species near the surface have different bond structures than those embedded in the bulk
- Quantum confinement effects, due to their inherently small size, that significantly influence charge transport, electronic band structure, and optical properties.

3.4 Strategies of Synthesis for Purposeful Use

The present progression of science and technology develops innovative strategies for material synthesis/fabrication that imparts fundamental importance toward properties of material. Controlled growth of functional ceramic materials can be integrated for assembled organization of primary nanoblocks to tailor the composition and morphology. The synthetic routes empower new bottom-up approaches for self-assembly of ceramic nanoparticles. In the current chapter, the diverse synthesis methodologies have been discussed in the context of individual applicability for each ceramic nanomaterial. However, wet-chemical methods (sol-gel, coprecipitation, solvothermal, hydrothermal, microemulsion, substrate growth, organic-group functionalized, template-based, and microwave) are exclusively used for generating isotropic/anisotropic crystallographic structures (Cao, 2011). Mostly, emphasis is put on material performance, flexibility in different aspects of the working environment (energy, environment, and human), scale-up potential, reproducibility, and cost effectiveness.

3.5 Promising Ceramic-Based Nanomaterials

3.5.1 Tungsten Trioxide

Tungsten oxide (WO_3) is an *n*-type transition metal oxide with wide extending applications. Researchers renewed their interest in WO_3 with the emergence of its efficient electrochromic (EC) effect. The synthesis and analysis of WO_3 nanostructures have become increasingly prominent with the advent of nanotechnologies. The tungsten oxides consist of WO_6 octahedra arranged in various sharing (corners, edges, planes) configurations. The widely reported crystal phase transitions for WO_3 in its bulk form occur in the following sequence: monoclinic II (ε-WO_3, <–43°C) to triclinic (δ-WO_3, –43°C to 17°C) to monoclinic I (γ-WO_3, 17°C to 330°C) to orthorhombic (β-WO_3, 330°C to 740°C) to tetragonal (α-WO_3, >740°C) and cubic WO_3. These transitions are partially reversible. Monoclinic I (γ-WO_3) is the most stable phase at room temperature, whereas cubic WO_3 is not commonly observed. The main differences between the phases are the position alteration of the W atom within the octahedra and variation in W–O bond lengths. A quite complex behavior is observed upon nanostructuring that depends on the morphology and dimension of the material. As proposed by the Gibbs–Thomson expression, the size reduction of WO_3 crystallite enhances the surface energy, which affects the material property by decreasing the melting and sublimation temperatures (Boulova et al., 2002; Adhikari, 2015). Thus, in nanostructured WO_3, phase transition occurs at a lower temperature than in bulk WO_3.

Many different approaches have been disclosed for the synthesis of WO_3 nanostructures via different methods of predefined processing. Most of the processing has been carried for preparation of thin films. The wet-chemical method specifically includes sol-gel, acid precipitation, hydrothermal, solvothermal, and combustion methods (Nishide et al., 1995; Gu et al., 2007). These methods offer better control of the material morphology through hydrolysis, condensation, etching, and oxidation during the reaction. Synthesis of WO_3 nanostructures from a peroxo-tungstic acid precursor is most commonly reported under the wet-chemical process using acid for precipitation (Zhang et al., 2009). The acid coprecipitation method has been used to synthesize spherical WO_3 and rod-shaped WO_3 nanopowders. An acid-catalyzed reaction occurs between tungstic acid and hydrogen peroxide (30% w/w) in the presence of concentrated nitric acid. The pH of the solution is maintained at ~1. The obtained precipitate is filtered and freeze-dried, followed by flash heating at 500°C for 5 min to obtain WO_3 nanopowders. The fabrication of films uses this base precursor for deposition of particles, but the resultant films prepared consist of large crystallites. Thus, control over the precursor preparation methodology can produce many interesting morphologies. Surfactants like cetyl-tri-methyl-ammonium-bromide ($C_{19}H_{42}BrN$, CTAB) or tetrabutylammonium bromide (TBABr) are also used as the directing agent during a sol-gel technique to prepare films with a plate-like morphology (Rajeswari et al., 2007; Kim et al., 2011). Similar to the above process, a clear solution of sodium tungstate and CTAB is acidified by conc. HNO_3 to obtain WO_3 nanopowders. Synthesis of nanoparticles with different morphologies is always fascinating as they act indifferently under diverse conditions, however, the matter of interest here is their reaction and formation mechanism. In an acid-catalyzed reaction, the two most important parameters are found to be temperature and pH of the reaction media in the presence of a hydrogen peroxide solution. The exothermic nature of the reaction could possibly be due to spontaneous decomposition of H_2O_2, leading to the formation of water vapor and nascent oxygen above the critical temperature of 95°C and pH > 5 (Nikitin et al., 1995). The preheated pot temperature of 90°C ± 5°C initiates the decomposition of H_2O_2 and accelerates the reaction between nascent oxygen atom and H_2WO_4 to form peroxotungstic acid ($WO_3 \cdot xH_2O_2$). Simultaneous addition of nitric acid favors dehydroxylation by the acid catalyst oxidation reaction, and $WO_3 \cdot xH_2O_2$ (x = 1–2) transforms to amorphous WO_3, which is annealed to form crystalline monoclinic spherical WO_3 nanopowders. On the other hand, an optimum amount of CTAB provides encapsulation to WO_4^{2-}, forming a confined reaction zone where the nuclei prefer an anisotropic growth leading to the development of a rod-like structure (Kiely et al., 1998). The probable reaction inside the reverse micelle formed by CTAB encapsulation is found similar to the nanorod formation (Shen et al., 2005). A distinct high current density of 1.8 mA/cm^2 is found for rod-shaped-WO_3-nanoparticle-coated indium tin oxide (ITO) glass compared to a 0.14 mA/cm^2 current density for spherical particles under ±0.5 V at a 50 mV/s scan rate. An elongated-rod-shaped nanoparticle preferentially favors more EC response from the counterpart spherical particles (Adhikari and Sarkar, 2013).

Another interesting hydrothermal process is single step, fast, and economical and produces high-yield powders. After annealing of WO_3 at temperatures less than 500°C, it retains its hydrated structure. Difficulties have been observed in producing different directional morphologies through variant physical and chemical methods. Investigation of the application potentiality depends on the development of a low-dimension morphology to enhance the electrical and optical performance for tuning modern device properties. A hydrothermal process takes place at an elevated temperature (120°C–300°C) for a certain period of time, allowing the nucleation and growth of the crystallites (Zhao et al., 2009). Herein, the selected hydrothermal method follows the expected development of WO_3 crystals from aqueous metal salt solutions under critical pressure at a pertinent temperature and solution pH through structure-directing reagents (SDRs). The WO_3 nanoparticle formation mechanism with a particular growth direction depends on the solubility of the metal oxides and the reaction kinetics during the synthesis. It significantly alters the critical point of temperature assistance to form high pressure and changes the dielectric constant of aqueous media (Hayashi et al., 2010). SDRs such as acids, surfactants, metal sulfates, chlorides, and sulfides are extensively used in this methodology. Sodium tungstate is one of the base precursors used in this method. Recently, monoclinic WO_3 nanocuboids (Fig. 3.2a) and hexagonal WO_3 nanofibers (Fig. 3.2b) have been developed using fluoroboric acid (HBF_4) and sodium chloride (NaCl) as SDRs with the base precursor sodium tungstate. Initially, fluoroboric acid addition to sodium tungstate forms tungstite (H_2WO_4) with a structure of WO_6 octahedra sharing four equatorial oxygen atoms, which are linked through hydrogen bonds derived from interaction between water molecules and oxygen present in the axial position of the octahedra. Plate-shaped particles are formed due to inhibition of the (010) crystal plane, which is the normal growth direction of the H_2WO_4 structure. Due to insufficient BF_4^- anion concentration, structural orientation of plates cannot be induced by adding BF_4^- anions to the hydrogen bond system to occupy apexes with a boron atom at the center. Thus, the growth axis is not restricted and rather prefers growth along this direction, forming cuboids (Adhikari and Sarkar, 2014a). On the other hand, formation of nanofibers is found to be a multiple-step process: it follows a somewhat sequential formation of spherical nuclei and their formation into a plate to a rod to fibers. Different states have different dispersion efficacies with respect to processing condition. The four coordinated window structures of WO_3 adsorb Na^+ ions, but the surface of WO_3 contains many hydroxyl groups, which have more interactions with H^+ ions than Na^+ ions due to hydrogen bonding at a lower pH. The layered structure of tungsten trioxide is assumed to self-assemble through an electrostatic link between the layers. Electrostatic repulsion between Na^+ and H^+ ions occur, which causes folding of the layered structure, forming a bundle of fibers. Na^+ ions can be presumed to act as a bonding bridge between the growth units to form crystal nuclei. Stabilization of the hexagonal phase only requires a certain amount of Na^+ ions, but morphology stability is strongly influenced by the number of Na^+ ions (Adhikari et al., 2014). Moreover, high ionic strength of the precursor solution also plays an important role in dispersion of particles. An increase in particle dispersion is found with a high ionic concentration. The current density of hexagonal-WO_3-nanofiber-coated WO_3 is 3.9 mA/cm^2 and that of monoclinic WO_3 nanocuboid is 1.2 mA/cm^2. There has been a large difference in the current densities between both fabricated electrodes as they use different morphology WO_3 along with different crystal structures. This difference in current densities is influenced by the difference in morphologies of the same crystal structure. But a limited comparative electrochemical representation of WO_3 is found for different morphologies having different crystal structures. The synthesized monoclinic WO_3 has high symmetry and high stability at room temperature, in which small ions can easily travel through the four coordinated tunnels formed upon layering of the octahedral, which eases the ion transportation and intercalation in the structure (Fig. 3.2c). Hexagonal structures have three possible locations for this ion insertion/extraction (Fig. 3.2d); they are the hexagonal window, the trigonal cavity, and the four coordinated square windows. The degree of openness and the ion-holding capacity in hexagonal crystal structure are higher than other layered structures of WO_3. The hexagonal WO_3 nanofibers exhibit high current density in comparison to the monoclinic WO_3 nanocuboids. Moreover, interconnectivity and existence of nanoscale pores have an additional advantage for fibrous morphology. The EC parameters such as coloration efficiency and reversibility were found best for hexagonal-WO_3-nanofiber-coated ITO glass substrates (Adhikari et al., 2015b).

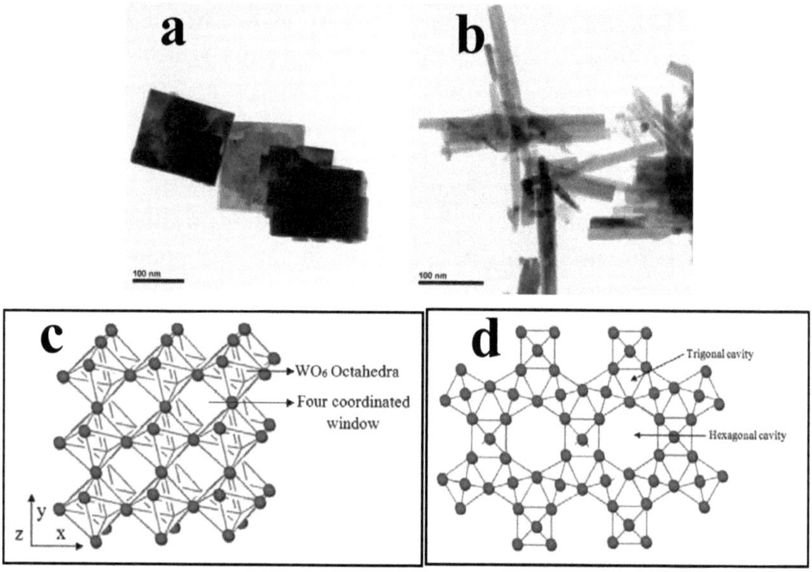

Figure 3.2 Transmission electron microscopy images of WO_3 (a) nanocuboids and (b) nanofibers and schemata of WO_3 (c) monoclinic and (d) hexagonal crystal structures. (a) Reproduced from Adhikari S, Sarkar D. (2014b). Hydrothermal synthesis and electrochromism of WO_3 nanocuboids. *RSC Adv.*, **4**: 20145–20153, with permission from the Royal Society of Chemistry. (b) Reprinted from *Electrochimica Acta*, **138**, Adhikari S, Sarkar D, High efficient electrochromic WO_3 nanofibers, 115–123, Copyright (2014a), with permission from Elsevier. (c, d) From Adhikari S, Sarkar D. (2015). Synthesis and electrochemical properties of nanocuboids and nanofiber WO_3. *J. Electrochem. Sci.*, **162**(1): H1–H7. Reproduced by permission of The Electrochemical Society.

WO_3 has high sensitivity toward different categories of gases, such as (i) flammable gases (H_2), (ii) toxic gases (CO, H_2S, NH_3, and NO_x), and (iii) environment-related gases (ozone O_3) and organic gases (ethanol, methanol, toluene, and others). Morphologies such as flower, nanoplate, sphere, octahedra have been synthesized for sensitizing organic gases (Long et al., 2015). Recently, porous WO_3 particulate films have been fabricated for effective dye-sensitized solar cell (DSSC) application. However, the efficiency achieved was only up to 0.75%. It is also found effective in hydrogen production from water splitting. WO_3 has also been explored for oxidization of organic compounds via an advanced oxidation process. Although it is a visible light harvester minimal photo-oxidization is observed with WO_3 alone. Thus, it is necessary to modify it with other metal oxide semiconductors to achieve better efficiency. In the recent past, a composite of WO_3-ZnO was designed for effective degradation of methylene blue and orange G in visible light (Adhikari et al., 2015b). Degradation took place via a charge transfer mechanism in the presence of 1:9 weight ratio of a WO_3 and ZnO mixture. The photocatalytic reaction was enhanced due to the reduction in the recombination of photogenerated electrons and holes. The high degree of degradation (90%) of both dyes is due to the activity of the mixed oxides, which is much higher than that obtained either with WO_3 or ZnO individually. When the composite is radiated by visible light, the activation of ZnO to produce the photogenerated electron-hole pairs is not possible due to its large absorption gap, while the narrow-bandgap WO_3 can effectively absorb the visible light and gets excited into generating electron-hole pairs. The photogenerated electrons in WO_3 transfer its electrons to the conduction band of ZnO, reducing the probability of recombination of photogenerated electron-hole pairs and increasing the number of active species for degradation. In this composite, ZnO plays the role of a cocatalyst that traps the electron from further recombination. The photoelectrons easily trap the dissolved O_2 to form a superoxide anion radicals, and photoinduced holes trap OH^- to form OH radicals to photodegrade dyes. The reactivity of formed radicals OH and O_2^- is sufficient to degrade the organic dye molecules. WO_3 does not impart any specific biological application, but its films acts as biosensors for certain biological compounds.

3.5.2 Zinc Oxide

Zinc oxide (ZnO) is one of the key technological semiconductor materials with a wide bandgap of ~3.2 eV, suitable for short wavelength optoelectronic applications. It is also transparent to visible light and can be made highly effective by doping/coupling. It possesses a hexagonal wurtzite structure. The crystal structure of ZnO can be described as a tetrahedral coordination of O^{2-} and Zn^{2+} ions with a number of alternating planes and stacked along *c* axis alternately. ZnO in tetrahedral coordination results in a noncentral symmetric structure that contributes to the piezoelectric and pyroelectric properties (Park et al., 2003; Ohshima et al., 2004). The interesting properties, like piezoelectricity, large photoconductivity, and high excitonic binding energy, make it suitable for various applications in the fields of photoelectrics, UV light emitters, window materials, solar blind photodetectors, transparent power electronics, displays, etc. (Fang et al., 2009; Liao et al., 2010). Moreover, its ability to harvest solar energy greater than that of TiO_2 makes it useful in the field of photocatalysis (Tryba et al., 2009). Methods such as precipitation (Mahmood et al., 2011), hydrothermal method (Li et al., 2000), template-based growth (Slama et al., 2011), sol-gel method (Yu et al., 2013), solvothermal method (Nagaraja et al., 2012), and combustion synthesis (CS) (Reddy et al., 2011) have been used for synthesis of nanosized ZnO. Other versatile methods include electrospinning, electrochemical deposition, and vapor-phase deposition. Many researchers followed these methods to achieve their desired goals as different methods are applicable for producing different morphologies with varying sizes of ZnO nanoparticles. Directional growth of nanoparticles is an important factor from the perspective of surface-based applications that provide a high surface area with active surface sites for reaction to take place.

In recent history, combustion of oxidizer zinc nitrate and fuel oxalic acid results in quasi-fibrous ZnO. The optimized processing parameters such as the oxidizer-to-fuel ratio, time, and temperature gave the resultant crystal structure and morphology. The hexagonal phase formation does not depend on the fuel ratio, but a stoichiometric ratio of oxidizer to fuel at 450°C produced highly crystalline hexagonal ZnO having a 3 μm length and a 0.5 μm width (Fig. 3.3). The quasi-fiber originates from the partial fusion of near-spherical (60 nm) particles during the rapid reaction rate in the combustion process. The rapid dissolution-recrystallization favors the formation of a neck within two particles and thereafter results in the formation of a continuous fiber-like structure. The porous particles showed high catalytic activity degradation of methyl orange under both UV and visible light irradiation (Adhikari et al., 2015a). CS has many advantages, such as a high production rate, energy efficiency, a low processing cost, and easy tailoring of the properties. The self-propagating exothermic redox reaction during combustion, which takes place with environmental friendly by-products, is a major advantage (Manjunath et al., 2014). Another major advantage of CS is that this method yields highly porous structures that could be beneficial for photo- and electrochemical applications. Various other fuels, like citric acid, dextrose, glycine, oxalyl dihydrazide, acrylamide, valine, and urea, have been used for the purpose to prepare high-surface-area porous particles.

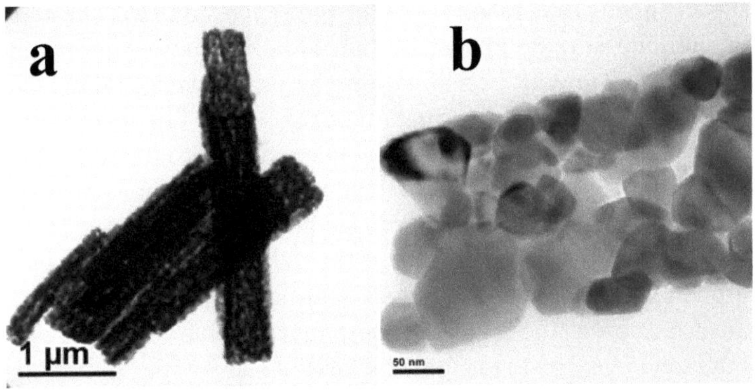

Figure 3.3 Transmission electron microscopy images of combustion-synthesized ZnO on (a) a 1 µm and (b) a 50 nm scale. Reproduced from Adhikari S, Sarkar D, Madras G. (2014). Synthesis and photocatalytic performance of quasi-fibrous ZnO. *RSC Adv.*, **4**: 55807–55814, with permission from the Royal Society of Chemistry.

One of the important characteristics of ZnO is polar surfaces. The basal plane is the most common polar surface. Spontaneous polarization along the *c* axis occurs due to the generation of positive charge from Zn and negative charge from O, diverging the surface energy. For structure stability, polar surfaces have facets or undergo massive surface reconstructions to form 1D morphologies. Researchers have been able to produce nanorings of ZnO by introducing small amounts of indium as impurity (Kong et al., 2004). The polar ZnO nanobelt is the building block for nanoring structures that grows along $[10\bar{1}0]$, with side surfaces $\pm(1\bar{2}10)$ and top/bottom surfaces $\pm(0001)$, and has a typical width of ~15 nm and a thickness of ~10 nm. The top and bottom of a nanobelt has polar charges. If the surface charges are not compensated during growth, it may get folded itself as its length gets longer, to minimize the area of the polar surface. One possible approach is to line the positively charged Zn-(0001) plane at the top surface, with the negatively charged O-(0001) plane at the bottom surface, for the neutralization of the local polar charges, reducing the surface area and thus forming a loop with an overlapped end. This type of folding is at 90° with respect to the folding direction for forming the nanospring or nanospiral, which may be due to the aspect ratio difference of the nanobelts and relative size of the polar surfaces. The radius of the loop may be known by the initial folding of the nanobelt in the initial growth period, but the loop size cannot be too small to reduce the elastic deformation energy. Thus, the polar charges, elastic deformation, and surface area contribute to the total energy involved in the process. The long-range electrostatic interaction is probably the initial driving force for nanobelt folding to form the first loop as the nucleation of the nanoring for the subsequent growth. As the growth proceeds, the nanobelt may be naturally attracted onto the rim of the nanoring due to electrostatic interaction and extends parallel to the rim of the nanoring to neutralize the local polar charge and reduce the surface area, resulting in the formation of a self-coiled, coaxial, uniradius, multilooped nanoring structure. Recently, hexagonal ZnO nanorings were designed via the hydrothermal route for effective sensing of trimethylamine (Li et al., 2015). ZnO is known to be an alternate photoanode material after TiO_2 due to a similar bandgap. Other research groups (Chang et al., 2003) prepared ZnO nanocones and nanorods with exposed $\{10\bar{1}1\}$ or $\{10\bar{1}0\}$ facets by the hot injection method for efficient photoconversion efficiency. The DSSC assembly having ZnO cones coated with C218 dye (organic photosensitizer) demonstrated three times higher conversion efficiency, of 4.36%, in comparison to ZnO nanorods. ZnO has also been exploited for hydrogen storage purpose. To improve the hydrogen storage property, it has been doped with metals like Sb and Al, which were synthesized by the controlled thermal evaporation method. Al-doped ZnO nanobelts under a pressure of 5 MPa showed maximum H_2 storage capacity of 2.94%, and 81.6% was released at 373 K under ambient pressure, proving it a promising candidate in this field (Wan et al., 2004; Hui et al., 2006). Nowadays, individual nanowires/nanobelts/nanotubes are most laterally bonded on a flat surface for field-effect transistor (FET) configuration, in which the substrate acts as the gate electrode and the current transported from the drain to the source along the nanowire is tuned by the gate voltage applied or the species adsorbed on the nanowire surface. In some cases, semiconductive and piezoelectric properties of ZnO are coupled for working as a nanodevice via utilization of the piezoelectric effect. The working principle is as follows: under strain, the piezoelectric potential is created inside a nanowire that serves as a gate voltage for fabricating a pioneering FET and diode. A freestanding ZnO nanowire cantilever upon mechanical impact creates a piezoelectric potential distributed across its width at roots, which simultaneously produces a reverse depletion layer locally of a high donor concentration that intensely changes the current flow from source to electrode to drain electrode when the device is under the influence of a fixed voltage bias. The device shows a prompt response in time less than 20 ms and exhibits high stable sensitivity of 2%/μN (Fei et al., 2009).

Although ZnO has the tendency to adsorb the bacteria and inhibit their growth, recently Ag-modified ZnO nanostructures have been explored for antibacterial activity toward the *E. coli* bacterium under natural solar light. The photocatalytic tests showed that Ag-impregnated ZnO (Fig. 3.4a) exhibits excellent bacterial inactivation. The porous and open structure of ZnO favored the effective inactivation by increasing the number of active sites, and effective charge separation was observed in the presence of Ag. Ag acts as an electron sink and the hole as a hydroxyl radical, the former for the photochemical killing of bacteria. The accepted photochemical theory predicts that

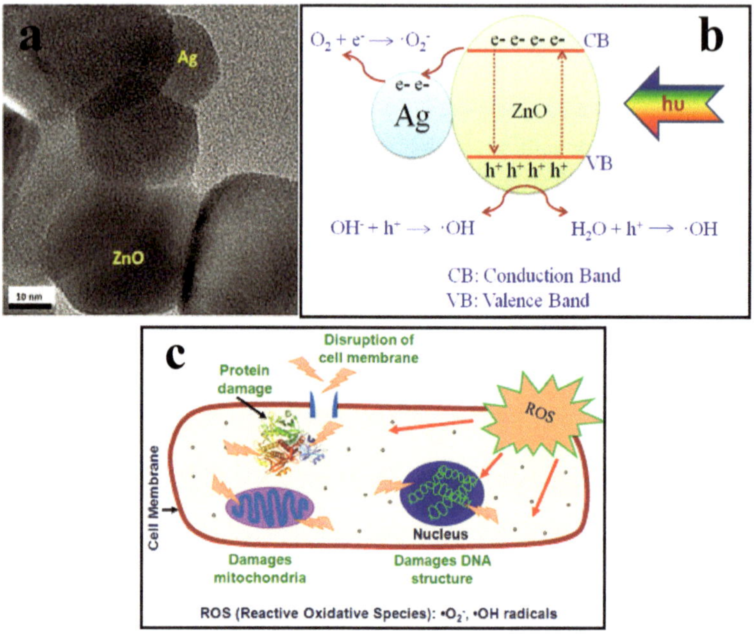

Figure 3.4 (a) Transmission electron microscopy image, (b) schematic photocatalytic mechanism, and (c) schematic cell disruption mechanism of Ag-impregnated ZnO nanoparticles. From Adhikari S, Banerjee A, Eswar NKR, Sarkar D, Madras G. (2015). Photocatalytic inactivation of *E. coli* by ZnO-Ag nanoparticles under solar radiation. *RSC Adv.*, **5**: 51067–51077. Published by The Royal Society of Chemistry.

electron excitation takes place from the valence band (VB) to the conduction band (CB), leaving behind a hole, when a photon having an energy of $h\nu$ equals or exceeds the energy bandgap of the semiconductor. However, these photogenerated electron and hole pairs can recombine and dissipate the input energy as heat. This recombination is prevented by suitable surface defect states or scavengers via trapping of the electron or hole. The hydroxyl groups on the surface also play an important role in trapping of a photogenerated electron or hole and actively participate in the reaction. The formation of superoxide anion experiences acid-base equilibrium to contend with the adsorbed water molecules for photogenerated holes and becomes the precursor for hydroxyl radicals (Adhikari et al., 2015a). These reactive oxidative species (Fig. 3.4b) such as peroxide and hydroxyl radicals inhibit the lag phase of bacterial growth via the disruption of the cell membrane. They invade the DNA structure to stop the functioning of minerals, proteins, and genetic materials that are leaked due to cell wall disruption, causing ultimate cell death, as shown in Fig. 3.4c. Recent explorations reveal its activity toward drug delivery and selective tumor destruction.

The food industry has found ZnO an essential antibacterial agent for food packaging against food pathogens. High reactivity, bioactivity, biocompatibility, and creative surface possession of nanoparticles favor food packaging technology (Sirelkhatim et al., 2015). A revolution has been witnessed for the food industry through nanotechnology implementation. For an antimicrobial purpose, ZnO nanoparticles possess excellent stability, with a long life span, in comparison to organic-based disinfectants. The significance of ZnO nanoparticles in various application fields has developed global interest. The documented actions of ZnO have stimulated a considerable range of versatile and novel applications.

3.5.3 Titanium Dioxide

The activity realization of titanium dioxide (TiO_2) was discovered 90 years ago. However, the celebrated breakthrough of its ability to split water by Honda–Fujishima got tremendous recognition (Fujishima et al., 1972). TiO_2 is one of those materials that have been exploited widely in energy-harvesting, environmental, and biological applications. The basic use of TiO_2 is in the form of pigment in paints due to a high refractive index and high optical absorption to impart brightness to the pigment. It also imparts high resistance toward UV light without any discoloration. Other commercial applications of TiO_2 are in toothpastes, sunscreens, self-cleaning, antifogging, lithography, sterilization, and prevention of metal corrosion via coating that includes glazes and enamels (Daghrir et al., 2013). Its hydrophilic property makes it suitable for medical device fabrication, air-conditioning filters, and sanitary ware. Industrially, it follows an economic trend as it contributes in major sectors like paper and paperboard, paints, and plastics. The increase in demand is due to its oxidative and hydrolysis properties, which improve the efficiency via different photo- and electrochemical processes. Thermodynamically stable phases of TiO_2 occur in three different forms, namely anatase (tetragonal structure), brookite (orthorhombic structure), and rutile (tetragonal structure). Alteration to the unique physicochemical properties is carried by modifying the intrinsic electronic structure through amendment of dopants and also fine-tuning the size and shape. The nanoparticles of TiO_2, less than 10–20 nm in size, are able to produce electricity via electrolytic water splitting. The methodology of photocatalytic degradation for environmental remedy is similar to that of the water-splitting process, whereby the only difference is the media under which the photochemical reaction proceeds (Tian et al., 2014). Taking into account the amount of solar spectrum impinging on earth, only 3%–4% of UV light can be beneficial for inducing the activity in TiO_2.

Apart from the above-mentioned synthesis processes for WO_3 and ZnO, methods such as flame reaction, melt-phase separation technique, thermal hydrolysis, and atmospheric pressure CVS have been followed to control the material properties (Zhou et al., 2010). Lately, the nucleation and growth kinetic study of nano-TiO_2 has gained attention. In a kinetic study of TiO_2 nanoparticles by Charbonneau et al., the growth rate was reported to be faster than the nucleation rate and tends to follow the Avrami model equation (Mehranpour et al., 2010). The formation of spherical nanoparticles follows two steps: the nucleation and aggregation of nanocrystalline particles (2–10 nm) take place followed by growth to self-assembled elongated nanofibers along (110) atomic planes of the rutile phase. The Avrami equation infers the kinetics of phase transformation, and the equation is

$$X = 1 - \exp(-kt)^n, \tag{3.1}$$

where X is the factor of transformation, k is the kinetic constant, t is the time, and n = 0.9–1.2 for TiO_2. The activation energy for the transformation can be calculated through the Arrhenius equation by incorporating the value of the kinetic constant. In another study, transition kinetics took place at a lower temperature for diffusion-controlled nanoparticles, which depicts nucleation and growth sites were more in these particles in comparison to others. The calculated nucleation activation energy for anatase to rutile phase transformation is 20 kJ/mol, which is the lowest activation energy found till date.

Some of the researched morphologies are nanotubes, nanofibers, honeycomb structure, hollow spheres, urchin-like structure, and etc. Hierarchical nanoarchitectural arrays of anatase TiO_2 that comprise a long nanowire trunk and short TiO_2 nanorods as branches on conductive fluorine-doped tin oxide (FTO) glass were fabricated by a hydrothermal method (Wu et al., 2013), and the power conversion efficiency observed for fabricated DSSC is 7.34%. Recently, vertically aligned anatase TiO_2 nanowire arrays on FTO glass were fabricated by a reactive pulsed DC magnetron sputtering method, but its application in DSSCs was still limited by its highly technical requirements as well as difficulty in increasing the lengths of nanowires (Su et al., 2012). In another study, 1D-nanorod/3D-nanotube-network-structure-coated FTO of rutile TiO_2 exhibited 7.68% efficiency, which is the highest reported value for rutile polymorph (Wang et al., 2015). There has been always a debate on the activity of the polymorphs of TiO_2. The anatase form of TiO_2 is considered photocatalytically active because of its chemical properties, enhanced charge carrier dynamics, and increased number of surface hydroxyl radicals. It has inherent surface band bending that forms impulsively in a deep region with a steep potential. Thus, its surface hole trapping dictates because spatial charge separation is achieved by the photogenerated hole transfer toward the surface of the particle via the strong upward band bending. However, in the rutile phase, bulk recombination of electrons and holes predictably occurs, and only the holes adjacent to the surface are trapped and transferred to the surface for reaction (Liu et al., 2011; Roy et al., 2011). For the purpose of energy conversion via DSSC, 1D and functionalized 3D nanostructures have been found fruitful. A schematic of DSSC using a TiO_2 photoanode is shown in Fig. 3.5. TiO_2 has also been experimented with for environmental remediation purposes, such as adsorption, degradation of organic compounds (phenol, acetaldehyde, textile dyes), bacterial killing, and wastewater purification. In the hunt for making TiO_2 a visibly active material, scientists manipulated TiO_2 by doping with transition metals and nonmetallic elements and also forming composites with other semiconductor materials. Particularly, doping has been found powerful enough to modify the electronic structure and construct heteroatomic structures for highly efficient photocatalysis under solar light (Rauf et al., 2011). Widespread research has been carried after the improvement in photocatalytic activity via nitrogen-doped TiO_2 for the photodegradation of methylene blue and gaseous acetaldehyde under illumination of visible light. To modify the electronic structure of TiO_2, almost all nonmetals and metal ions have been explored from the perspective of photodegradation (Bouras et al., 2007).

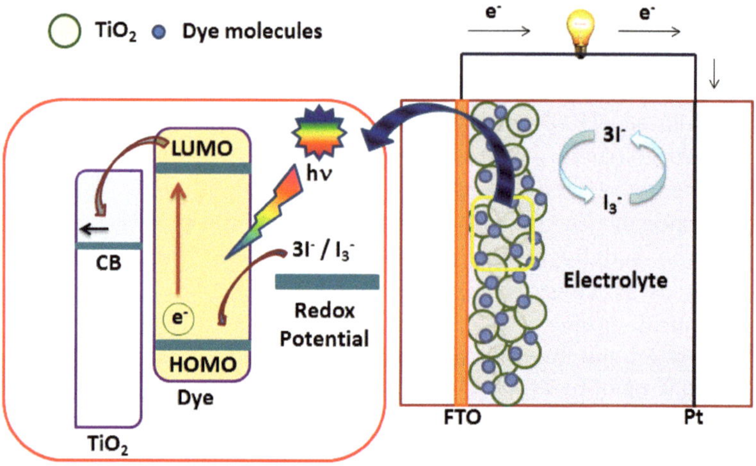

Figure 3.5 Schematic of dye-sensitized solar cells for a TiO_2 fabricated photoanode.

The inherent oxygen vacancies in a TiO_2 crystal are capable of absorbing gas that releases electrons into TiO_2, changing the resistance to higher or lower, explaining a typical sensing mechanism of TiO_2-based gas sensor. Nanostructures such as 3D dendrites, opal structures, nanobelts, branched nanorods, and hierarchical nanowires have been exploited for the detection of oxidative and reductive gases, for the monitoring of the chemical oxygen demand (COD), and as biosensors for a variety of food samples, body fluids, and cell cultures. Apart from the numerous factors that affect the sensing mechanism, the basic mechanism follows two processes—receptor process and transducer process—as shown in Fig. 3.6. The receptor process occurs at the TiO_2 surface via physisorption and chemisorption whereas the transducer process takes place through the transportation of electrons and their transformation to outer signals. The electron transfer pattern, such as surface, grain, and neck control, affect the process differently (Bai et al., 2014).

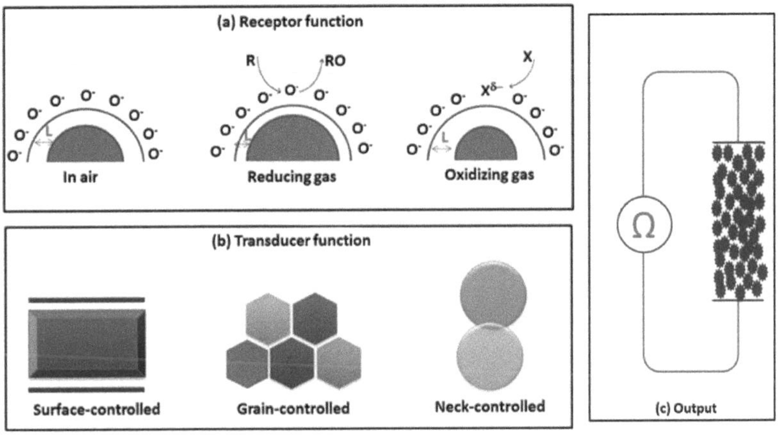

Figure 3.6 Schematic of gas sensing under a different mode.

Apart from the active nature of TiO_2, its biocompatibility and environment friendliness are a prospective interface for immobilization of biomolecules. The classified research fields on the basis of functional activities from the perspective of biological applications are the drug delivery system, cell imaging, photodynamic therapy for cancer, biosensors for biological assay, and genetic engineering. In the context of maximizing the therapeutic activity with minimal side effects, the ideal drug distribution takes place when the target is specified. For the purpose of drug delivery, TiO_2 nanoparticles of various shapes, such as whiskers, porous shapes, and capsules, have been developed. TiO_2 is studied as the carrier material and is used for drugs like valproic acid (Uddin et al., 2011), temozolomide (Lopez et al., 2006), sodium phenytoin (Cervera et al., 2009), and daunorubicin (DNR). The release rates of drugs are initially higher, which become constant over a period of time. Scientists have prepared 1D titania whiskers to enhance the antitumor intracellular efficiency of DNR by increase in the concentration of DNR inside SMMC-7721 (human hepatocarcinoma cell line), while relieving the toxic side effects and making ultimate use out of DNR. In addition, UV radiation further enhances the growth inhibition of cancerous cell by photocatalysis (Li et al., 2009). Another study revealed that the loading mode between TiO_2 and drug can also significantly affect the therapeutic efficacy (Qin et al., 2011). It is essential to direct the carrier drug to the target cell for effective therapeutic outcome. Recently, silica/titania hollow nanoparticles were designed, followed by modification with herceptin and a high amount of camptothecin loading. The drug showed highly efficient anticancer activity, with low toxicity toward normal cells (Kim et al., 2012).

Cell imaging via TiO_2 nanoparticles acts as a benchmark for noninvasive and targeted detection of molecules. The nanosized material prolongs the circulation time in the host body in comparison to small molecules and penetrates and accumulates in the tumor tissue. The insolubility of TiO_2 nanoparticles in an aqueous medium hampers its utility in the photodynamic therapy for cancer. This therapy administers a tumor by localizing a photosensitizing agent that activates by light of a specific wavelength. Another researcher (Seo et al., 2007) was able to design water-soluble, short, rod-shaped TiO_2 nanoparticles through a high-temperature nonhydrolytic method that showed higher toxicity toward human melanoma cells (A375) under UV light than commercial P-25 TiO_2 nanoparticles. Recognition of cancer stem cells remains a great challenge. Thus, a monoclonal antibody (mAb) Nilo1 with TiO_2 nanoparticles was successfully coupled, which was able to deplete cancer stem cells under UV illumination, in vitro (Elvira et al., 2012). Another biological assay of TiO_2 is electrochemical biosensors for disease diagnosis, drug discovery, environmental monitoring, and food safety. Nanoparticles, nanotubes, 3D macroporous matrices, nanosheets, and other sol-gel matrices prove to be effective as biosensors (Yin et al., 2013). High-performance nanosheet-based TiO_2 microspheres with a hollow core-shell structure with unique trumpet-shaped pores between the nanosheets were fabricated by Xie et al. (2011). The inner core of the microspheres easily adsorbs enzyme, and the substrate has easy access to the immobilized enzyme. Thus the substrate concentration and enzyme in this region will have increased chances of collision. This particular morphology showed good performance against H_2O_2 detection, with a low detection limit of 0.5 mM and at a linear range of 0.4–140 mM. The fast response and excellent long-term stability make this type of structure a promising biosensor for biological analysis. Recombination of DNA for direct genetic modification of organisms or populations of organisms is a technique being used in genetic engineering. Recently, detection of TiO_2-DNA nanocomposites was carried out by Gd (III)-modified particles that were able to retain specific locations inside cells by the conjugated DNA oligonucleotides hybridizing to intracellular targets while being monitored by MR imaging for in vivo detection of tumors (Endres et al., 2007). Another researcher modified the nanocomposites with Alizarin Red S (ARS), followed by transfection of ARS-labeled conjugates into PC-3M cells, for successful visualization (Thurn et al., 2009). More effort is required by researchers for advancement in these applications that finds new pathways for higher-end implementation.

3.5.4 Barium Titanate

Barium titanate ($BaTiO_3$) is the oldest ceramic material belonging to the perovskite family, having the general chemical formula ABO_3. It is highly infamous for its ferroelectric properties at or above room temperature. It is often used in multilayer ceramic capacitors for its high dielectric constant and low loss characteristics. Doping in $BaTiO_3$ has extended its application in semiconductors, piezoelectric devices, and thermistors. At Curie temperature T_c = 130°C, $BaTiO_3$ transforms from the paraelectric to the ferroelectric phase, resulting in internal stress at room temperature, affecting the ferroelectric property of $BaTiO_3$ (Jona, 1993; Koelzynski et al., 2005). It is generally proved to be a strong candidate for field-induced piezoelectric transducers due to its large polarization and large permittivity and also the large induced strains achievable in this material. Long ago, it was demonstrated as pyroelectric through infrared detection via a ferroelectric sensing element. It is used in ferroelectric-based pyroelectric detectors. In tetragonal $BaTiO_3$, the potential of Ti^{4+} along the *c* axis can be reversed by an electric field toward the polarizing direction and defeat the barrier, resulting in a reversible polarization. Till date, barium strontium titanate has been the most commercially successful pyroelectric (Akcay et al., 2006). Among the ferroelectric ceramics, $BaTiO_3$ can be potentially used to develop high-performance energy harvesters due to its lead-free environmental friendliness. Initially, aligned ZnO nanowire was used for this purpose, but its low electromechanical coupling coefficient limits its application toward energy harvesting (Kumar et al., 2012). In this context, vertically aligned ferroelectric $BaTiO_3$ nanowire arrays were deposited directly onto conductive FTO glass, and their efficient applicability toward nanoelectromechanical system (NEMS)-based energy harvester driven by base vibrations is explored. The design of the NEMS energy harvester was to have resonance below 1 kHz. However, for effective energy harvesting of ambient mechanical vibrations, the resonance resides in the range 1 Hz to 1 kHz. The vertically aligned nanowire arrays of ferroelectric $BaTiO_3$ showed a 16 times higher power density than the extensively used ZnO (Koka et al., 2014). With the discovery of high dielectric constant and urge to enhance the dielectric constant by ferroelectricity of $BaTiO_3$ ceramics, ABO_3 crystals are explored with respect to ferroelectricity. The electrical poling of $BaTiO_3$ ceramic provided the complete orientation of all the domains in the single crystals. The relatively low T_c of $BaTiO_3$ limits its practice in high-power transducers. It also exhibits low electromechanical coupling factor (k_p = 0.35) in regard to piezoelectric materials. In spite of having a low piezoelectric constant (d_{33} = 191 pC/N), the piezoelectric property of $BaTiO_3$ is considered to be closely related to both grain and domain sizes.

In the early 1940s, $BaTiO_3$ nanopowders were commercially prepared via a hydrothermal process. The stoichiometry of the starting solution employed with a Ba/Ti = 1.05–1.10:1 ratio drives the completion of the reaction. Although it is difficult to control precise stoichiometry via a hydrothermal method, the low cost, easy handling of the reagents, and fast reaction rate under low temperature ensure narrow size distribution of small deagglomerated powders (Lencka et al., 1993). The dissolution-precipitation models have been proposed in the context of homogeneous and heterogeneous reactions. In the heterogeneous preparation model, nuclei of $BaTiO_3$ form on the TiO_2 surface, followed by an in situ transformation mechanism. However, in the homogeneous model, low-concentration TiO_2 dissolves in the form of intermediate hydroxytitanium complexes that react with Ba^{2+} ions in the solution for precipitation of $BaTiO_3$ (Hertl et al., 1988). A typical synthesis process includes inorganic

$BaCl_2 \cdot 2H_2O$ salt and food-grade ammonium bicarbonate (NH_4HCO_3) to be dissolved separately, followed by the addition of clear ammonium bicarbonate solution to $BaCl_2$ solution at room temperature, and kept for 1 h to complete the formation of $BaCO_3$ nanorods. Subsequently, cosmetic-grade commercially available TiO_2 nanopowder (15 nm) was added within as-synthesis $BaCO_3$ nanorods and mixed for another 1 h to obtain homogeneous mixed powder precursors. The resultant white powder slurry was filtered, washed till the removal of chloride ions, and dried at 45°C for thermal treatment. Thermal treatment of the dried precursor powder was carried out in air to study the reaction mechanism of precursor $BaCO_3$ and TiO_2 mixed powders: 10 g of a dry precursor ($BaCO_3$ + TiO_2) mixture was placed in a 60 cm^3 Pt crucible and heated up to 1100°C with isothermal treatment for 2 min. Following that the furnace door was quickly opened and the Pt crucible was taken out and cooled down to 5°C through ice-cooled water. The cooling time from 1100°C to 5°C was ~5 s. This work was aimed to produce highly tetragonal, pure, and spherical $BaTiO_3$ nanopowder through a combined wet-chemical and rapid calcination process (Sarkar, 2010). The c/a ratio was 1.00908, and rattling Ti ions displaced an amount of 0.1 Å along the *c* axis (Fig. 3.7). The optimum content of sintering additive and sintering profile consolidated ~99% relative dense disks with a 200 nm average grain size. The ferroelectric domain exhibited a hysteresis loop. The remnant polarization and coercive field measured from a *P–E* hysteresis curve were 1.64 $\mu C/cm^2$ and 4.91 kV/cm at 20 Hz, respectively. The measured dielectric constant and tan δ were 3700 and 0.05 at 1 kHz, respectively.

$BaTiO_3$ is considered an important semiconductor. There have been few studies about the visible light–induced photocatalytic activity of $BaTiO_3$ because its large bandgap (E = 3.25 eV) restricts its use only to the narrow light response range of the UV spectrum. A relatively high recombination rate of electron-hole pairs leads to poor efficiency of the photocatalytic reaction for this material. In a recent study, a series of $BaTiO_3$-graphene nanocomposites have been synthesized by a facile one-pot hydrothermal approach, during which reduced graphene oxide and an intimate interfacial contact between $BaTiO_3$ nanoparticles and the graphene nanosheets were obtained. The role of graphene in the $BaTiO_3$-graphene nanocomposites is to act as an organic dye-like photosensitizer for large-bandgap $BaTiO_3$. The photosensitization process of $BaTiO_3$ by graphene transforms the wide-bandgap $BaTiO_3$ semiconductor into visible light photoactivity in dye degradation. The photocatalytic activity of the obtained

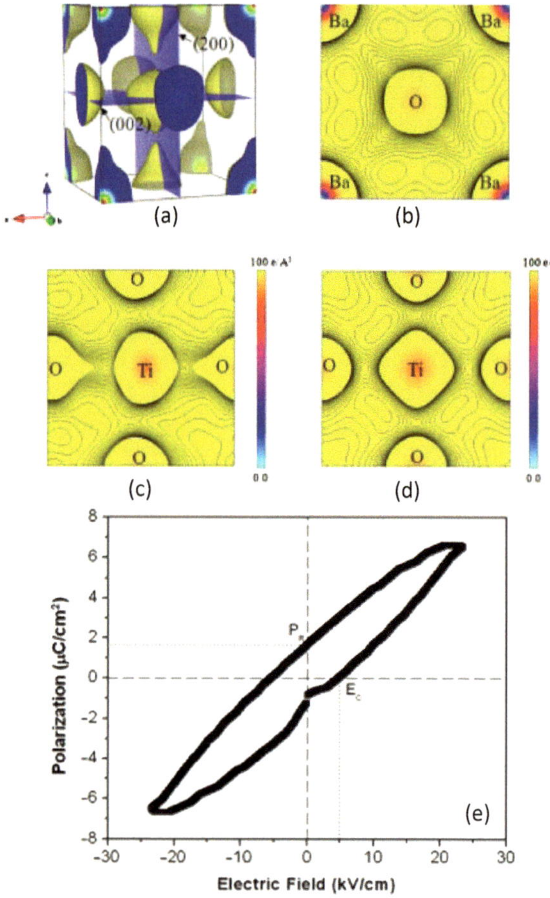

Figure 3.7 Electron density distributions obtained by the maximum entropy method (MEM) for the tetragonal phase in $BaTiO_3$: (a) the crystal structure of $BaTiO_3$; (b, c, d) (001), (200), and (002) planes, respectively (the isosurface level is 100 e/A^3, represented by the color scale bar); and (e) *P–E* characteristics of the sintered disk (ϕ, 15 mm; thickness, 100 mm) at 20 Hz; the hysteresis loop is observed as a signature the ferroelectric domain. Reproduced from Sarkar D. (2011). Synthesis and properties of $BaTiO_3$ nanopowders. *J. Am. Ceram. Soc.*, **94**(1): 106–110, with permission from Wiley.

(Wang et al., 2015). In another approach, surface sensitization was done by photodeposition of silver onto $BaTiO_3$ to produce a $BaTiO_3$-graphene composites for the degradation of methylene blue under visible light irradiation was investigated and a higher photocatalytic activity than that of pure $BaTiO_3$ was observed modified $BaTiO_3$. The higher tetragonal material content shows the highest photocatalytic efficiency even though there is a reduced surface area due to heat treatment to the material. The change in the catalytic performance of the catalyst is associated with improvements in charge carrier separation due to the internal space charge layer formed in a ferroelectric due to the internal dipole (Cui, 2015).

Recently, ceramic materials based on perovskite-like oxides have grabbed interest in nanotechnology research. However, evidence of any biological applications and investigations is rare in the literature. In this context, recently, an efficient method was obtained to stabilize $BaTiO_3$ nanoparticle dispersion using a noncovalent wrapping with poly-L-lysine (Ciofani et al., 2010). An in vitro testing of the obtained $BaTiO_3$ dispersions on the H9C2 cell line showed that the metabolism and viability are not affected up to a concentration of 5 g/ml in the culture medium. No apoptotic phenomenon was noticed at such concentration, and also differentiation of cells in myotubes occurred without much qualitative and quantitative differences from the control cultures. A self-assembling approach favored the formation of $BaTiO_3$ nanoparticles labeled with fluorescent bovine serum albumin (BSA) that considerably increases the protein internalization by the cells. It is, of course, the polymeric coating that influenced the behavior of nanoparticles concerning the interaction between the biological matter and particles. The intracellular environment causes the polymeric wrapping to undergo dissolution to achieve an actual material–cell interface, and other physical properties, such as a high dielectric constant and piezoelectricity, are rendered by the bulk material. The obtained results suggest that $BaTiO_3$ nanoparticles are able to retain a high potential in several biomedical applications, possibly ranging from sensors and transducers for biomolecule detection to "sensible" substrates for tissue engineering and to cellular nanostimulators.

3.5.5 Zirconium Dioxide

Zirconium dioxide (ZrO_2) is popularly known as zirconia, which is a white crystalline oxide of zirconium. It is a polycrystalline ceramic material without a glassy phase, and its existence is in several forms. The mechanical properties of zirconia are very similar to those of metals, and its color resembles tooth color (Manicone et al., 2007; Vagkopoulou et al., 2009). Recently, zirconia has emerged as a versatile and promising material among dental ceramics due to its excellent mechanical properties owing to the transformation-toughening mechanism. The polymorphic material transforms into three temperature-dependent forms: monoclinic (from room temperature to 1170°C), tetragonal (1170°C–2370°C), and cubic (2370°C up to melting point) (Denry et al., 2008). Stabilizing oxides, such as magnesia, calcium, ceria, and yttria, are added to zirconia to retain the tetragonal phase, which occurs as a metastable phase at room temperature, enabling a phenomenon called transformation toughening to take place. The partially stabilized crystalline tetragonal zirconia, in response to mechanical stimuli, such as tensile stress at crack tips, transforms to the more stable monoclinic phase with a local increase in volume of approximately 4%. This increase in volume closes the crack tips, effectively blunting crack propagation. It is this transformation-toughening process that gives zirconia its strength and toughness, exceeding all currently available sintered ceramics (Piconi et al., 1999). Compared to alumina, zirconia has twice the flexural strength, partly because of its grain size and the transformation-toughening mechanism. Zirconia-toughened alumina, magnesium-doped partially stabilized zirconia (Mg-PSZ), and 3 mol% yttrium-doped tetragonal zirconia polycrystal (3Y-TZP) are well-accepted bioceramics because of their biocompatibility and moderate mechanical properties (Christel et al., 1989).

Mg-PSZ is not a popular choice because of less sinterability and spontaneous transformation from t $\rightarrow$ m, leading to reduced mechanical properties and machining problem. The important and commercial 3Y-TZP is, however, less stable above a critical grain size and more susceptible to spontaneous t $\rightarrow$ m transformation, whereas a smaller grain size (<0.2 μm) is restricted to spontaneous transformation rate and favors high toughness and mechanical properties (Cottom et al., 1996; Chevalier et al., 2006). In the recent literature, equiaxed (0.2–0.5 μm) sintered 3Y-TZP microstructure provided improved mechanical properties for the application of bioimplants (Guazzato et al., 2004). However, the hardness of 3Y-TZP is inversely proportional to grain size but fracture toughness depends on transformation-toughening phenomenon of the zirconia grain size, which is predominantly controlled either by yttria content or by the sintering atmosphere. Hence, synchronization of such properties is favorable for the development of zirconia bioceramics. Since the last decade, different zirconia powder preparation processes, fabrication of identified shapes, and classic explanation on phase-transformation-dependent mechanical properties are documented for the development and application of zirconia structural ceramics (Tan et al., 2003). Several researchers have prepared doped zirconia nanoparticles through a hydroxide precipitation technique called the sol-gel process from inorganic precursor zirconium oxychloride and yttrium nitrate and subsequent calcination at a relatively high temperature, say 1200°C (Zalga et al., 2011). However, the hydroxide synthesis route has the tendency to form hard agglomeration, as reported elsewhere (Sarkar et al., 2007). In recent past, an approach of the direct rapid calcination (DPC) technique without any hydroxide precipitation has been implemented to synthesize narrow-particle-size-distributed tetragonal zirconia nanopowders from such inorganic salts. Following that, green cuboid blanks are fabricated through centrifuge casting, sintered at a prerequisite temperature, their mechanical properties evaluated, and nanoscale hydroxyapatite (HAp) particle deposition mechanism on sintered nanostructured zirconia in the presence of a simulated body fluid (SBF) solution proposed (Sarkar et al., 2013). DPC of inorganic salt solutions at a relatively low temperature of 600°C develops narrow-size-distributed 20 nm spherical tetragonal zirconia nanoparticles without formation of any hard agglomeration. Nanostructured grains (100–400 nm) are developed after sintering at 1350°C for 2 h. A combination of such high hardness and moderate fracture toughness is a suitable material for tribological application. Discontinuous and spherical HAp particles are deposited and adhered after 28 days' incubation within the SBF solution. This nanostructured zirconia could be used as a bioactive ceramic.

Solid oxide fuel cells (SOFCs) are electrochemical reactors that are efficient in converting chemical fuels to electrical power (Fig. 3.8). Low cost and stability in a redox environment explain the potentiality of yttria-stabilized zirconia (YSZ) as one of the most popular electrolytes. In one of the studies (Wei et al., 2008), asymmetric YSZ hollow-fiber membranes were prepared at relatively lower sintering temperatures and shorter sintering durations following a combined phase inversion and sintering method. The designed YSZ hollow-fiber membranes with an asymmetric structure show a high mechanical strength of 296.7 MPa and a gastight property. The inner porous structure of the YSZ hollow-fiber membranes provides good enhancement in nickel coating that results in low resistance between the electrolyte (YSZ) and the electrode (nickel). In current past, flake-shaped NiO-YSZ with nanocrystalline YSZ grains was synthesized using the sucrose-concentrated H_2SO_4 dehydration reaction. The properties of flake-shaped NiO-YSZ as anode materials were evaluated for catalytic activity on methane steam/CO_2 reforming reactions. The catalytic performance of the flake-shaped NiO-YSZ particles was found to show better activity than that of the mixed commercial NiO-YSZ in both steam reforming and CO_2 reforming of methane. The flake-shaped NiO-YSZ presented a much lower polarization resistance at an operating temperature below 800°C than the mixed commercial NiO-YSZ (Wu et al., 2012).

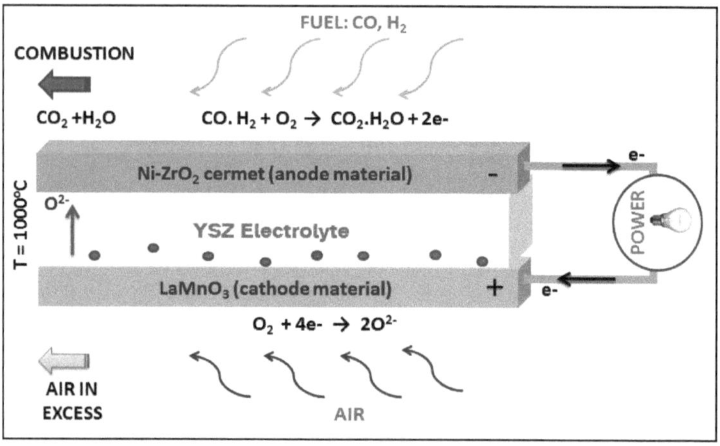

Figure 3.8 Solid oxide fuel cell model with yttria-stabilized zirconia as the electrolyte.

3.5.6 Hydroxyapatite

Nowadays artificial biomaterial use has increased due to limited accessibility of autografts, and a significant disadvantage of allografts is the risk of induction of transmissible diseases in the recipient (Deijkers et al., 1997; Daculsi et al., 2003). As a consequence, the interest shifts toward the use of synthetic implantable biomaterials for repair of bone defects, bone fracture repairs, alveolar ridge augmentation, dental implants, ear implants, and spine fusion that simulates the bone-bonding properties and morphology of natural bone. HAp is a calcium-phosphate-based ceramic and known to be one of the most important implantable biomaterials due to its biocompatibility and osteoconductive properties (Geesink et al., 2002; Shi et al., 2009). Natural bone is made of HAp (~70 wt%) and organic substances such as collagen, proteins, and water. Bone skeleton has HAp nanocrystals with an average size less than 50 nm and collagen through the formation of an interconnected net like open porous structure. HAp is one of the most biocompatible ceramics because of its significant chemical and physical resemblance to the mineral constituents of human bones and teeth (Hench, 1998). It is a bioactive ceramic and widely used as powder or in particulate forms in various bone repairs and as coatings for metallic prostheses to improve their biological properties (Jaffe et al., 1996). HAp is a thermodynamically stable calcium phosphate ceramic compound nearest to the pH, temperature, and composition of the physiological fluid. Recently, HAp has been used for a variety of biomedical applications, including matrices for drug release control (Palazzo et al., 2005).

In the recent past, different morphology HAp nanoparticles were synthesized using $(CH_3COO)_2Ca$ and KH_2PO_4 (Swain et al., 2012). In aqueous media, two inorganic salt solutions were separately prepared and slowly mixed to maintain the Ca:P molar ratio of 1.67. The solution pH was controlled through NH_4OH and a tris-buffer (tri-methylhydroxy aminomethane) solution for spherical and rod morphology, respectively. Spherical HAp nanoparticles, rod HAp nanoparticles, and fibroid HAp nanoparticles were developed at pH = 12.25, temperature = 298 K; pH = 9.5, temperature = 303 K; and pH = 5.25, temperature = 353 K, respectively, as shown in Fig. 3.9. Low-temperature synthesized different grade nanometric HAp nanoparticles were further studied for the protein BSA adsorption and release (Swain and Sarkar, 2013).

Figure 3.9 Transmission electron microscopy image of (a) spherical, (b) rod, and (c) fibroid HAp nanoparticles. Reprinted from *Applied Surface Science*, **286**, Swain SK, Sarkar D, Study of BSA protein adsorption/release on hydroxyapatite nanoparticles, 99–103, Copyright (2013), with permission from Elsevier.

The particle morphology, particle size, Ca:P molar ratio, and crystallinity assist the BSA adsorption characteristic on HAp nanoparticles. The rod morphology exhibits a relatively higher BSA adsorption capacity compared to the counterpart spherical and fibroid HAp nanoparticles. Preferably, optimized 1 g of HAp nanorod adsorbs 28 mg of BSA in 48 h and around 75% release within 96 h. Fast adsorption rate and slow but steady release kinetic are favorable for futuristic implantation and delivery media. Furthermore, the combination of freeze casting and cryogenic curing for 5 h in liquid nitrogen develops HAp-gelatin nanobiocomposite scaffolds with interconnected porosity and excellent protein adsorption/delivery property. Such a scaffold exhibits an 80 vol% porosity with pores of 50–200 μm, a moderate compressive strength of 5 MPa, and large strain failure up to 24%, as shown in Fig. 3.10 (Basu, Swain, and Sarkar, 2013). Both the structure and the compression response are similar to that of natural cancellous bone. A HAp gelatin scaffold has a higher affinity toward the adherence of BSA on surfaces due to its porous structure. A higher surface area and an interconnected porous structure of the scaffold facilitate adherence of more amounts of BSA compared to the synthesized HAp nanopowders. One gram of the HAp scaffold adsorbed 50 mg of BSA in 48 h and 60 wt% of release is observed in 24 h of the loaded scaffold.

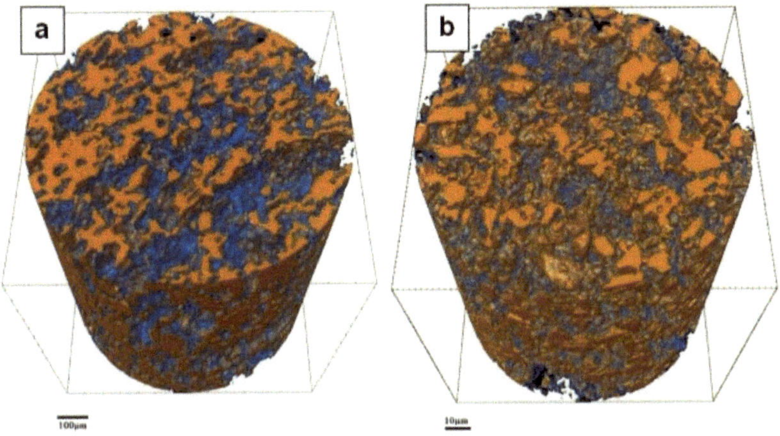

Figure 3.10 Representative 3D micro-CT scan image of the as-synthesized cryo-treated hydroxyapatite-gelatin macroporous scaffold (CHAMPS) (a) and BSA protein-loaded CHAMPS (b). Reproduced from Basu B, Swain SK, Sarkar D. (2013). Cryogenically cured hydroxyapatite–gelatin nanobiocomposite for bovine serum albumin protein adsorption and release. *RSC Adv.*, **3**(34): 14622–14633, with permission from the Royal Society of Chemistry.

HAp was typically considered to exhibit antiferroelectric order. HAp was considered to be centrosymmetric until the discovery of two polar symmetries of HAp with no center of symmetry, that is monoclinic P21 and hexagonal P63 (Haverty et al., 2005). These noncentrosymmetric and polar structures originated from ferroelectric ordering of the hydroxyl ions (OH) along the crystallographic [001] direction of the *c* axis. The possibility of antiferroelectric to ferroelectric transition was hypothetically marked possible. Not recently, synthetic HAp nanocrystalline films were found to exhibit strong piezoelectric and pyroelectric properties. A prediction of spontaneous polarization in HAp was made since 2005, but experimental evidence to reversible polarization, that is, ferroelectricity, was required (Tofail et al., 2005). Piezoresponse force microscopy was used to demonstrate ferroelectricity in nanocrystalline HAp, which strengthened the investigations regarding electrical polarization in biomineralization and diseases related to bone density. Thus, a systematic exploration toward lead and rare-metal-free ferroelectric devices could be stimulated for potential application in diverse areas. It has been found that HAp indeed reveals a ferroelectric order that stabilizes at nanoscale dimensions. This innovation stimulated a new pathway to the biomineralization mechanism and also formation of the characteristic hierarchical structure of bone. Reversible spontaneous polarization in HAp also opens up a number of novel technical applications, such as in electronics, in vivo energy harvesting, biosensors, and microelectromechanical systems (MEMS) (Lang et al., 2013). It has been explored as a biocompatible sorbent, namely nano-HAp-embedded gelatin (n-HAp-Gel) composite, prepared by an in situ precipitation method. The synthesized biocomposite was imperiled due to the removal of hexavalent chromium from the aqueous solution. The sorption of chromium(VI) onto n-HAp-Gel biocomposite follows the Langmuir isotherm, and the mechanism of sorption was mainly due to electrostatic-attraction-coupled reduction (Gopalakannan et al., 2015). In some cases, paramagnetic iron containing HAp nanoparticles are designed for efficient sorption properties. It was observed that an increase in the surface oxygen content and the formation of iron phases on the surface strongly enhance the Cu^{2+} adsorption capacity of HAp-based nanomaterials. The maximum sorption value was found to be 550–850 mg Cu^{2+} per gram of nanoparticles suspended in an aqueous solution at pH 7. It is almost 10 times the maximum values observed for HAp nanoparticle suspensions under the same conditions (Mercado et al., 2014).

3.5.7 Silicon Carbide

Silicon carbide (SiC) is an industrial material, explored since 1885, when Cowles and Cowles introduced its synthesis method. The development in the synthesis process by Acheson overtook in 1892 (Cowles, 1885; Acheson, 1892). It is known to be one of the important structural ceramic materials due to its properties such as retention of strength at high temperatures, excellent oxidation resistance, high wear resistance, thermal conductivity, and also good thermal shock resistance. Such properties are imparted due to strong covalent chemical bonding between silicon and carbon atoms. Despite the strong bonding and extremely low self-diffusion coefficient, densification of SiC requires high pressure and a temperature exceeding 2100°C (Ledoux et al., 1990). Thus, sintering aids are required to densify SiC. The crystal structure of SiC exhibits polytypism, where there are a number of different 1D ordering sequences without stoichiometric variation. SiC polytypes with a cubic structure are known as β-SiC. However, noncubical structures like hexagonal and rhombohedral are collectively known as α-SiC. Additives such as Al_2O_3 or Al_2O_3 + Y_2O_3 are added to SiC for liquid phase formation during pressureless sintering for higher densification. This approach possibly reduces the sintering temperature to 1850°C (Heimann, 2010). Among the densification methods, pressureless sintering in a solid state, infiltration of reaction-bonded SiC by molten silicon, and hot pressing are a few common methods. SiC formed from a solid state has a fine-grained equiaxed microstructure that shows good high-temperature creep and oxidation resistance but exhibits poor fracture toughness (3–4 MPa $m^{1/2}$) and has high flaw-sensitive strength. The prolonged soaking under high temperature during sintering causes abnormal grain growth, resulting in anisotropic α-SiC crystal phase formation. The other limitations are related to the melting point of silicon and densification via the hot pressing technique. Considering the case of SiC, methods such as in situ toughening, weight loss control by means of controlled sintering atmosphere, variation of sintering aids, and secondary phase crystallization are some of the processing and sintering possibilities. A large number of crystal growth experiments have been conducted by various scientists in an attempt to define stability regions for these basic SiC polytypes obtained from both vapors and liquids. In most of the cases, the resultant polytype was observed to be dependent on both the growth rate and temperature. Under high supersaturation (i.e., nonequilibrium conditions) the cubic polytype was initially formed in sublimation and growth experiments over the temperature range of 1000°C–2750°C. However, in some cases, this polytype subsequently disappears as a result of transformation and recrystallization processes, resulting in α-SiC formation. Under identical growth conditions of temperature and purity, but with lower supersaturations, it leads to lower growth rates and other basic SiC structures were observed to have formed. The stabilization of the polytype is important; thus the effect of impurity addition and nonstoichiometry should be taken into account. The impurity elements that have a pronounced influence on polytype formation are elements in groups III and V of the periodic table. In the context of impurities, electronic donors such as N and P and electronic acceptors such as Al and B are fruitful with respect to SiC. The first notation to stabilize the 4H structure was carried out through the influence of Al. A similar stabilizing effect was also attributed to boron, whereas, on the other hand, electron donors like N and P stabilize the cubic structure of SiC (Izhevskyi et al., 2005).

β-SiC is an attractive narrow-bandgap p-type semiconductor (2.2 eV), known for its interesting optical performance under visible region and chemical stability during the photochemical process. Thus, in recent past nanocomposites of β-SiC-TiO_2 heterojunction were formed by the sol-gel process. A 20 wt% SiC-loaded TiO_2 heterojunction nanocomposite was used as an effective photocatalyst for efficient degradation of rhodamine B in the presence of solar light. The enhanced degradation can be attributed to the high surface area (85.12 m^2/g) and the heterojunction between β-SiC and TiO_2 (Mishra et al., 2014). A new approach to synthesize a medium-surface-area and nonmicroporous TiO_2-β-SiC photocatalytic material was carried out via the shape memory synthesis (SMS) method, which follows a three-step process. The first step involves the sol-gel precipitation of titanium tetraisopropoxide in the presence of activated charcoal powder to produce high-surface-area TiO_2 supported on activated charcoal. The second step involves a gas-solid reaction between the above precursor and vapors of SiO at 1300°C under dynamic vacuum, which forms TiC-β-SiC by carboreduction, and the final step is air oxidation at a temperature ranging from 420°C to 600°C, producing a TiO_2-β-SiC material. The ratio of TiO_2 to SiC could be directly altered by adjusting the content of the TiO_2 precursor, whereas the anatase-to-rutile ratio could be modified by changing the postsynthesis calcination temperature. According to the SMS concept, the gross morphology of activated charcoal was taken care of during the carburization and the subsequent oxidation processes (Keller et al., 2004).

SiC is also used in the development of UV photodetectors. SiC-based photodetectors can achieve large gains, solar-blind operation, and high signal-to-noise ratios. These features make them ideal photodetectors that are certainly applicable for monitoring the UV spectrum without the need for solar rejection filters. Moreover, SiC also offers exceptional long-term stability even after operation under high-intensity UV radiation (up to $1000\,W/m^2$) and at a high operating temperature. Recently, an extended state of the art was formulated with a goal to develop a small, low-cost, highly reliable UV detector with high responsivity and repeatability. Ultrathin nano-Schottky contact with high efficiency and responsivity of 0.18 A/W with a very stable baseline and repeatability was achieved. Even at a high operating temperature of 300°C, the fabricated detectors displayed good performance in responsivity, stability, and repeatability (Fan, 2014; Aldalbahi et al., 2016).

Apart from being a structural material, SiC has been explored as a bioinspired material due to its properties such as charge-driven selective localization in cells for bioimaging applications. The SiC nanoparticle's surface can be functionalized by various functional groups to govern the intracellular fate (Alekseev et al., 2007). The surface of SiC nanoparticles possesses carboxylic groups that are deprotonated and negatively charged at certain pH. The reaction then transforms the carboxylic moieties into a positively charged aminoethylcarboximide functional group and protonated under physiological conditions. Thus, the negatively charged SiC nanoparticles target the cell nuclei. However, the positively charged SiC nanoparticles get accumulated in the cytosol because of unsuccessful nuclei penetration. The excellent fluorescence of both negatively and positively charged nanoparticles helps in visualization of cells treated by nanoparticles (Serdiuk et al., 2013). For the same purpose, recently, SiC-based nanoparticles were delivered into living eukaryotic cells, facilitated in the presence of cationic and anionic cell-penetrating peptides (CPPs). The cationic "sweet arrow peptide" (SAP) was found suitable for delivery of surface containing carboxylic group SiC nanoparticles, whereas amine-grafted SiC nanoparticles get delivered into cell nuclei by glutamate-based anionic SAP-E (an analog of SAP) (Serdiuk et al., 2015). The functional-group-modified SiC nanoparticle combined with cell-penetrating peptide introduces an additional mode of delivery control. A similar study through optimized carrier CPPs can be done for other nanoparticles.

3.6 Conclusions

The interesting and new genre of nanocrystals can be innovated by proper tuning of chemical reactions. Though the above ceramic materials are vastly explored for respective application fields, their versatility, combined with technological advances, makes them fit for multipurpose use in terms of energy, the environment, and the well-being of humanity. The unlimited research opportunity in many laboratories and industries around the globe provides a platform to understand the basic science behind ceramic nanomaterials. The future is undoubtedly bright for ceramic nanomaterials that can serve functionally in countless ways.

References

Acheson AG. (1892). England Patent No. 17911.

Adhikari S, Sarkar D. (2013). Electrochemical response for spherical and rod-shaped WO_3 nanoparticles. *ISRN Nanotechnol.*, 2013: 2793928 (1–9).

Adhikari S, Sarkar D. (2014a). High efficient electrochromic WO_3 nanofibers. *Electrochim. Acta*, **138**: 115–123.

Adhikari S, Sarkar D. (2014b). Hydrothermal synthesis and electrochromism of WO_3 nanocuboids. *RSC Adv.*, **4**: 20145–20153.

Adhikari S, Sarkar D. (2015). Synthesis and electrochemical properties of nanocuboids and nanofiber WO_3. *J. Electrochem. Sci.*, **162**(1): H1–H7.

Adhikari S, Banerjee A, Eswar NKR, Sarkar D, Madras G. (2015a). Photocatalytic inactivation of *E. coli* by ZnO-Ag nanoparticles under solar radiation. *RSC Adv.*, **5**: 51067–51077.

Adhikari S, Sarkar D, Madras G. (2014). Synthesis and photocatalytic performance of quasi-fibrous ZnO. *RSC Adv.*, **4**: 55807–55814.

Adhikari S, Sarkar D, Madras G. (2015). Highly efficient WO_3-ZnO mixed oxides for photocatalysis. *RSC Adv.*, **5**: 11895–11904.

Adhikari S, Sarkar D, Maiti HS. (2014). Synthesis and characterization of WO_3 spherical nanoparticles and nanorods. *Mater. Res. Bull.*, **49**: 325–330.

Adhikari S. (2015). *Nanostructured WO_3 for Electrochromic and Photocatalytic Applications*. e-Thesis, NIT Rourkela, India.

Akcay G, Zhong S, Alpay SP, Mantese JV. (2006). Dynamic pyroelectric enhancement of homogeneous ferroelectric materials. *Solid State Commun.*, **137**: 589–594.

Aldalbahi A, Li E, Rivera M, Velazquez R, Altalhi T, Peng X, Feng PX. (2016). A new approach for fabrications of SiC based photodetectors. *Sci. Rep.*, **6**: 1–10.

Alekseev SA, Zaitsev VN, Botsoa J, Barbier D. (2007). Fourier transform infrared spectroscopy and temperature-programmed desorption mass spectrometry study of surface chemistry of porous H-SiC. *Chem. Mater.*, **19**(9): 2189–2194.

Bai J, Zhou B. (2014). Titanium dioxide nanomaterials for sensor applications. *Chem. Rev.*, **114**: 10131–10176.

Basu B, Swain SK, Sarkar D. (2013). Cryogenically cured hydroxyapatite–gelatin nanobiocomposite for bovine serum albumin protein adsorption and release. *RSC Adv.*, **3**(34): 14622–14633.

Boulova M, Lucazeau G. (2002). Crystallite nanosize effect on the structural transitions of WO_3 studied by raman spectroscopy. *J. Solid State Chem.*, **167**: 425–434.

Bouras P, Stathatos E, Lianos P. (2007). Pure versus metal-ion-doped nanocrystalline titania for photocatalysis. *Appl. Catal. B: Environ.*, **73**: 51–59.

Cao G, Wang Y. (2011). *Nanostructures and Nanomaterials: Synthesis, Properties and Applications*. World Scientific, Singapore.

Cervera BEH, González-Azcorra SA, Gattorno GR, López T, Ortiz-Islas E, Oskam G. (2009). Controlled release of phenytoin from nanostructured TiO_2 reservoirs. *Sci. Adv. Mater.*, **1**: 63.

Chang CE, Tran VH, Wang JB, Fuh YK, Lin LW. (2010). Direct-write piezoelectric polymeric nanogenerator with high energy conversion efficiency. *Nano Lett.*, **10**: 726–731.

Chang J, Ahmed R, Wang H, Liu H, Li R, Wang P, Waclawik ER. (2003). ZnO nanocones with high-index {1011} facets for enhanced energy conversion efficiency of dye-sensitized solar cells. *J. Phys. Chem. C*, **117**: 13836–13844.

Chevalier J. (2006). What future for zirconia as a biomaterial? *Biomaterials*, **27**: 535–543.

Christel P, Meunier A, Heller M, Torre J, Peille C. (1989). Mechanical properties and short-term in vivo evaluation of yttrium-oxidepartially-stabilized zirconia. *J. Biomed. Mater. Res.*, **23**: 45–61.

Ciofania G, Dantib S, Moscatoc S, Albertazzid L, D'Alessandro D, Dinuccie D, Chiellini F, Petrini M, Menciassia A. (2010). Preparation of stable dispersion of barium titanate nanoparticles: potential applications in biomedicine. *Colloids Surf. B*, **76**: 535–543.

Cottom BA, Mayo MJ. (1996). Fracture toughness of nanocrystalline ZrO_2-3 mol percent Y_2O_3 determined by vickers indentation. *Scr. Mater.*, **34**: 809–814.

Cowles AH, Cowles EH. (1885). Electric smelting furnace. US Patent No. 319945.

Cui Y. (2015). *Ferroelectric Barium Titanate for Semiconductor Photocatalytic Application*. PhD thesis, Queen Mary University of London, London.

Daculsi G, Laboux O, Malard O, Weiss P. (2003) Current state of the art of biphasic calcium phosphate bioceramics. *J. Mater. Sci. Mater. Med.*, **14**: 195–200.

Daghrir R, Drogui P, Robert D. (2013). Modified TiO_2 for environmental photocatalytic applications: a review. *Ind. Eng. Chem. Res.*, **52**: 3581–3599.

Deijkers RM, Bloem RM, Petit PC, Brand R, Vehmeyer SW, Veen MR. (1997). Contamination of bone allografts: analysis of incidence and predisposing factors. *J. Bone Joint Surg.*, **79**: 161–166.

Denry I, Kelly J. (2008). State of the art zirconia for dental applications. *Dent. Mater.*, **24**: 299–307.

Elvira G, Moreno B, Valle ID, Garcia-Sanz JA, Canillas M, Chinarro E, Jurado JR, Silva A. (2012). Targeting neural stem cells with titanium dioxide nanoparticles coupled to specific monoclonal antibodies. *J. Biomater. Appl.*, **26**: 1069–1089.

Endres PJ, Paunesku T, Vogt S, Meade TJ, Woloschak GE. (2007). DNA-TiO_2 nanoconjugates labeled with magnetic resonance contrast agents. *J. Am. Chem. Soc.*, **129**: 15760–15761.

Fan J, Chu PK. (2014). *General Properties of Bulk SiC in Silicon Carbide Nanostructures: Fabrication, Structure, and Properties.* Springer International, Switzerland.

Fang X, Bando Y, Gautum UK, Zhai T, Zeng H, Xu XJ. (2009). ZnO and ZnS nanostructures: ultraviolet light emitters, lasers and sensors. *Crit. Rev. Solid State Mater. Sci.*, **34**, 190–223.

Fei P, Yeh PH, Zhou J, Xu S, Gao Y, Song J, Gu Y, Huang Y, Wang ZL. (2009). Piezoelectric potential gated field effect transistor based on a free standing ZnO wire. *Nano Lett.*, **9**(10): 3435–3439.

Fujishima A, Honda K. (1972). Electrochemical photolysis of water at a semi-conductor electrode. *Nature*, **238**: 37–38.

Geesink RGT. (2002). Osteoconductive coating for total joint arthroplasty, *Clin. Orthop. Relat. Res.*, **395**: 53–65.

Gopalakannan V, Viswanathan N. (2015). Development of nano-hydroxyapatite embedded gelatin biocomposite for effective chromium(VI) removal. *Ind. Eng. Chem. Res.*, **54**: 12561–12569.

Gu Z, Li H, Zhai T, Yang W, Xia Y, Ma Y, Yao J. (2007). Large-scale synthesis of single-crystal hexagonal tungsten trioxide nanowires and electrochemical lithium intercalation into the nanocrystals. *J. Solid State Chem.*, **180**: 98–105.

Guazzato M, Albakry M, Ringer SP, Swain MV. (2004). Strength, fracture toughness and microstructure of a selection of all-ceramic materials. Part II: zirconia-based dental ceramics. *Dent Mater.*, **20**: 449–456.

Haverty D, Tofail SAM, Stanton KT, McMonagle JB. (2005). Structure and stability of hydroxyapatite: density functional calculation and rietveld analysis. *Phys. Rev. B*, **71**: 94103.

Hayashi H, Hakuta Y. (2010). Hydrothermal synthesis of metal oxide nanoparticles in supercritical water. *Materials*, **3**: 3794–3817.

Heimann RB. (2010). *Classic and Advanced Ceramics: From Fundamentals to Applications*. Wiley, Weinheim.

Hench LL. (1998). Bioceramic. *J. Am. Ceram. Soc.*, **81**: 1705–1728.

Hertl W. (1988). Kinetics of barium titanate synthesis. *J. Am. Ceram. Soc.*, **71**: 879–883.

Hui P, Jizhong L, Han S, Yuanping F, Cheekok P, Jianyi L. (2006). Hydrogen storage of ZnO and Mg doped ZnO nanowires. *Nanotechnology*, **17**: 2963–2967.

Izhevskyi VA, Bressianiw AHA, Bressian JC. (2005). Effect of liquid phase sintering on microstructure and mechanical properties of Yb_2O_3-AlN containing SiC-based ceramics. *J. Am. Ceram. Soc.*, **88**(5): 1115–1121.

Jaffe WL, Scott DF. (1996). Current concepts review: total hip arthroplasty with hydroxyapatite coated prostheses. *J. Bone Joint Surg.*, **78A**: 1918–1934.

Jona F, Shirane G. (1993). *Ferroelectric Crystals*. Dover, New York.

Keller N, Keller V, Barraud E, Garin F, Ledoux MJ. (2004). Synthesis and characterization of a new medium surface area TiO_2-b-SiC material for use as photocatalyst. *J. Mater. Chem.*, **14**: 1887–1895.

Kiely CJ, Fink J, Brust M, Bethell D, Schiffrin DJ. (1998). Spontaneous ordering of bimodal ensembles of nanoscopic gold clusters, *Nature*, **396**: 444–446.

Kim C, Kim S, Oh WK, Choi M, Jang J. (2012). Efficient intracellular delivery of camptothecin by silica/titania hollow nanoparticles. *Chem. Eur. J.*, **18**: 4902–4908.

Kim SJ, Hwang IS, Choi JK, Lee JH. (2011). Gas sensing characteristics of WO_3 nanoplates prepared by acidification method. *Thin Solid Films*, **519**: 2020–2024.

Koelzynski A, Tkacz-Smiech K. (2005). From molecular picture to the band structure of cubic and tetragonal barium titanate. *Ferroelectrics*, **314**: 123–134.

Koka A, Zhoub Z, Sodano HA. (2014). Vertically aligned $BaTiO_3$ nanowire arrays for energy harvesting. *Energy Environ. Sci.*, **7**: 288–296.

Kong XY, Ding Y, Yang RS, Wang ZL. (2004). Single-crystal nanorings formed by epitaxial self-coiling of polar-nanobelts. *Science*, **303**: 1348–1351.

Kulik OP. (1999). Current state of development of new ceramic materials (review of foreign literature). *Powder Metall. Met. Ceram.*, **38**: 93–101.

Kumar B, Kim SW. (2012). Energy harvesting based on semiconducting piezoelectric ZnO nanostructures. *Nano Energy*, **1**: 342–355.

Lang SB, Tofail SAM, Kholkin AL, Wojtas M, Gregor M, Gandhi AA, Wang Y, Bauer S, Krause M, Plecenik A. (2013). Ferroelectric polarization in nanocrystalline hydroxyapatite thin films on silicon. *Sci. Rep.*, **3**(2215): 1–7.

Ledoux MJ, Guille J, Hantzer S, Dubots D. (1990). US Patent, US 4914070.

Lencka MM, Riman RE. (1993). Thermodynamic modeling of hydrothermal synthesis of ceramic powder. *Chem. Mater.*, **5**: 61–70.

Leonov V, Vullers RJM. (2009). Wearable electronics self-powered by using human heat: the state of the art and the perspective. *Renewable Sustainable Energy Rev.*, **1**: 062701–062714.

Li C, Lin Y, Li F, Zhu L, Sun D, Shen L, Chen Y, Ruan S. (2015). Hexagonal ZnO nanorings: synthesis, formation mechanism and trimethylamine sensing properties. *RSC Adv.*, **5**: 80561–80567.

Li Q, Wang X, Lu X, Tian H, Jiang H, Lv G, Guo D, Wu C, Chen B. (2009). The incorporation of daunorubicin in cancer cells through the use of titanium dioxide whiskers. *Biomaterials*, **30**: 4708–4715.

Li Y, Meng GW, Zhang LD, Phillip F. (2000). Ordered semiconductor ZnO nanowires arrays and their photoluminescence properties. *Appl. Phys. Lett.*, **76**: 2011–2013.

Liao Y, Li H, Liu Y, Zou Z, Zeng, Xie C. (2010). Characterization of photoelectric properties and composition effect of $TiO_2/ZnO/Fe_2O_3$ composite by combinatorial methodology. *J. Comb. Chem.*, **12**: 883–889.

Liu S, Yu J, Jaroniec M. (2011). Anatase TiO_2 with dominant high-energy {001} facets: synthesis, properties, and applications. *Chem. Mater.*, **23**: 4085–4093.

López T, Sotelo J, Navarrete J, Ascencio JA. (2006). Synthesis of TiO_2 nanostructured reservoir with temozolomide: structural evolution of the occluded drug. *Opt. Mater.*, **29**: 88–94.

Long H, Zeng W, Zhang H. (2015). Synthesis of WO_3 and its gas sensing: a review. *J. Mater. Sci. Mater. Electron.*, **26**: 4698–4707.

Mahmood MA, Baruah S, Dutta J. (2011). Enhanced visible light photocatalysis by manganese doping or rapid crystallization with ZnO nanoparticles. *Mater. Chem. Phys.*, **130**: 531–535.

Manicone PF. (2007). An overview of zirconia ceramics: basic properties and clinical applications. *J. Dent.*, **35**: 819–826.

Manjunath K, Ravishankar TN, Kumar D, Priyanka KP, Varghese T, Naika HR, Nagabhushana H, Sharma SC, Dupont J, Ramakrishnappa T, Nagaraju G. (2014). Facile combustion synthesis of ZnO nanoparticles using cajanus cajan (L) and its multidisciplinary applications. *Mater. Res. Bull.*, **57**: 325–334.

Mehranpour H, Askari M, Ghamsari MS, Farzalibeik H. (2010). Study on the phase transformation kinetics of sol-gel derived TiO_2 nanoparticles. *J. Nanomater.*, **2010**: 626978(1–5).

Mercado DF, Magnacca G, Malandrino M, Rubert A, Montoneri E, Celi L, Prevot AB, Gonzalez MC. (2014). Paramagnetic iron-doped hydroxyapatite nanoparticles with improved metal sorption properties. *ACS Appl. Mater. Interfaces*, **6**: 3937–3946.

Mishra G, Parida KM, Singh SK. (2014). Solar light driven rhodamine B degradation over highly active b-SiC–TiO_2 nanocomposite. *RSC Adv.*, **4**: 12918–12928.

Nagaraja R, Kottam N, Girija CR, Nagabhushana BM. (2012). Photocatalytic degradation of Rhodamine B dye under UV/solar light using ZnO nanopowder synthesized by solution combustion route. *Powder Technol.*, **215**: 91–97.

Nikitin ED, Pavlov PA, Popov AP, Nikitina HE. (1995). Critical properties of hydrogen peroxide determined from direct measurements. *J. Chem. Thermodyn.*, **27**: 945–952.

Nishide T, Mizukami F. (1995). Crystal structures and optical properties of tungsten oxide films prepared by a complexing-agent-assisted sol-gel process. *Thin Solid Films*, **259**: 212–217.

Ohshima E, Ogino H, Niikura E, Maeda K, Sato M, Ito M, Fukuda T. (2004). Growth of the 2-in-size bulk ZnO single crystals by the hydrothermal method. *J. Cryst. Growth*, **260**: 166–170.

Palazzo B, Sidoti MC, Roveri N, Tampieri A, Sandri M, Bertolazzi L, Galbusera F, Dubini G, Vena P, Contro R. (2005). Controlled drug delivery from porous hydroxyapatite grafts: an experimental and theoretical approach. *Mater. Sci. Eng. C*, **25**: 207–213.

Park S, Lee DW, Lee JC, Lee JH. (2003). Photocatalytic silver recovery using ZnO nanopowders synthesized by modified glycine-nitrate process. *J. Am. Ceram. Soc.*, **86**: 1508–1512.

Piconi C, Maccauro G. (1999). Zirconia as a ceramic biomaterial. *Biomaterials*, **20**: 1–25.

Qin Y, Sun L, Li X, Cao Q, Wang H, Tang X, Ye L. (2011). Highly water dispersible TiO_2 nanoparticles for doxorubicin delivery: effect of loading mode on therapeutic efficacy. *J. Mater. Chem.*, **21**: 18003- 18010.

Rajeshwari J, Kishore PS, Vishwanathan B, Varadarajan TK. (2007). Facile hydrogen evolution reaction on WO_3 nanorods. *Nanoscale Res. Lett.*, **2**: 496–503.

Rauf MA, Meetani MA, Hisaindee S. (2011). An overview on the photocatalytic degradation of azo dyes in the presence of TiO_2 doped with selective transition metals. *Desalination*, **276**: 13–27.

Reddy AJ, Kokila MK, Nagabhushana H, Rao JL, Shivakumara C, Nagabhushana BM, Chakradhar RPS. (2011). Combustion synthesis, characterization and Raman studies of ZnO nanopowders. *Spectrochim. Acta Part A*, **1**: 53–58.

Roy P, Berger S, Schmuki P. (2011). TiO_2 nanotubes: synthesis and applications. *Angew. Chem. Int. Ed.*, **50**: 2904–2939.

Sarkar D, Mohapatra D, Ray S, Bhattacharyya S, Adak S, Mitra N. (2007). Synthesis and characterization of sol-gel derived ZrO_2 doped Al_2O_3 nanopowder. *Ceram. Int.*, **33**: 1275–1282.

Sarkar D, Swain SK, Adhikari S, Reddy BS, Maiti HS. (2013). Synthesis, mechanical properties and bioactivity of nanostructured zirconia. *Mater. Sci. Eng. C*, **33**: 3413–3417.

Sarkar D. (2010). Synthesis and properties of $BaTiO_3$ nanopowders. *J. Am. Ceram. Soc.*, **94**(1): 106–110.

Seo JW, Chung H, Kim MY, Lee J, Choi IH, Cheon J. (2007). Development of water-soluble single-crystalline TiO_2 nanoparticles for photocatalytic cancer-cell treatment. *Small*, **3**: 850–853.

Serdiuk T, Bakanovich I, Lysenko V, Alekseev SA, Skryshevsky VA, Afonin S, Berger E, Gélöen A, Komarov IV. (2015). Delivery of SiC-based nanoparticles into live cells driven by cell-penetrating peptides SAP and SAP-E. *RSC Adv.*, **5**: 20498–20502.

Serdiuk T, Lysenko V, Mognetti B, Skryshevsky VA, Gélöen A. (2013). Impact of cell division on intracellular uptake and nuclear targeting with fluorescent SiC-based nanoparticles. *J. Biophotonics*, **6**: 291–297.

Shen M, Du Y, Yang P, Jiang LJ. (2005). Morphology control of the fabricated hydrophobic gold nanostructures in W/O microemulsion under microwave irradiation. *J. Phys. Chem. Solids*, **66**: 1628–1634.

Shi Z, Huang X, Cai Y, Tang R, Yang D. (2009). Size effect of hydroxyapatite nanoparticles on proliferation and apoptosis of osteoblast-like cells. *Acta Biomater.*, **5**: 338–345.

Sirelkhatim A, Mahmud S, Seeni A, Kaus NHM, Ann LC, Bakhori SKM, Hasan H, Mohamad D. (2015). Review on ZnO nanoparticles: antibacterial activity and toxicity mechanism. *Nano-Micro Lett.*, **7**(3): 219–242.

Slama R, Ghribi F, Houas A, Barthou C, Mir LE. (2011). Visible photocatalytic properties of vanadium doped zinc oxide aerogel nanopowder. *Thin Solid Films*, **519**: 5792–5795.

Su I. (2012). Low temperature synthesis of transparent, vertically aligned anatase TiO_2 nanowire arrays: application to dye sensitized solar cells. *Bull. Korean Chem. Soc.*, **33**: 1989–1992.

Swain SK, Sarkar D. (2013). Study of BSA protein adsorption/release on hydroxyapatite nanoparticles. *Appl. Surf. Sci.*, **286**: 99–103.

Swain SK, Dorozhkin SV, Sarkar D. (2012). Synthesis and dispersion of hydroxyapatite nanopowders. *Mater. Sci. Eng. C*, **32**: 1237–1240.

Tan Q, Zhang Z, Tang Z, Luo S, Fang K. (2003). Influence of polyelectrolyte dispersant on slip preparation of nano-sized tetragonal polycrystals zirconia for aqueous-gel-tape-casting process. *Mater. Chem. Phys.*, **80**: 615–619.

Thurn KT, Paunesku T, Wu A, Brown E, Lai B, Vogt S, Maser J, Aslam M, Dravid V, Bergan R, Woloschak GE. (2009). Labeling TiO_2 nanoparticles with dyes for optical fluorescence microscopy and determination of TiO_2-DNA nanoconjugate stability. *Small*, **5**: 1318–1325.

Tian J, Zhao Z, Kumar A, Boughton RI, Liu H. (2014). Recent progress in design and applications of one-dimensional TiO_2 nanostructured surface heterostructures: a review. *Chem. Soc. Rev.*, **43**: 6920–6937.

Tofail SAM, Haverty D, Stanton KT, McMonagle JB. (2005). Structural order and dielectric behaviour of hydroxyapatite. *Ferroelectrics*, **319**: 117–123.

Tryba B, MIszcz M, Morawski AW. (2009). Photocatalytic activity of TiO_2-WO_3 composites. *Int. J. Photoenergy*, 1–7.

Uddin MJ, Mondal D, Morris CA, Lopez T, Diebold U, Gonzalez RD. (2011). An in vitro controlled release study of valproic acid encapsulated in a titania ceramic matrix. *Appl. Surf. Sci.* **257**: 7920–7927.

Vagkopoulou T, Koutayas SO. (2009). Zirconia in dentistry: Part 1. Discovering the nature of an upcoming bioceramic. *Euro. J. Esthet. Dent.*, **2**: 2–23.

Wachtmann JB. (1983). Ceramic fever: advanced ceramics in Japan. *Ceram. Int.*, **121**(6): 24–33.

Wan Q, Lin CL, Yu XB, Wang TH. (2004). Room-temperature hydrogen storage characteristics of ZnO nanowires. *Appl. Phys. Lett.*, **84**: 124–126.

Wang H, Wang B, Yu J, Hu Y, Xia C, Zhang J, Liu R. (2015). Significant enhancement of power conversion efficiency for dye sensitized solar cell using 1D/3D network nanostructures as photoanodes. *Sci. Rep.*, **9305**: 1–5.

Wang RX, Zhu Q, Wang WS, Fana CM, Xu AW. (2015). $BaTiO_3$–graphene nanocomposites: synthesis and visible light photocatalytic activity. *New J. Chem.*, **39**: 4407–4413.

Wei CC, Li K. (2008). Yttria-stabilized zirconia (YSZ)-based hollow fiber solid oxide fuel cells. *Ind. Eng. Chem. Res.*, **47**: 1506–1512.

Wu WQ, Lei BX, Rao HS, Xu YF, Wang YF, Su CY, Kuang DB. (2013). Hydrothermal fabrication of hierarchically anatase TiO_2 nanowire arrays on FTO Glass for dye-sensitized solar cells. *Sci. Rep.*, **1352**: 1–7.

Wu Y, Wang W, Wang K, Zeng Y, Dong D, Shao Z, Wang H. (2012). Morphology and catalytic performance of flake-shaped NiO-yttria-stabilized zirconia (YSZ) particles with nanocrystalline YSZ grains. *Ind. Eng. Chem. Res.*, **51**: 6387–6394.

Xie Q, Zhao Y, Chen X, Liu H, Evans DG, Yang W. (2011). Nanosheet-based titania microspheres with hollow core-shell structure encapsulating horseradish peroxidase for a mediator-free biosensor. *Biomaterials*, **32**: 6588–6594.

Yin Q, Zhu B, Zeng H. (2010). *Microstructure, Property and Processing of Functional Ceramics*. Metallurgical Industry Press, Springer, Berlin, Heidelberg.

Yin ZF, Wu L, Yang HG, Su YH. (2013). Recent progress in biomedical applications of titanium dioxide. *Phys. Chem. Chem. Phys.*, **15**: 4844–4858.

Yu C, Yang K, Xie Y, Fan Q, Yu JC, Shu Q, Wang C. (2013). Novel hollow Pt-ZnO nanocomposite microspheres with hierarchical structure and enhanced photocatalytic activity and stability. *Nanoscale*, **5**: 2142–2151.

Zalga A, Abakeviciene B, Zarkov A, Beganskiene A, Kareiva A, Tamulevicius S. (2011). On the properties of yttria-stabilized zirconia thin films prepared by sol-gel method. *Mater. Sci.*,**17**: 91–196.

Zhang Y, Chen Y, Liu H, Zhou Y, Li R, Cai M, Sun X. (2009). 3D Hierarchical structure of single crystalline tungsten oxide nanowires:

construction, phase transition and voltammetric behavior. *J. Phys. Chem. C*, **113**: 1746–1750.

Zhao Z, Miyauchi M. (2009). Shape modulation of tungstic acid and tungsten oxide hollow structures. *J. Phys. Chem. C*, **113**: 6539–6546.

Zheng H, Ou, JZ, Strano MS, Kaner RB, Mitchell A, Zadeh KK. (2011). Nanostructured tungsten oxide-properties, synthesis, and applications. *Adv. Funct. Mater.*, **21**: 2175–2196.

Zhou W, Liu H, Boughton RI, Du G, Lin J, Wang J, Liu D. (2010). One-dimensional single-crystalline Ti-O based nanostructures: properties, synthesis, modifications and applications. *J. Mater. Chem.*, **20**: 5993–6008.

Chapter 4

Fabrication of Porous Nanoceramic Materials Based on Sol-Gel Chemistry

Neetu Talreja and Dinesh Kumar

Department of Chemistry, Banasthali University, Banasthali 304022, Rajasthan, India

dsbchoudhary2002@gmail.com

Porous nanostructured ceramics are a very attractive class of materials that find potential for various end applications, ranging from simple to socialite, such as bioimaging, sensors, paints, pigments, optics, and electronics. Their real-time application is restricted due to their incapability to economically produce nanorange ceramic structures on an industrial scale and simultaneously achieve precise control of their size. Nevertheless, sol-gel is an efficient process that can fabricate nanoceramics on an industrial scale. There has been remarkable progress in the fabrication of ceramic nanomaterials by the sol-gel process. The ceramic nanomaterial is becoming useful and is used in several applications owing to its surface- and size-dependent properties. In this chapter, the various ceramic nanomaterial systems that

Smart Ceramics: Preparation, Properties, and Applications
Edited by Ajay Kumar Mishra

ISBN 978-981-4774-30-7 (Hardcover), 978-1-315-16359-8 (eBook)
www.panstanford.com

have been fabricated so far are presented and the physical and chemical property enhancements due to the nanosize and the various end applications have been discussed in the chapter.

4.1 Background

The chapter explores the possibility of using the synthesis route, especially the sol-gel process, for the preparation of nanoceramic materials. Ceramic nanostructures have various end applications and are marked for their stability and their properties. In spite of various properties significant research has been made on process development of nanomaterials by which their structure and properties can be controlled. A number of issues still need to be resolved relating to their industrial production, synthesizing desired functionality, and understanding their novel structural process. Being used in huge applications nanostructured ceramic materials could almost replace all bulk ceramics in applications such as catalysis, fuel cells, solar cells, membranes, hydrogen storage batteries, structural applications requiring high mechanical strength, and biological applications such as tissue engineering, biomolecular machines, and biosensors, due to their high value addition (Vakifahmetoglua et al., 2016). Besides, nanostructured ceramics have potential applications in nanoelectronics and optical, sensor, resonator, and opto- and magnetoelectronic devices due to the physical and chemical properties (Jiang et al., 2016).

Nanoceramic materials exhibit diverse and extensive electronic properties, such as ferroelectricity, piezo- and pyroelectricity, and superconductivity. In the electronics industry, nanocrystal ceramics can bring about dramatic changes (Goel et al., 2013; Li et al., 2014). For example, the high dielectric strength of ferroelectric thin films composing ceramic nanocrystals compared to corresponding single-crystal thin films allows fabrication of small interconnects in metal oxide semiconductors. Ferroelectric thin films composing these ceramic nanocrystals have a much higher dielectric (Ramaseshan et al., 2007). Nanocomposite ceramics have high modulus and low gas permeability and have superior chemical and physical properties, such as enhanced barrier properties, higher mechanical strength, thermal stability, chemical stability, recyclability, dimensional stability, light and heat resistance, and optical clarity (Motloung et al., 2016). These properties enable the nanocomposite ceramics to be used in numerous applications ranging from simple coatings to functionality (Al-Hazmi et al., 2016).

The ceramic materials are a major field of research as they have exceptional mechanical properties. It is expected that mechanical properties can be enhanced by reducing the size of the composite to the nonmetric range. However, there are various methods available for synthesizing nanoshaped materials (<100 nm), but sintering is the issue for the nanometric range as the sintering process increases the particle size due to the high temperature provided for a long time. Hence there should be proper control on the sintering process as the properties can only be controlled by controlling the parameters (Mukhopadhyay et al., 2006). Moreover, the development and surface functionalization aspect of enhanced properties of nanoceramic nanoparticles has been studied in various area of research, such as silica, alumina, titania, zirconia, silicon nitride, and silicon carbide. The progression and development of nanotechnology has led to the synthesis of nanosized silica, SiO_2, which has been extensively used as in various applications (Nenova et al., 2016).

4.2 What Are Nanoceramics?

Ceramic nanocomposites that have excellent mechanical properties were first synthesized in Japan by K. Nihara, mainly by nanocomposite technology. These ceramic nanocomposites have two main categories: (i) nanocomposites that are prepared using the distribution of nanosized particles within micron-size matrix grains and (ii) matrix of nano/nanocomposites, together with nanosized particles within nanosized nanoceramics, that can be classified as follows: (a) intergranular nanocomposites, (b) intragranular, and (c) inter-/intragranular. In intergranular nanocomposites the nanomaterials are present at the particles' boundaries and triple junctions of the matrix, which results in better creep resistance. Intragranular nanocomposites have better mechanical strength and toughness, which results in nanoscaled reinforcements within the matrix grains. In inter-/intragranular nanocomposites, both inter- and intragranular nanocomposite properties exist. These nanocomposites have several enhanced properties at higher temperature, such as toughness, hardness, strength, fracture, resistance against creep and fatigue, and thermal shock fracture resistance. In nano/nanocomposites, both particulates and the matrix are nanosized. These materials also have properties of inter-/intragranular nanocomposites due to the presence of a large volume fraction of grain boundaries and show superplasticity and machinability due to improved sliding of the grain boundary. In a nano/micron composite, the particle size is the deciding factor of mechanical and physical strength. Figure 4.1 shows the schematic representation of the microstructural features of the various nanocomposites.

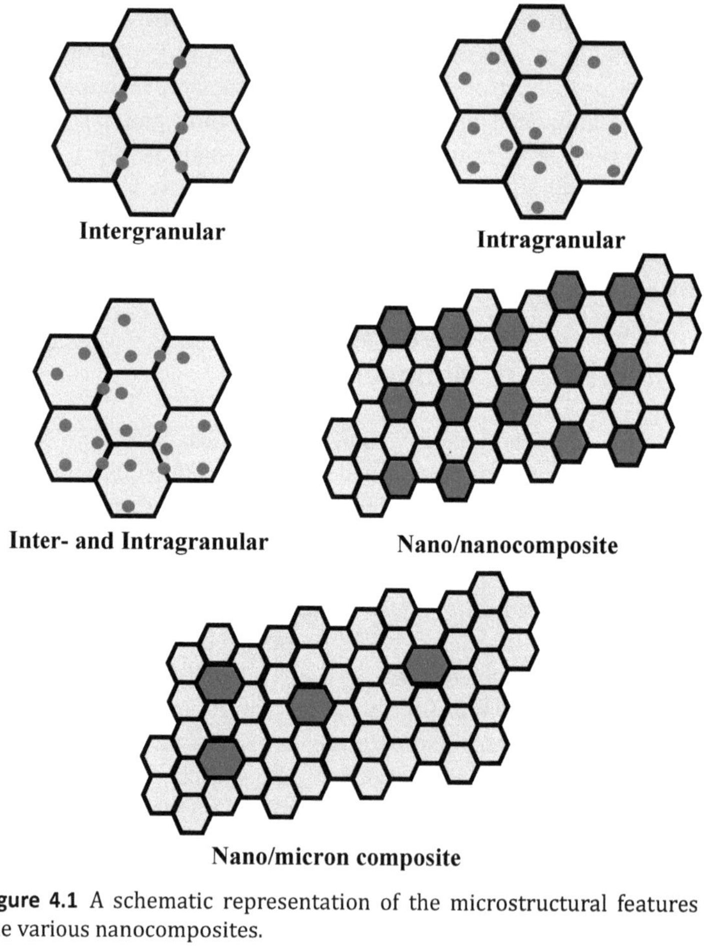

Figure 4.1 A schematic representation of the microstructural features of the various nanocomposites.

4.3 Synthesis of Nanoceramic Composites

The synthesis of a nanoceramic composite is quite necessary to improve various physicochemical characteristics, thereby enhancing applicability in end applications. However, the nanoceramic synthesis process is not easy; there are several issues to overcome to make the process easier and also benefit the materials industry.

4.3.1 Challenges Involved in Processing

Synthesis of nanoceramic composites on an industrial scale has continued to be a challenge for the materials community. An appropriate process of fabrication of nanoceramics can improve mechanical properties in nanocomposites. While designing a composite structure, besides considering the chemical composition of the matrix and the second phase, the amount, distribution, and morphology of the reinforcement phase should be properly evaluated. A key point is achieving a homogeneous distribution of the second phase in the matrix material, but this is still challenging when nanocrystalline particles are used, due to their extremely high specific surface and, hence, intrinsically high tendency toward agglomeration (Kumar et al., 2015). This shows that all manufacturing steps toward the elaboration of nanocomposite ceramics should be carried out with particular care, from the synthesis of the composite powders to the forming of green bodies and their sintering. Although the processing and sintering routes of nanoceramics have been revised in some previous papers, the role of synthesis in the microstructure and properties of sintered composite ceramics has hardly been discussed. The most investigated structural ceramic nanocomposites are Al_2O_3/SiC and Si_3N_4/SiC. However, many other different phases, such as TiN, TiC, TiO_2, ZrO_2, Cr_3C_2, and yttrium aluminum garnet (YAG), can be used as nanoreinforcements in Al_2O_3, Si_3N_4, MgO, mullite, or SiAlON ceramic matrix. In the following, the mainly used synthesis routes for the above composite systems are reviewed. First, the traditional mixing and milling method of single-phase ceramic powders to produce a composite feedstock is reviewed:

- Conventional powder route
- Mechanochemical synthesis
- Polymer precursor route
- Vapor-phase reaction technique
- Self-propagating high-temperature synthesis (SHS)
- Combustion synthesis
- Solution techniques:
    - o Coprecipitation
    - o Spray decomposition
    - o Solution combustion
    - o Sol-gel
- Surface modification methods
- Industrial production of ceramic composite powders

4.3.2 Coprecipitation

In the process evaporation of the aqueous phase or a chemical agent is used for increasing the solubility of the metal hydroxide in the solution. The particle size of composites and their nucleation and growth depend on pH and solution temperature. On the other hand, a common problem for coprecipitation is the achievement of a suitable condition for the simultaneous precipitation of all the species present in the solution. Nucleation and growth both factors are essential in precipitation as a large nucleation rate and small growth are essential for a small particle size. Various synthesis methods are available for precipitation as these processes are the factors that affect the crystalline temperature and phase evolution of the synthesized material at the time of calcination. These processes are precisely gel precipitation (GPT), in which the powder becomes an agglomerate after washing, and washed precipitation (WPT), in which a less agglomerated powder is formed and washing involves first the separation of the precipitation phase, and then, without the drying process, it undergoes washing using alcohol and hot water and precipitation. The washing procedure in powder precipitate (PPT) is simple as first it is separated from the aqueous phase and then washed after drying. The lower agglomeration rate requires a lower temperature for the sintering process. The powder developed using the WPT method has the lowest sintering temperature requirement, whereas GPT requires a high sintering temperature.

Various studies show the preparation method of alumina using the coprecipitation method in which an aqueous solution of aluminum nitrate hydrate and zirconium acetate is taken for the synthesis of a composite then dried at a certain temperature on a Teflon sheet to produce an amorphous powder. This study states the process temperature and composition as required for the synthesis of small grains. Tetragonal zirconia and γ-Al_2O_3 phases with smaller grain sizes, ranging from 40 and 100 nm, were observed at a temperature lower than 1100°C. Another study was conducted on the lamellar morphology, which is formed by the strain energy and diffusion phenomena of the $\gamma \rightarrow \alpha$ phase transformation.

4.3.3 Pyrolysis or Spray Decomposition

Pyrolysis is a technique that produces various materials, including ceramics, in which a gas (carrier) flows through an aerosol, where these aerosols are formed within the furnace. The various synthesis processes are (i) synthesis of the precursor of an aerosol, (ii) solvent evaporation and precipitation of the solute, and (iii) decomposition of the solute to produce a ceramic powder. The main advantages of this method are low cost, safety, easy availability of water-soluble salts, highest-purity nanosized particles, and homogeneity, while the disadvantages of this process are mainly difficulty in scaling up production or bulk production, the high cost of pure solvents, and required proper disposal. Studies are replete with Al_2O_3-/ZrO_2-based materials. Data suggest that lower agglomeration, controlled stoichiometry powder, and poorly controlled porous particles are observed.

4.3.4 Solution Combustion

Solution combustion is a process between nitrates ions and fuels due to an exothermal redox reaction. The process generally requires high-temperature combustion of metal nitrate, in which the particle size decreases significantly. The higher-volume gases formed during the reaction promote the conversion of the precursor gel into nanocrystalline form and subsequently undergo the calcination process. Some studies show nanocrystalline Al_2O_3/ZrO_2 powders are formed using a combustion process in which a metal nitrate salt is an oxidizing agent and urea is a source of fuel. The size of the produced nanopowder (α-Al_2O_3) was found to be 40 nm. The data suggest that the powder produced using this method has smaller-size particles, and controlled tetragonal zirconia with the highest purity. Some other processes, such as microwave heating technology in the solution combustion process, involving Al_2O_3/ZrO_2 powder permit a decrease in the particle size as compared to the traditional process, and smaller particles (<20 nm) are observed with a uniform distribution of ZrO_2 particles in the Al_2O_3 matrix.

4.3.5 The Sol-Gel Method

Ceramics prepared using the sol-gel method have several benefits compared to other methods, such as better uniformity of dissolved composites, highest purity, and lower stoichiometric losses (Verma et al., 2013). Moreover, researchers have an interest in transparent lithium/mica/glass-ceramic-based materials having luminescent properties and sol-gel synthesis for various end applications. Recently, Secu et al. (2016) used the sol-gel process for the synthesis of porous nanocrystalline Eu^{3+}-doped LaOCl nanoglass ceramic and identified its crystallization and luminance properties. Motloung et al. (2016) synthesized $ZnAl_2O_4$:1.5% Pb nanocrystals using the sol-gel process. Hazmi et al. (2016) used the sol-gel process for the synthesis of $Ca_{(10-x)}$ Bi_x $(PO_4)_6(OH)_2$ nanopowders hydroxyapatite-based ceramics. B_2O_3-doped MN nanoceramics were synthesized through the sol-gel method, and their microwave dielectric properties and microstructures were investigated by Wu et al. (2015). Nano-apatite-zirconia glass ceramics was synthesized by Montazerian et al. (2015) using the sol-gel process and their sintering behavior and bioactivity tested. Wang et al. (2015) fabricated 3D oxide fiber-reinforced Al_2O_3-SiO_2-SiOC composites of alumina and SiO_2 using the sol-gel process.

Being a simple, economical process, sol-gel has attracted attention over the past 30 years. Different ceramics are prepared using this process. This method requires low-temperature-based chemical synthesis, under a specific pH, and a concentration condition sol containing a chemical precursor, a binder, and a cross-linking agent. Gels are a complex network, which after drying and subsequent heating, at a particular temperature, are modified into a glassy material. In general, sol-gel-derived materials exhibit uniformity, purity, and greater chemical modification than other techniques (Fig. 4.2). There are several advantages and disadvantages of the process, which are given in subsequent sections.

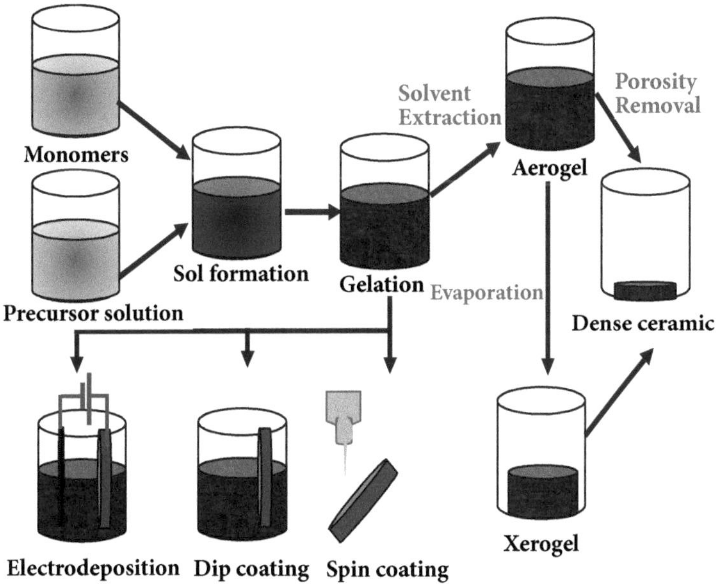

Figure 4.2 A schematic representation of the sol-gel process.

4.3.5.1 Advantages of sol-gel

Being a simple, cost-effective, and viable process, sol-gel also provides homogeneity of the solution. The greater adhesion between substrate and top coat is the major advantage. The material can be modified to any shape as the physical state is gel based. Due to mechanical agitation between precursor and solvent, the highest-purity product possible is obtained. The low temperature required is also a cause of it being a simple, commercial, and operative method to produce high-quality application products.

4.3.5.2 Disadvantages of sol-gel

There is still a lack of industrial potential due to some limitations, one of which is that the porosity cannot be controlled by this method. Due to thermal mismatch and its surface dependency the technology suffers a drawback presently. Also, it is a multistep process and time consuming, which are also drawbacks of the process. Moreover, the main drawback of sol-gel synthesis is that it is hard to achieve crack-free materials (glass monolith) with more than 1 cm diameter during the melting process. The solution is partially sintered glass gel powder of a desirable density for several applications, mainly scaffolds, coatings, and bone grafts. The crystallization might be encouraging for future end applications. However, little research has so far addressed the sintering of bioactive glass gel. The lack of such studies is due to the highly porous nature and a tendency toward crystallization, which negates the sintering ability of gel glass powder.

4.4 Chemistry of the Sol-Gel Process

The chemical route of the process can be commonly described by the following steps: initial component homogenization, their conversion into a sol and then into a stable sol, and the final drying. The most important step involves alkoxide hydrolysis and reverse precipitation of the metal salts. The first approach includes water addition to the alcohol solutions of metal and nonmetal alkoxides with the general formula $M(OR)_n$ (M = Si, Al, Ti, V, Cr, W, Zr; R = CH_{3^-}, $C_2H_{5^-}$, $C_3H_{7^-}$, etc.). This pathway results in hydrolysis, with further polycondensation at low temperatures, according to the following reactions (Danks et al., 2015):

$$M(OR)n + H_2O \longrightarrow [M(OR)_{n-1}(OH)] + ROH$$

$$M\text{-}OH + M\text{-}OX \longrightarrow M\text{-}O\text{-}M + XOH \ (X = H \text{ or } M)$$

Consequently, hydroxilated M-OH groups and hydroxide monomers are formed; the synthesized group evidence themselves as active centers in the polycrystallization reaction that probably occurs according to the alkoxylation mechanism.

4.5 Properties of Sol-Gel Ceramics

4.5.1 Thermal Resistance

Ceramic material shows better thermal shock compared to alumina due to the lower thermal expansion coefficient or thermal shock resistance, which depends on multiple parameters such as size, shape, thermal, physical, and mechanical properties of the material. Thermal stock resistance can be increase by strong filament monocrystals or be strengthened by other methods. The thermal shock resistance of ordinary alumina ceramics cannot withstand significant quench into a cold airflow.

4.5.2 Mechanical Properties

It may be possible to understand the mechanical properties of nanoceramics and their composites such as higher strength, hardness, and modest increase in fracture toughness as a result of their ultrafine grain sizes and nanoscaled reinforcements and derive the benefits due to success in developing such materials. These materials exhibit outstanding properties compared to their counterparts. These materials have high toughness, good chemical stability, and biocompatibility.

4.6 Some Common Examples of Nanoceramics via the Sol-Gel Process

4.6.1 Silica-Based Ceramics

Silica-based ceramics prepared by the sol-gel process are widely used in many industrial as well as scientific research studies. These materials have many applications in almost all fields, from complicated biomedicals to simple coatings. The sol-gel is a widely used method for the preparation of silica-based ceramic materials as it provides homogeneity of composites as well as purity at the lowest reaction temperature. The process involves five major steps for the synthesis, including hydrolysis followed by polycondensation of an alkoxide precursor ($Si(OR)_4$) in the presence of an acid or a base, which catalyzes the reaction as well.

Commonly used alkoxides are inorganic salts, tetraethylorthosilicate (TEOS, $Si(OC_2H_5)_4$). Two major steps are involving in the process of silica particle development, nucleation followed by a growth mechanism.

The proposed models basically describe the process of silica chain growth. One is monomer addition and the other is controlled aggregation of silica. As per the first model after the addition of hydrolyzed monomer on the particle, chain growth occurs on early eruption of nucleation. On the other hand, according to the aggregation model nucleation occurs continuously during the reaction and it aggregates to form dimers, trimers, and larger particles (secondary particles). The final product of both models is the formation of a complex gel network or spherical particles, which depends on reaction conditions (Pan et al., 2016).

4.6.2 Zirconium-Based Ceramics

Zirconia-based ceramics exhibit outstanding properties. They have high toughness, good chemical stability, and biocompatibility. Zirconia and its composite-based solid solutions are widely applied in various industrial applications, such as solid electrolytes (Khartona et al., 2004) for fuel cells (Shao et al., 2013) and electrochemical sensors (Montazeriana et al., 2015) due to the high ionic conductivity of these materials. At the same time, it is possible to use them in the exploitation regimes under the conditions that eliminate electronic conductivity harmful for fuel cell and electrochemical sensor applications.

4.6.3 Alumina-Based Ceramics

Being a porous ceramic, alumina finds versatile applications in the field of polymeric foams for packaging and porous ceramics for water purification and is used for making porous ceramics because most of the alumina-based ceramics possess relatively high strength along with improved thermal and chemical stability. Because of their inherently brittle nature, pores have been traditionally avoided in ceramic components because pores reduce the material's strength. However, applications of porous ceramics have increased in the last decades, especially for environments where high temperatures, excessive wear, and corrosiveness are involved. Such applications include the filtration of molten metals, high-temperature thermal insulation, filtration of particulates from diesel engine exhaust gases, support for catalytic reactions, and filtration of hot, corrosive gases in various industrial operations (Motloung et al., 2016). Due to several advantages porous ceramics are used in various applications. The major advantages include high melting point, high corrosion resistance, and wear resistance, along with the features gained by the replacement of solid material by voids in the material. Such features include low thermal mass, low thermal conductivity, controlled permeability, high surface area, low density, high specific strength, and low dielectric constant. All these properties can be altered for each specific application by controlling the composition and microstructure of the porous ceramic. Fluctuations in open and closed porosity, pore size distribution, and pore morphology can have a major effect on a material's properties. These microstructural features are, in turn, greatly influenced by the processing route used for the production of the porous material. Alumina exists both in the transition or metastable and stable forms. Because of this alumina is used mainly in catalysis and in absorbent applications. There are different crystal structures of α-alumina, mostly round and hexagonal. The crystals also have small surface areas.

Alumina is one of the inert biomaterials used in implants. It is, therefore, a biodegradable material, well tolerated by the biological environment. In literature, there is information on obtaining Al_2O_3 by the sol-gel method using different precursors: aluminum triisopropylate in a hydrolysis system consisting of octanol and acetonitrile, aluminum nitrate in an aqueous medium, and aluminum secondary butoxide in an alcoholic medium (Odaka et al., 2008, 2009; Hafez et al., 2012).

4.7 Other Advanced Nanoceramics Composites

4.7.1 YAG-/SIC-Based Composites

The mechanical strength and other properties of nanoceramics can be enhanced by incorporation of nanosized ceramics such as SiC particles into YAG. These composites are well known due to significant improvement in features. The nanocomposites show extraordinary band strength compared to pure YAG, consolidated using spark plasma sintering (SPS). The reported bond strength value is less than that of hot-pressed pure YAG. The nanosized SiC dispersions were in the intragranular regions (Fedyk et al., 2007).

4.7.2 SiC-/TiC-Based Composites

The other reported nanocomposite is the combination of SiC-based nanocomposites containing dispersions of nanosized TiC. The composites show excellent properties, such as high fracture toughness and high bend strength, at a sintering temperature of around 1800°C. Investigation of the fracture mode revealed that SiC grains contain transgranular fractures, which are responsible for improvement of bend strength, while intergranular fracture of TiC resulted in the improvement of fracture toughness, as weak grain/interface boundaries enhance crack-bridging intensities.

4.7.3 SI_3N_4/SIC Nanocomposites

There are various other nanocomposites, including the mixture of Si_3N_4 and SiC nanocomposites synthesized by hot-pressing a mixture of Si_3N_4 and SiC at a particular temperature for some hours. The experiment involves heat treatment of powder mixtures at an optimized temperature (1400°C–1500°C). It was reported that the heat treatment resulted in the development of a microstructure characterized by the presence of a significant fraction of nanocrystalline SiC particles at the grain boundaries of the fine-grained Si_3N_4 matrix. The developed nanocomposite with SiC nanocrystalline reinforcements exhibited high flexural strength at room temperature, which was maintained up to 1200°C. Hot-pressed Si_3N_4/SiC nanocomposites are characterized by inter-/intragranular nanocrystalline SiC. Such nanocomposites also exhibited similarly high strength at room temperature and maintained the strength up to a high temperature of 1400°C (Nishimura et al., 2007; Ebrahimpour et al., 2014).

4.8 Applications of Sol-Gel-Derived Ceramics

The use of ceramic materials explores the possibilities of the application of these materials in various fields. There are various fields, including gas turbine engines, where high temperatures are required and hence ceramic materials are advantageous to use. Some common advantages are higher operating temperatures, improved fuel efficiency, reduced weight, and greater thrust-to-weight ratio. Using ceramic materials also decreases the dependence on strategic materials. A metallic alloy is less dense than ceramics and maintains its strength even in the hottest places. Because of their bonding, most monolithic ceramics are brittle and lack fracture toughness. In ceramic engine parts, these characteristics have led to terrible failure due to point contact stresses. Recently developed fiber-reinforced ceramic matrix composites offer the promise of overcoming these problems. The fracture toughness for these composites is significantly higher than for monolithic materials. Figure 4.3 shows the schematic representation of various applications of sol-gel. Further applications are given in subsequent sections.

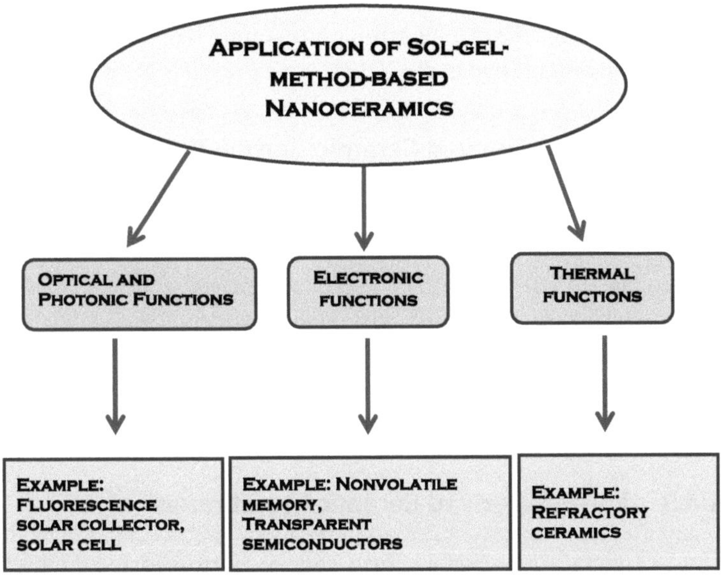

Figure 4.3 A schematic representation of various applications of sol-gel.

4.8.1 Biomedical Applications

The sol-gel-derived bioceramics have a great potential for application as coatings on metallic substrates to provide a high degree of biocompatibility and promote a rapid healing response with minimal adverse biological events. Without a doubt one of the main benefits of applying sol-gel approaches to the production of bioactive coatings is the relatively low temperatures under which synthesis can be achieved. However, compatibility is but one facet of sol-gel-derived biomedical applications. Low-temperature synthesis also allows for the inclusion of materials that would otherwise be destroyed by excessively high temperatures (Hong et al., 2010; Abrishamchian et al., 2013; Wu et al., 2015).

4.8.1.1 Nanoceramics for dental applications

The success of all-ceramic crowns and patient demand for metal-free, tooth-colored restorations have led to the development and introduction of restorative systems for all-ceramic fixed partial dentures (FPDs). These systems continue to be evaluated in clinical studies for their predictability and long-term success. Since then, developments in dental ceramics have led to the introduction of new high-strength ceramic core materials for all-ceramic FPDs (Shao et al., 2013).

4.8.2 Sol-Gel-Derived Ceramic-Carbon Composite Electrodes

The recently developed ceramic composite carbon composite electrodes (CCEs) have various advantages, such as a chemically modified surface, hardness, and high porosity. These electrodes are much more stable than their counterparts. On the basis of the sensing application the surfaces of these electrodes can be easily modifiable (Abrishamchian et al., 2013; Robak et al., 2016).

4.8.3 Sol-Gel-Derived Ceramic Membranes

The gas separation performance can be improved by modifying mesoporous alumina using the sol-gel process.

Ceramic membranes have played an important role in fields such as particulate filtration, gas separation, and analytical application (sensors for gases and chemicals). For alumina membranes in particular, they have made a successful entry into the market as in-line gas filters. In microelectronic packaging, food, and pharmaceutical industries, where clean air is required, removal of fine particles from gas streams can be achieved by means of porous ceramic membranes (Yang et al., 2015; Bechervaisea et al., 2016; Shao et al., 2016). They are selected due to the following aspects:

- These materials have thermal stability as well as chemical inertness.
- They do not react with the process gases or generate impurities, nor do they produce a hydrogen outgassing problem.
- Their pore size distributions are narrow and can retain an extremely high percentage of particulates.
- They can be easily regenerated periodically by back flushing when membrane fouling occurs.
- They can be steam sterilized without changing their microstructure and chemical characteristics.

4.8.4 Nanocomposite Electrodes

Ceramics have also been used for the fabrication of nanoceramic-based electrode materials. Due to various properties (explained above) these materials behave as excellent electrode materials in the form of nanocomposites. Recent reports show the fabrication of nanocomposite electrodes made of ceramic composite. Wang et al. (2016) developed $(Pb_{0.87}Ba_{0.08}Sr_{0.02}La_{0.02})(Zr_{0.65}Sn_{0.27}Ti_{0.08})\ O_3$ electrodes. $Sr_2FeSi_2O_7$ was developed by Pu et al. (2016). These materials show excellent dielectric properties and are used as electrode materials in various applications (Pu et al., 2016; Wang et al., 2016).

4.8.5 Nanoceramic Sensors

Various metal oxide-based ceramics are used as CO detectors, oxygen sensors, and alcohol breath analyzers. The sensing mostly occurs at a high temperature due to the semiconductor nature of these metallic-oxide-based ceramics. The temperature ranges typically between 100° and 400°. Depending on whether the semiconducting is n-type, it will oxidize or reduce. The electrical resistance of the sensor may increase or decrease with respect to its value in air in the absence of the gas, that is, its baseline resistance. Electromagnetic radiation is also detected by using ceramic sensors. A hydrogen-sensing ability was determined by Chen et al. (2016) using Pt-TiO_2 nanocomposite ceramic as a sensor. B_2O_3 nanoceramics were used as humidity sensors (Xu et al., 2012; Chen et al., 2016; Senturk et al., 2016; Sharma et al., 2016).

4.9 Future Aspects

Nanoceramic materials with improved mechanical strength still need to be developed, as the basic causes of imperfections in mechanical properties are still not clearly understood after such extensive research on nanocomposites. After achieving a significant rise in strength and hardness, the current concern is on fracture toughness in ceramic nanocomposites as it is insufficient. Hence, extensive research of nanoceramics is still needed to develop newer approaches to enhance their mechanical properties. Future studies should focus on nanocomposites with high mechanical and chemical functionality. Although nanoceramic composites cover a variety of applications they are not a priority in some fields due to a lack of large-scale production.

4.10 Conclusion

Nanoceramics synthesized through the sol-gel process have high purity, good homogeneity, and high mechanical behavior. This technique can control the shape and size of the material. This technique has an advantage over other techniques in terms of lower-temperature requirement to prepare nanocomposites. The ceramics developed using the process are used in various applications such as film coatings, electrodes, sensors, electronics, different medical applications, and fiber pulling. Overall, the sol-gel technique is the better option for the development of nanoceramic materials due its wider applicability toward various end applications.

References

Abrishamchian A, Hooshmand T, Mohammadi M, Najafi F. (2013). Preparation and characterization of multi-walled carbon nanotube/hydroxyapatite nanocomposite film dip coated on Ti-6Al-4V by sol-gel method for biomedical applications: an in vitro study. *Mater. Sci. Eng. C*, **33**: 2002–2010.

Al-Hazmi F. (2016). Synthesis and electrical properties of Bi-doped hydroxyapatite ceramics. *J. Alloys Compd.*, **665**: 119–123.

Bechervaisea P, Carrb D, Bird MR. (2016). Removal of thermophilic spores from gum Arabicstreams using ceramic alumina microfiltration membranes. *Food Bioprod. Process.*, **99**: 147–155.

Chen WP, Xiong Y, Li YS, Cui P, Guo SS, Chen W, Tang ZL, Yan Z, Zhang Z. (2016). Extraordinary room-temperature hydrogen sensing capabilities of porous bulk PteTiO_2 nanocomposite ceramics. *Int. J. Hydrogen Energy*, **41**: 3307–3312.

Danks AE, Hallb SR, Schnepp Z. (2015). The evolution of "sol-gel" chemistry as a technique for materials synthesis. *Mater. Horiz.*, **3**: 91–112.

Das G, Ferraioli L, Bettotti P, De Angelis F, Mariotto G, Pavesi L, Di Fabrizio E, Soraru GD. (2008). Si-nanocrystals/SiO_2 thin films obtained by pyrolysis of sol-gel precursors. *Thin Solid Films*, **516**: 6804–6807.

Ebrahimpour O, Dubois C, Chaouki J. (2014). Fabrication of mullite-bonded porous SiC ceramics via a sol-gel assisted in situ reaction bonding. *J. Eur. Ceram. Soc.*, **34**: 237–247.

Fedyk R, Hreniak D, Łojkowski W, Strek W, Matysiak H, Grzanka E, Gierlotka S, Mazur P. (2007). Method of preparation and structural properties of transparent YAG nanoceramics. *Opt. Mater.*, **29**: 1252–1257.

Goel P, Biradar AM. (2013). A typical dielectric behavior in sol-gel derived fine grain PZT/CeO_2 nanocomposites. *J. Phys. Chem. Solids*, **74**: 854–861.

Hafez HS, El-fadaly E. (2012). Synthesis, characterization and color performance of novel Co^{2+}-doped alumina/titania nanoceramic pigments. *Spectrochim. Acta Part A*, **95**: 8–14.

Hong Y, Fan H, Li B, Guo B, Liu M, Zhang X. (2010). Fabrication, biological effects, and medical applications of calcium phosphate nanoceramics. *Mater. Sci. Eng. R*, **70**: 225–242.

Jiang Y, Fan J, Jiang B, Mao X, Zhou C, Zhang L. (2016). Structure and optical properties of transparent Er^{3+}-doped CaF_2-silica glass ceramic prepared by controllable sol-gel method. *Ceram. Int.*, **42**: 9571–9576.

Khan MA, Alam MM, Naushad M, Alothman AZ, Kumar M, Ahamad T. (2015). Sol-gel assisted synthesis of porous nano-crystalline $CoFe_2O_4$ composite and its application in the removal of brilliant blue-R from aqueous phase: An ecofriendly and economical approach. *Chem. Eng. J.*, **279**: 416–424.

Khartona, VV, Marquesa, FMB, Atkinson A. (2004). Transport properties of solid oxide electrolyte ceramics: a brief review. *Solid State Ionics*, **174**: 135–149.

Kumar A, Yadav N, Bhatt M, Mishra NK, Chaudhary P, Singh R. (2015). Sol-gel derived nanomaterials and its applications: a review. *Res. J. Chem. Sci.*, **5**(12): 98–105.

Li WL, Cao WP, Xua D, Wanga W, Fei WD. (2014). Phase structure and piezoelectric properties of NBT–KBT–BT ceramics prepared by sol-gel flame synthetic approach. *J. Alloys Compd.*, **613**: 181–186.

Montazeriana M, Schneiderb, JF, Yektaa B, Marghussiana VK, Rodriguesc AM, Zanottoc ED. (2015). Sol-gel synthesis, structure, sintering and properties of bioactive and inert nano-apatite–zirconia glass–ceramics. *Ceram. Int.*, **41**: 11024–11045.

Motloung SV, Tsega M, Dejene FB, Swart HC, Ntwaeaborwa OM, Koao LF, Motaung TE, Hato MJ. (2016). Effect of annealing temperature on structural and optical properties of $ZnAl_2O_4$:1.5% Pb nanocrystals synthesized via sol-gel reaction. *J. Alloys Compd.*, **677**: 72–79.

Mukhopadhyay A, Basu B. (2006). Bulk nanoceramic composites for structural applications: a review. *Proc. Indian Natl. Sci. Acad.*, **72**: 97–111.

Nenova Z, Kozhukharov S, Nenov T, Nedev N, Machkova M. (2016). Combined influence of titania and silica precursors on the properties of thin film humidity sensing elements prepared via a sol-gel method. *Sens. Actuators B*, **224**: 143–152.

Nishimuraa T, Xub X, Kimotoa K, Hirosakia N, Tanaka H. (2007). Fabrication of silicon nitride nanoceramics: powder preparation and sintering: a review. *Sci. Technol. Adv. Mater.*, **8**: 635–643.

Odakaa A, Yamaguchi T, Fujita T, Taruta S, Kitajima K. (2008). Densification of Ca-doped alumina nanopowders prepared by a new sol-gel route with seeding. *J. Eur. Ceram. Soc.*, **28**: 2479–2485.

Odakaa A, Yamaguchi T, Fujita T, Taruta S, Kitajima K. (2009). Densification of rare-earth (Lu, Gd, Nd)-doped alumina nanopowders obtained by a sol-gel route under seeding. *Powder Technol.*, **193**: 26–31.

Pan Y, Yina J, Yaoa D, Zuo K, Xia Y, Liang H, Zeng Y. (2016). Effects of silica sol on the microstructure and mechanical properties of $CaSiO_3$ bioceramics. *Mater. Sci. Eng. C*, **64**: 336–340.

Pu Y, Dong Z, Zhang P, Wu Y, Zhao J, Luo Y. (2016). Dielectric, complex impedance and electrical conductivity studies of the multiferroic $Sr_2FeSi_2O_7$-crystallized glass-ceramics. *J. Alloys Compd.*, **672**: 64–71.

Ramaseshan R, Sundarrajan S, Jose R. (2007). Nanostructured ceramics by electrospinning. *J. Appl. Phys.*, **102**: 111101.

Robaka J, Burnata B, Andrzej Leniarta L, Kisielewskab A, Brychta M, Skrzypek S. (2016). The effect of carbon material on the electroanalytical determination of 4-chloro-3-methylphenol using the sol-gel derived carbon ceramic electrodes. *Sens. Actuators B*, **236**: 318–325.

Secu M, Secu CE, Bartha C. (2016). Crystallization and luminescence properties of a new Eu^{3+}-doped LaOCl nano-glass-ceramic. *J. Eur. Ceram. Soc.*, **36**: 203–207.

Senturk E, Duman S, Bagsi S, Soykan HS, Aslanoglu Z. (2016). Humidity sensing properties of steatite ceramic containing B_2O_3. *Sens. Actuators A*, **240**: 80–84.

Shao L, Jiang D, Gong J. (2013). Nanoindentation characterization of the hardness of zirconia dental ceramics. *Adv. Eng. Mater.*, **15**: 707–708.

Shao X, Wanga Z, Xua S, Xie K, Hua X, Donga D, Parkinsona G, Li CZ. (2016). Microchannel structure of ceramic membranes for oxygen separation. *J. Eur. Ceram. Soc.*, **36**: 3193–3199.

Sharma SK, Chauhan VS, Kumar A. (2016). Detection of electromagnetic radiation in ferroelectric ceramics for non-contact sensing applications. *J. Alloys Compd.*, **662**: 534–540.

Vakifahmetoglua C, Zeydanlib, D, Colomboc P. (2016). Porous polymer derived ceramics. *Mater. Sci. Eng. R*, **106**: 1–30.

Verma K, Sharma S, Sharma DK, Kumar R, Rai R. (2012). Sol-gel processing and characterization of nanometer sized $(Ba,Sr)TiO_3$ ceramics. *Adv. Mater. Lett.*, **3**(1): 44–49.

Wang X, Yang Tn, Shen J. (2016). Unique dielectric behavior of $(Pb_{0.87}Ba_{0.08}Sr_{0.02}La_{0.02})(Zr_{0.65}Sn_{0.27}Ti_{0.08})O_3$ antiferroelectric ceramics. *Mater. Lett.*, **181**: 220–222.

Wang Y, Liu, H, Cheng H, Wang J. (2015). Fabrication and properties of 3D oxide fiber-reinforced Al_2O_3–SiO_2–SiOC composites by a hybrid technique based on sol-gel and PIP process. *Ceram. Int.*, **41**: 1065–1071.

Wu C, Xia L, Han P, Xu M, Fang B, Wang J, Chang Ja, Xiao Y. (2015). Graphene-oxide-modified b-tricalcium phosphate bioceramics stimulate in vitro and in vivo osteogenesis. *Carbon*, **93**: 116–129.

Wu HT, Mei QJ. (2015). Effects of B_2O_3 addition on sintering behavior and microwave dielectric properties of $Mg_5Nb_4O_{15}$ ceramics prepared by the sol-gel method. *J. Alloys Compd.*, **651**: 393–398.

Xu B, Zhang J, Tian Z, Yuan SL. (2012). Synthesis, dielectric and nonlinear electrical properties of $Na_1/2Bi_1/2Cu_3Ti_4O_{12}$ ceramics by a sol-gel technique. *Mater. Lett.*, **75**: 87–90.

Yang F, Zhao H, Yang J, Fang M, Lu T, Du Za, Świerczek K, Zheng K. (2015). Structure and oxygen permeability of $BaCo_{0.7}Fe_{0.3\text{-}x}In_xO_{3\text{-}\delta}$ ceramic membranes. *J. Membr. Sci.*, **492**: 559–567.

Chapter 5

Technology of Refractory Materials Based on SHS in Metal Oxide Systems

S. M. Fomenko, E. E. Dilmuhambetov, and Z. A. Mansurov

Institute of Combustion Problems, 172 Bogenbay Batyr Str., Almaty City, Republic of Kazakhstan

ZMansurov@kaznu.kz

The work presents the physicochemical and technological principles of obtaining refractory materials based on SHS. The macrokinetic regularities of aluminothermal combustion in SHS mode of chromium oxides, silicon, and iron, constituting the active components of SHS refractories, are considered. The associative role of silica sols and the impact of nanosized silicon on the SHS process in oxide systems are shown. It is shown that the burning of these oxides is a multistage factor and is accompanied by the formation of intermediate compounds with a lower state of metal oxidation. The features of SHS in oxides in the presence of carbon and the field of application of carbonaceous SHS refractories are considered. The results concerning successful implementations

Smart Ceramics: Preparation, Properties, and Applications
Edited by Ajay Kumar Mishra

ISBN 978-981-4774-30-7 (Hardcover), 978-1-315-16359-8 (eBook)
www.panstanford.com

of SHS refractories during repair of shaft furnaces for limestone calcining and induction furnaces for melting of nonferrous metals in recent years are presented.

5.1 Introduction

The most attractive aspect of self-propagating high-temperature synthesis (SHS), which determines the constant interest in this phenomenon, is the ability to produce valuable products of chemical reactions in the form of powders or finished moldings directly by the SHS process. Currently, on the basis of SHS various combined processes are being developed. One of the promising directions that have been developed in recent years is the combination of SHS and metallothermy (Merzhanov, 1975). Metallothermic oxidation-reduction reactions, occurring in combustion wave, are characterized by high temperatures and thermal effects. The synthesis temperatures reach 1600°C–3000°C. At such temperatures, the formation of hard-melting oxide compounds and salts occurs, representing refractory materials of practical interest. For the first time, the possibility to obtain refractory materials and products by the SHS method was shown by Nersesyan et al. (1984, 1985). Systematic investigations concerning solid-phase metallothermic combustion of oxide systems in order to develop new production technology of refractories were made at the Institute of Combustion Problems under the guidance of Ksandopulo and Ismailova (Ksandopulo et al., 1986; Ksandopulo et al., 1987; Ksandopulo and Ismailov, 1987; Ismailov, 1992). The result of these research studies was the creation of SHS technology refractories "FURNON," which now have widespread practical use during the repair of lining of high-temperature thermal units—metal furnaces for limestone and cement, preheat furnaces for coke and anode paste, arches of electric steel furnaces, package boilers, thermal power plants (TPPs), hydroelectric power plants (HEPPs), boiler houses, etc. The industries where the SHS refractories are very popular are, for example, metallurgy, chemicals, petrochemicals, fire, construction, aluminum, and combined heat and power (Mansurov, 2010).

5.2 Macrokinetics of SHS Porous Oxides

Modern concepts of rate theories of gasless combustion in an SHS regime typically use an assumption concerning a one-step chemical reaction. In this case, the mathematical models of solid-phase combustion with a weak braking action and narrow wave front of SHS lead to a unique dependence of the burning rate u on the combustion temperature T_Γ (Wolpert and Barzykina, 1988):

$$u^2 \cong (RT_\Gamma^2/EQ)\cdot\exp(-E/RT_\Gamma), \tag{5.1}$$

where E is the effective energy of activation and Q is the heat of the process. Linear dependence, $\ln(u/T_\Gamma)$, from the reciprocal temperature allows you to determine the effective energies of the activation SHS process.

Aluminothermal combustion of simple oxides generally is a multistep process, and regularities of SHS significantly deviate from the findings of a thermal combustion theory. In general the aluminothermal burning of oxides can be represented by the chemical reaction:

$$4/3\text{Al} + 2/m\cdot\text{Me}_n\text{O}_m \rightarrow 2n/m\cdot\text{Me} + 2/3\text{Al}_2\text{O}_3 \tag{5.2}$$

For the purpose of a detailed study of such processes the investigations of various systems are conducted, including aluminum and metal oxides. In the case of the Al-Cr_2O_3 system (Ismailov and Fomenko, 1992) it was shown that at the molar ratio of reactants, $8 \leq [\text{Al}]/[\text{Cr}_2\text{O}_3] \geq 14$, the samples that are prepared in the following manner are resistant to ignition. In the ratio range of $[\text{Al}]/[\text{Cr}_2\text{O}_3] \geq 0.8{:}12$ and 10:14 there is a pulsating combustion.

The combustion temperature of the system Al–Cr_2O_3 (Fig. 5.1) has three strongly pronounced maximums at a molar aluminum content: 2, 4, and 6. At the same time, the linear velocity of combustion wave propagation is characterized by one maximum at the molar content of aluminum, which is equal to three. These results indicate the absence of one-to-one correspondence between velocity and combustion temperature, which suggests a multistage combustion mechanism, which is subjected to the following general principles: the recovery of chromium occurs in accordance with the following scheme:

$$Cr^{3+} \rightarrow Cr^{2+} \rightarrow Cr^{0} \quad (5.3)$$

The oxidation of aluminum occurs in stages:

$$Al \rightarrow Al^{+} \rightarrow Al^{2+} \rightarrow Al^{3+} \quad (5.4)$$

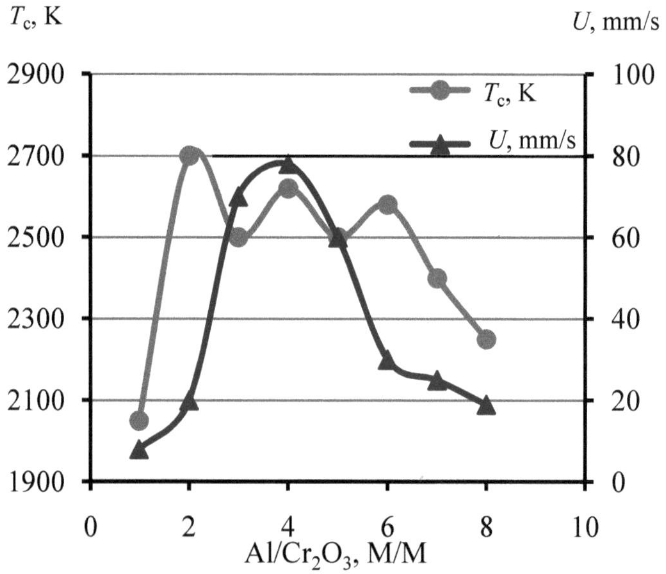

Figure 5.1 The temperatures and combustion rates in the system Cr_2O_3 + nAl. Reproduced from Dilmukhambetov EE, Mansurov ZA, Fomenko SM. (2011). Aluminothermic SHS: the effect of silica sol added as a binder. *Int. J. Self-Propag. High-Temp. Synth.*, **20**(2): 82–87.

Adiabatic temperature calculations of aluminothermic reduction of chromium oxide, based on empirical temperature dependence and average atomic enthalpy of the process (Dubrovin and Pliner, 1962), establish a single maximum temperature at a molar ratio of 2. It corresponds to the stoichiometry of complete chrome reduction. A strong consideration of thermal effects at SHS and heating products (Novikov et al., 1975) provides a more complex temperature dependence of the process on the molar ratio Al/Cr_2O_3. The calculation results of adiabatic temperatures under the assumption of the formation of intermetallics of aluminum and chromium are shown in Fig. 5.2.

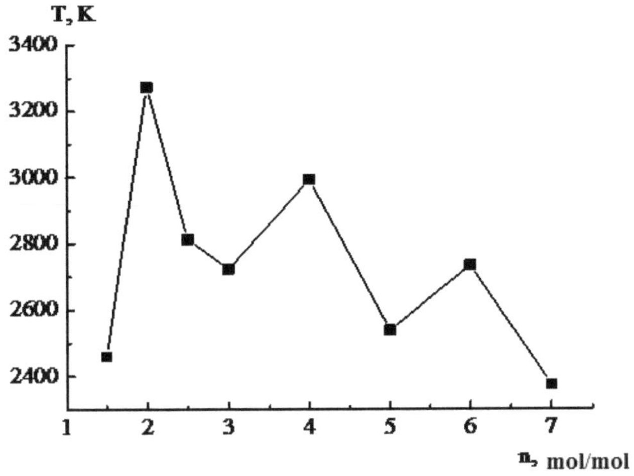

Figure 5.2 Assumed combustion temperatures of the Al-Cr_2O_3 system. Reproduced from Dilmukhambetov EE, Mansurov ZA, Fomenko SM. (2011). Aluminothermic SHS: the effect of silica sol added as a binder. *Int. J. Self-Propag. High-Temp. Synth.*, **20**(2): 82–87.

In the calculations, known thermochemical characteristics for aluminum and chromium compounds were used (Glushko, 1979). As can be seen from the figure, the predicted combustion temperatures reproduce qualitatively the observed dependences of combustion temperatures in accordance with the molar ratio Al/Cr_2O_3. The calculation results of adiabatic temperatures point at the important role of the formation of intermetallics in the change of combustion temperatures in the Al-Cr_2O_3 system. The lack of correlation between the temperature and combustion rate in varying the exothermic mixture composition demonstrates the inapplicability of the simple model of solid-phase combustion with a narrow wave front of SHS.

The main products of high-temperature synthesis at different ratios of aluminum to chromium oxide are determined by x-ray diffraction analysis as presented in Table 5.1. From this table it can be seen that small ratios of Al Al/Cr_2O_3 do not reach the stoichiometry of complete recovery of chromium oxide, so the main product is a solid solution of the high-melting compound $Al_2O_3 \cdot Cr_2O_3$. There is also reduced chromium. In the case of the hyperstoichiometric ratio with plenty of aluminum, a complete reduction of chromium oxide occurs; at the same time the reduced chromium and aluminum form a chemical compound (intermetallic metal) Cr_5Al_8.

Table 5.1 The basis phases of SHS products

Al/Cr_2O_3, mole/mole	Solid solution $Al_2O_3 \cdot Cr_2O_3$, %	Cr, %	Al_2O_3, %	Cr_5Al_8, %
1.0 (bulk-free)	81.3	18.7	—	—
1.5 (bulk-free)	82.8	17.2	—	—
4.0 (c 30% Al_2O_3)	—	4.2	70.2	25.6
6.0 (c 30% Al_2O_3)	—	—	73.2	26.7

Source: Reproduced from Dilmukhambetov EE, Mansurov ZA, Fomenko SM. (2011). Aluminothermic SHS: the effect of silica sol added as a binder. *Int. J. Self-Propag. High-Temp. Synth.*, **20**(2): 82–87.

In samples that were obtained after SHS, the separation of metallic and metal-ceramic phases was performed, each of which was analyzed. X-ray fluorescence analysis (XFA) results showed that the metallic phase represents a monophase connection like Cr_5Al_8 with a hexagonal crystalline structure. The metal-ceramic component of SHS products has a complex composition. Table 5.2 shows XFA results of the metal-ceramic component.

Table 5.2 The composition of the metal-ceramic phase of SHS products

Compound	Content, %
Christobalite	1.3
Quartz	0.1
Aluminum oxide	62.7
Intermetallic metal Cr_5Al_8	35.9

Source: Reproduced from Dilmukhambetov EE, Mansurov ZA, Fomenko SM. (2011). Aluminothermic SHS: the effect of silica sol added as a binder. *Int. J. Self-Propag. High-Temp. Synth.*, **20**(2): 82–87.

These details help to realize the qualitative explanation of temperature–speed characteristics of SHS in the field of molar ratios smaller than the stoichiometry of full recovery of chromium oxide—in a combustion wave the processes of reorganization of crystal structures of chromium oxides and aluminum in the solid phase dominate. These processes are thermally activated, and they determine the speed of SHS, which shows a clear dependence on the combustion temperature. As a result of the reorganization of crystalline structures the solid solutions of chromium oxide and aluminum are formed. In the field of a high superstoichiometric ratio of aluminum to chromium oxide (4:6), combustion occurs substantially in the liquid phase of melted aluminum, which is accompanied by dissolving of reaction products. Dissolution processes do not require high activation energies, and therefore the combustion rate shows weak temperature dependence.

The speed measurements and combustion temperatures of silicon dioxide of various crystalline modifications and aluminum show that the combustion rate and temperature pass through the maximum at different molar ratios of reactants. The maximum combustion rate is achieved at a ratio of reagents $[Al]/[SiO_2] = 2$, but the maximum combustion temperature is achieved at $[Al]/[SiO_2] = 1.6$. The effective activation energies of the combustion process were determined at molar ratios of aluminum to silicon oxide from 0.5 to 1.5 and from 1.5 to 4.5, which were 27 and 23 kJ/mol for amorphous silicon dioxide, 105 and 40 kJ/mol for quartz, and 126 and 45 kJ/mol for christobalite. Research results show that the combustion of silicon oxide in crystalline modification at low content of aluminum is determined by diffusion processes that play an insignificant role during burning of amorphous silica. In the case of larger aluminum content, kinetic factors dominate. There are two characteristic regions: The first corresponds to the low temperatures at low molar ratios of reactants, $05 \leq [Al]/[SiO_2] \leq 1.5$, where the combustion rate increases monotonically. The second corresponds to the correlation of reactants $1.5 \leq [Al]/[SiO_2] \leq 4.5$, where the combustion rate decreases monotonically. This indicates the transition of the combustion process from kinetic field to diffusion. The combustion of silicon oxide in various modifications is approximately equal.

From these experimental studies of the system SiO_2–Al it can be assumed that SHS is carried out stepwise:

$$2SiO_2 + 2Al \rightarrow 2SiO + Al_2O_2 \quad (5.5)$$

$$2SiO + 2Al \rightarrow 2Si + Al_2O_2 \quad (5.6)$$

First of all, intermediate products such as monoxide silicon (SiO) and aluminum suboxides (AlO, Al_2O_2) are formed. These products were found with the help of transmission electron microscopy on electron-emitting thermal treatment of nanoscale films of silicon oxide with aluminum particles (Fomenko et al., 2003). Nonequilibrium defects at the surface of amorphous SiO_2 can serve as active centers, which were introduced into the surface layer as a result of heating or produced during combustion process. An interaction of aluminum with the surface SiO_2 begins at the site of the valence bond break, at the peaks of embryonic microcracks. At the time of the initial formation of burning centers the mixture SiO_2 + nAl reaction rate is small, but as they grow, the rate of reaction increases. Growing interaction centers are expanded independently of each other and cross with the formation of a united reaction front. This step of combustion of the exothermic mixture corresponds to the maximum burning rate.

In an air atmosphere and simultaneously in a combustion wave, oxidation reactions occur:

$$2Al + 2O_2 \rightarrow Al_2O_2 \quad (5.7)$$

$$2Al_2O_2 + O_2 \rightarrow 2Al_2O_3 \quad (5.8)$$

It is evident that at low temperature (heat penetration), the stage process that is initially started between reactants is in the solid state and passes into a reaction with the participation of the liquid phase. The appearance of the liquid phase accelerates mass-exchanging processes, so the kinetic characteristics of the reaction rise sharply, which is reflected in the Arrhenius plot of rate versus temperature. The reactions that lead to the final products are as follows:

$$SiO + Al_2O_2 \rightarrow Al_2O_3 + Si \quad (5.9)$$

$$3Al_2O_2 \rightarrow 2Al_2O_3 + 2Al \quad (5.10)$$

Disproportionating products SiO and Al_2O_2 condense in the reaction zone, forming the phases SiO_2, Al_2O_3, and silicon-aluminum fire beads. XFA shows that at the stoichiometric ratio of reactants, the reaction products are Al_2O_3, metallic Si, and Al.

The temperature and the combustion rate depend on the molar ratio of aluminum to iron oxide (III) for the samples with 20% of the inert filler (aluminum oxide). The presence of two combustion temperature maximums is notable. The stoichiometric ratio for the first is 2:1 (2600°C) and for the second is 3.5:1 (2400°C). The minimum value is about 1650°C corresponding to a ratio of 3:1.

The presence of two combustion temperature peaks of the considered system was registered in the literature earlier (Podergin et al., 1974), but the temperature values remained uncertain. Due to the addition to the system of an inert filler it was possible to reduce the general temperature dependence in the field, available for recording tungsten-rhenium thermocouples, and get a clear indication of the peaks of combustion temperatures.

It should be noted that the first maximum of combustion temperature corresponds to the stoichiometric equation of the complete recovery of iron:

$$Fe_2O_3 + 2Al \rightarrow Al_2O_3 + 2Fe \qquad (5.11)$$

This reaction is highly exothermic and the heat of the reaction is equal to 854 kJ/mol.

The second maximum may correspond to the formation of aluminide iron FeAl at a high aluminum content. Aluminide iron in the system Fe–Al with the formula FeAl is formed at the temperatures below 1150°C and mass concentrations of aluminum more than 60%. Apparently, these conditions are implemented in the SHS wave. The heat of the appropriate reaction (Eq. 5.12) is far less than the heat of the reaction in Eq. 5.3 and is equal to 750 KJ/mol:

$$Fe_2O_3 + 4Al \rightarrow Al_2O_3 + 2FeAl \qquad (5.12)$$

Temperature-speed characteristics of combustion of investigated systems point at the complex, multistage character of high-temperature reduction of iron oxide: there are intermediate compounds that are responsible for combustion temperature and compounds that are responsible for burning rate (Dilmuhambetov et al., 2005). Indeed, the x-ray phase analysis of SHS products shows the availability of a number of partial recovery products of iron oxide (III), iron oxide (II), magnetite, and iron aluminate.

Obtained results of investigation concerning aluminothermic combustion with participation of these oxides were the basis for the creation of SHS refractory materials—aluminosilicate, chrome-magnesia, and carbonaceous.

5.3 Influence of Silica Sols on Aluminothermal Combustion of Oxide Systems in an SHS Regime

Currently, on the basis of SHS various combined processes are developed, including SHS and metallothermy, SHS and sol-gel technology, and SHS and mechanoactivation (Merzhanov and Mukasyan, 2007; Dilmukhambetov et al., 2009). Investigations of SHS with the participation of nanodispersed components have significant scientific and practical interest. Among nanodispersed systems for various purposes, silicon in the form of a colloid solution or sol has a major expansion (Shabanova and Sarkisov, 2004). The most important factor is the establishment of macrokinetic features of aluminothermal combustion of silicon oxides, iron, and chromium (III) in an SHS regime in the presence of nanodispersed silica; there is an influence of heterocoagulation of silica sol on temperature-speed characteristics of SHS and the nature of end products (Dilmukhambetov et al., 2011). Measurements of temperatures and combustion rate were carried out on cylindrical samples with a diameter of 2 cm and a height of 5 cm, which are manufactured by compression of powder mixtures of aluminum and corresponding oxides. During manufacture of cylindrical specimens, as a moistening liquid of dry mixes (binder) silica sols with different contents of SiO_2 were used. Heterocoagulation of silica sol in the presence of fine powder materials leads to the formation of nanodispersed silicon oxide at the surface of particles of a heterogeneous environment.

5.3.1 Coagulation of Silica Sol in Heterogeneous Environments

Heterocoagulation of silica sol in the presence of fine powder materials leads to the formation of nanodispersed silicon oxide at the surface of particles of a heterogeneous environment. Figure 5.3a shows a microphotograph of particles of the initial silica sol, but Fig. 5.3b shows the heterocoagulation of sol on particles of aluminum. Almost all sample particles are coated by the sol. Electron diffraction patterns, which were obtained from such particles, indicate the lack of a crystal structure. Thus, the surface of aluminum particles is blocked by particles of nanodispersed silica. This circumstance leads to an interesting phenomenon—the powdered aluminum is placed in an alkaline medium of silica sol and does not react with water on hydrogen evolution. In the absence of silica sol, any alkaline solution with pH = 10–11 immediately leads to hydrogen evolution.

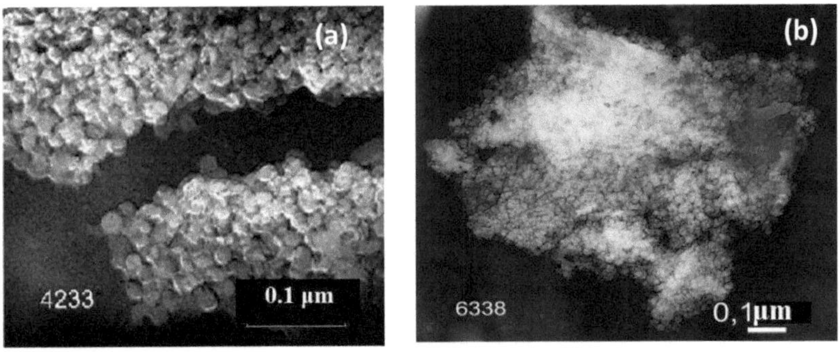

Figure 5.3 Aggregative silica sol particles (a) and particles of aluminum powder surrounded by an aggregative silica sol (b). Reproduced from Dilmukhambetov EE, Mansurov ZA, Fomenko SM. (2011). Aluminothermic SHS: the effect of silica sol added as a binder. *Int. J. Self-Propag. High-Temp. Synth.*, **20**(2): 82–87.

Figure 5.4 shows the image of the heterocoagulation of silica sol on silica particles. The sample was presented as plenty of translucent structures of different densities and sizes. Often the fragmentary particles have loose zones that are formed during silica exposure. Probably this is due to silica impact. There is a small number of crystallites that are coated by rounded sol particles.

The picture of heterocoagulation of silica sols in a medium of powdered iron oxides—magnetite and hematite—in Fig. 5.5 clearly shows the presence of aggregated sol particles at the surface of oxides.

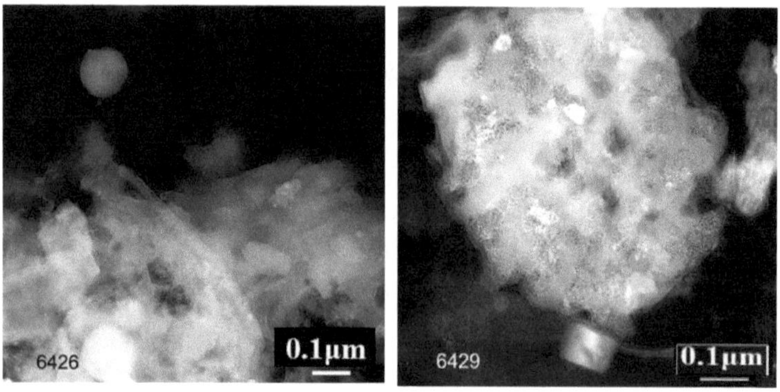

Figure 5.4 Electron micrograph of sol particles on quartz particles. Reproduced from Dilmukhambetov EE, Mansurov ZA, Fomenko SM. (2011). Aluminothermic SHS: the effect of silica sol added as a binder. *Int. J. Self-Propag. High-Temp. Synth.*, **20**(2): 82–87.

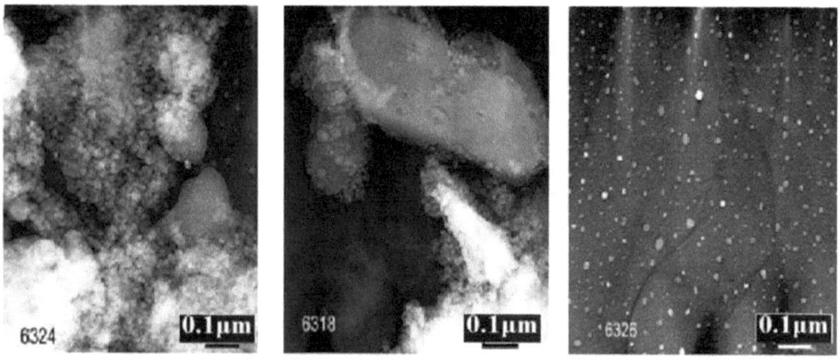

Figure 5.5 Heterocoagulation of silica sol on hematite. Reproduced from Dilmukhambetov EE, Mansurov ZA, Fomenko SM. (2011). Aluminothermic SHS: the effect of silica sol added as a binder. *Int. J. Self-Propag. High-Temp. Synth.*, **20**(2): 82–87.

Under the influence of electron beam heat the aggregative sol particles fall into smaller formations up to certain nanosized spherical particles. Sizes of sol particles range from 10 to 30 nm. Independent determination of the particle size of silica sol by the optical turbidity method in a colloidal solution shows size variation from 20 to 30 nm.

Figure 5.6 shows the pattern of heterocoagulation of silica sol on the particles of amorphous chromium oxide. In the sample there are loose rounded particles with a dominant size of 20–30 nm, forming clusters around the dense areas, transforming under the electron beam (neg. № 8716). Microdiffraction patterns show that there are several metallic phases (neg. № 8723).

Metallic phases, dense formations, and loose masses represent the various forms of amorphous chromium oxide, but the spherical particles represent particles of silica sol.

The results of electron microscopic studies indicate that regardless of the nature of the heterogeneous environment the heterocoagulation of silica sol occurs at the surface of amorphous particles; particles of the substance having crystal ordering do not interact with sol particles.

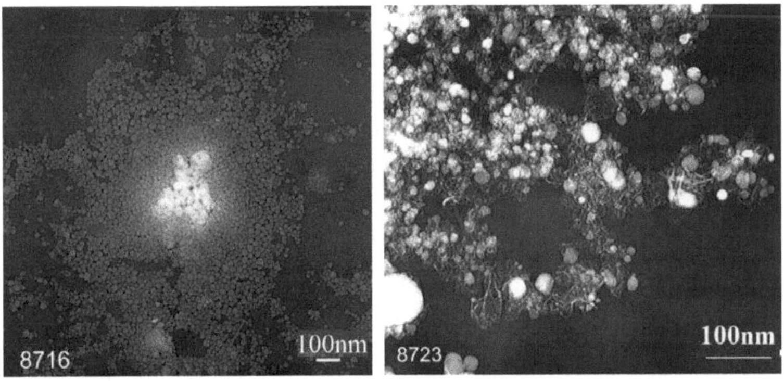

Figure 5.6 Heterocoagulation of silica sol on amorphous chromium oxide. Reproduced from Dilmukhambetov EE, Mansurov ZA, Fomenko SM. (2011). Aluminothermic SHS: the effect of silica sol added as a binder. *Int. J. Self-Propag. High-Temp. Synth.*, **20**(2): 82–87.

5.3.2 The Influence of Silica Sol on SHS in the Al–Silicon Oxide System

In a powder composition of aluminum and silicon oxide the addition of silica sol does not alter the chemical composition of aluminothermal system. In this case, the influence of the nanodispersed component on the macrokinetic of SHS will appear in "pure" form. The temperature and speed characteristics in SHS systems like silicon oxide and aluminum were considered above.

The observed displacement of peak temperature and combustion rate is relative to each other and indicates the deviation of experimental data from the theory. Other large deviations from the conclusions of thermal theory of solid-phase combustion are observed in the presence of nanodispersed silica.

Figure 5.7a–c shows the temperature-velocity functions of SHS in the presence of nanodispersed silica on the heterocoagulation of sols of different concentrations. The effect of nanodispersed silica appears generally in the form of a noticeable reduction of temperatures and combustion rates in comparison with net silicon oxide. The increase of the silica sol concentration leads to a weak dependence, resulting in a poor combustion rate depending on the temperature regardless of the molar ratio of aluminum to silicon oxide. Combustion temperatures in the presence of nanodispersed silica can be compared with the lower temperature limit for the combustion of pure oxide, and this fact is conditioned by the reduction of the effective energy of SHS activation.

Obtained results testify that the nanodispersed silica in an aluminothermal system encourages activation-less processes in the combustion wave with weak temperature dependence, which determines the wave propagation velocity. Destruction and dissolution of aggregative nanoparticles in liquid aluminum may act as such processes because the aluminum is only a liquid phase at recorded temperatures. The combustion temperatures are determined by destruction processes of the amorphous and crystalline structures of silicon oxide and the following multistage reduction oxidation reaction, with the release of a large amount of heat.

The results of x-ray phase analysis show that the products consist of aluminum oxide, free silicon, and silicon oxide. For the system of stoichiometric ratio $4Al/3SiO_2$ the SHS products are determined depending on the concentration of the silica sol. The results are shown in Table 5.3. As can be seen from the table, along with the basic compounds, the aluminum nitride is present almost always, the content of which decreases with an increase of the sol concentration. The concentration of silica sol equals 35–40 wt%, and in view of this, a complete reduction of silicon oxide is observed. Thus, nanodispersed silica at aluminothermy of silicon oxide contributes to the full restoration of the oxide at moderate temperatures of the SHS process.

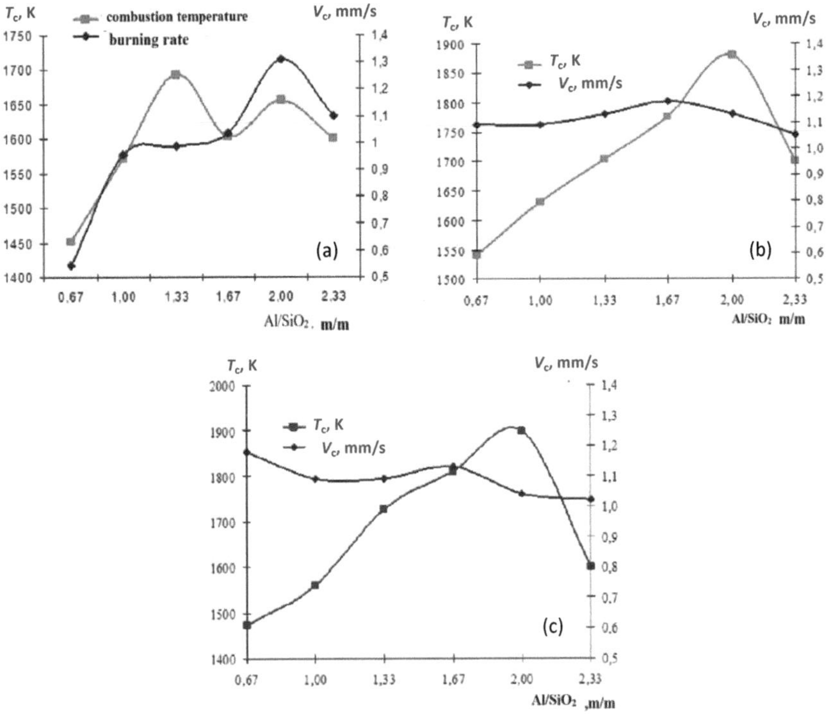

Figure 5.7 Temperatures and combustion rates of the Al-SiO_2 system in the presence of (a) 20% silica sol, (b) 30% silica sol, and (c) 40% silica sol. Reproduced from Dilmukhambetov EE, Mansurov ZA, Fomenko SM. (2011). Aluminothermic SHS: the effect of silica sol added as a binder. *Int. J. Self-Propag. High-Temp. Synth.*, **20**(2): 82–87.

Table 5.3 The XRF data of SHS products for amorphous silicon oxide in the presence of a sol

Sol concentration	Quartz, %	Christobalite, %	Corundum, %	Si, %	AlN, %
5%	17.2	1.8	38.7	15.6	9.7
10%	19.9	1.0	40.0	18.2	5.6
15%	9.2	0.6	48.7	18.7	3.4
20%	13.8	0.5	46.8	18.3	4.7

(*Continued*)

Table 5.3 (*Continued*)

Sol concentration	Quartz, %	Christobalite, %	Corundum, %	Si, %	AlN, %
25%	12.8	0.8	48.3	21.7	6.2
30%	10.3	0.9	53.8	20.6	2.5
35%	5.7	1.6	67.1	22.4	1.2
40%	1.9	1.4	60.7	23.7	1.7

Source: Reproduced from Dilmukhambetov EE, Mansurov ZA, Fomenko SM. (2011). Aluminothermic SHS: the effect of silica sol added as a binder. *Int. J. Self-Propag. High-Temp. Synth.*, **20**(2): 82–87.

5.3.3 The Influence of Silica Sol on SHS in Al–Iron Oxide and Al–Chrome Oxide Systems

The measurements of temperatures and the linear burning rate in the Al-Fe_3O_4 system on cylindrical specimens made by pressing aluminum powder mixtures and magnetite with the addition of silica sol were carried out. To eliminate combustion in the thermal explosion mode, aluminum oxide, which serves as an inert filler, was added to the system. Figure 5.8 shows the temperature–speed characteristics of SHS, depending on the molar ratio Al/Fe_3O_4.

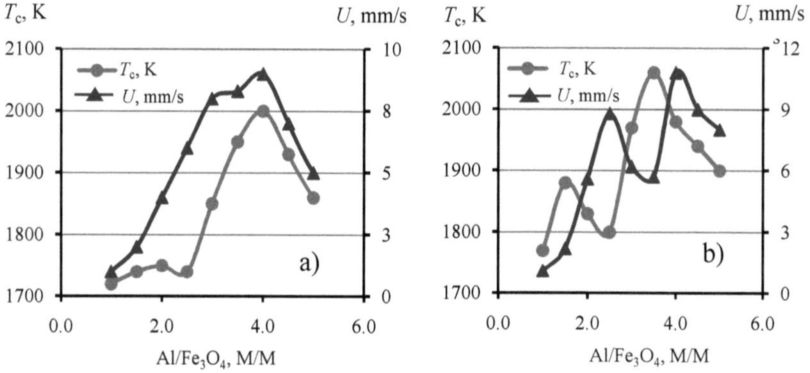

Figure 5.8 Temperatures and combustion rates of the Al-Fe_3O_4 system (a) in the absence of silica sol and (b) in the presence of silica sol. Reproduced from Dilmukhambetov EE, Mansurov ZA, Fomenko SM. (2011). Aluminothermic SHS: the effect of silica sol added as a binder. *Int. J. Self-Propag. High-Temp. Synth.*, **20**(2): 82–87.

Temperature dependence in Fig. 5.7a has a poorly resolved maximum at n = 2, corresponding to 1750 K; a pronounced maximum at n = 2, corresponding to 1750 K; and a distinct maximum at n = 4, corresponding to 2000 K. The linear burning rate of samples shows the maximum value at n = 4. The effect of silica sol on the temperature and speed of SHS is presented in Fig. 5.3b. In all variations of the molar ratio of aluminum to iron oxide the presence of nanodispersed silica causes a slight increase in the combustion temperature, with two well-defined maxima at a ratio of 1–1.5 mol/mol and 3.5–4 mol/mol. The combustion rates reach a maximum value with ratios of 2.5 mol/mol and 4 mol/mol.

For the Al-Fe_2O_3 system in the presence of 20% wt. of the inert filler Al_2O_3, the following regularities of SHS are found: in the absence of silica the combustion temperatures reach the maximum values of 2900 K and 2700 K at molar ratios Al/Fe_2O_3 equal to 2 and 4 and the combustion rate has one maximum at a molar ratio equal to 3 (Fig. 5.9).

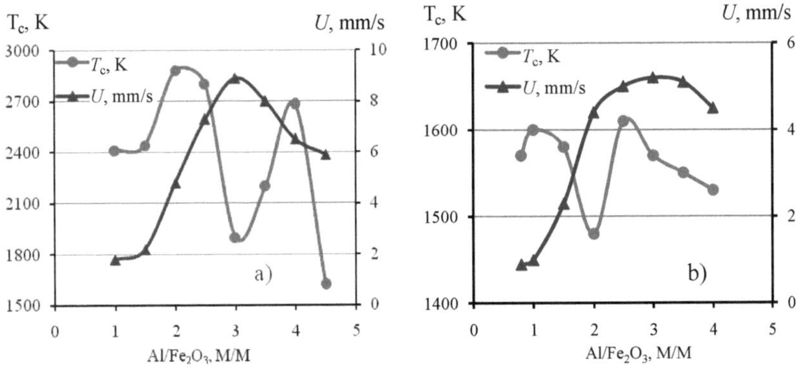

Figure 5.9 Temperatures and combustion rates of the Al-Fe_2O_3 system (a) in the absence of silica sol and (b) in the presence of silica sol. Reproduced from Dilmukhambetov EE, Mansurov ZA, Fomenko SM. (2011). Aluminothermic SHS: the effect of silica sol added as a binder. *Int. J. Self-Propag. High-Temp. Synth.*, **20**(2): 82–87.

An introduction of nanodispersed silica into the system leads to a massive reduction in the temperatures and combustion rates (Fig. 5.9b) as well as a shift of the peak temperature in the region of the smallest molar ratios. The rate takes the maximum value in a sufficiently broad range of molar ratios from 2.0 to 3.5.

The obtained results testify to the fact that there is no correlation between temperature and combustion rate in an SHS regime for considered systems with iron oxides, established by the simple model of solid-phase combustion with a narrow wave front.

The absence of a one-to-one relation between temperature and the rate of SHS is determined by the complex nature of a large number of physical and chemical processes during the combustion of iron oxides, which occur essentially in a liquid phase from reaction products: the formation of iron aluminates and aluminum alloys with iron, decomposition of oxides with oxygen separation, disproportionating of iron oxide (II), etc.

X-ray fluorescence (XRF) results of SHS products confirm the complex nature of combustion processes of iron oxides. Tables 5.4 and 5.5 show XRF results of SHS products at different molar ratios of aluminum oxide.

Table 5.4 The results of semiquantative x-ray phase analysis of SHS products of iron oxide (III)

n(Al)/ $n(Fe_2O_3)$	Fe_3O_4, %	Fe_2O_3, %	FeO, %	Fe, %	$FeAl_2O_4$	Al_2O_3	FeAl
1.0	19.2	11.7	0.7	5.4	40.2	18.0	—
1.5	16.4	7.2	0.8	11.4	48.1	16.0	—
4.5	0.7	5.6	0.8	26.4	11.6	39.0	8.7

Source: Reproduced from Dilmukhambetov EE, Mansurov ZA, Fomenko SM. (2011). Aluminothermic SHS: the effect of silica sol added as a binder. *Int. J. Self-Propag. High-Temp. Synth.*, 20(2): 82–87.

The main conclusions of XRF data are as follows: in the region of molar ratios below the stoichiometry, iron aluminate, a low-melting compound, is the main product. This indicates the active interaction of the inert filler, aluminum oxide, with the product of partial recovery, oxide Fe(II), with a lack of a reducing agent. At superfluous, hyperstoichiometric content of aluminum in products there are dominating aluminum alloys with iron and aluminum oxide.

Table 5.5 XRF results of SHS products of Al-Fe_3O_4 system

n(Al)/ n(Fe_2O_3)	$FeAl_2O_4$, %	FeO, %	Quartz, %	Fe_2O_3, %	Al_2O_3, %	$Al_{13}Fe_4$, %	$Al_{0.5}Fe_3Si_{0.5}$, %
1.5	89.4	0.9	1.8	1.4	6.4	—	—
2.5	80.7	—	—	1.4	17.8	—	—
4.0	—	—	—	—	15.0	Much	In significant quantity
5.0	—	—	—	—	17.0	Much	In significant quantity

Source: Reproduced from Dilmukhambetov EE, Mansurov ZA, Fomenko SM. (2011). Aluminothermic SHS: the effect of silica sol added as a binder. *Int. J. Self-Propag. High-Temp. Synth.*, **20**(2): 82–87.

The rates and temperatures of combustion of silica sol added to chromium oxide in a wide range of molar ratios Al/Cr_2O_3 in SHS mode are presented in Fig. 5.10. The specific differences between the combustion of pure chromium oxide are a noticeable decrease of temperature and rate and the lack of a temperature peak in the molar ratio Al/Cr_2O_3, which is equal to 2.

Thus, heterocoagulation of silica sols in dispersion mediums of different natures (oxides of silicon, iron, chromium, and powdered aluminum) is always accompanied by the formation of nanodispersed silicon oxide on the surfaces of amorphized particles of the medium with poorly resolved crystalline order. In studied oxides the macrokinetic regularities of aluminothermic combustion in a regime of SHS in the presence of silica sol significantly different from the regularities are observed in the absence of nanodispersed silica.

The effect of silica sols on temperatures and the rate of aluminothermic combustion of oxides is manifested primarily in a noticeable reduction of these characteristics measured depending on the molar ratio of the reducing agent (aluminum)/oxidizer. The temperature–velocity dependence of SHS, as usually a one-to-one correlation is absent, is predicted by a simple model of solid-phase combustion with a narrow wave front. Results are conditioned by the transition of a combustion regime to a solid phase with an increasing of molar ratio aluminum/oxide and reduction of the effective activation energy of SHS. At a molar ratio of aluminum-iron oxide with less stoichiometry of the complete recovery of iron in SHS products, there are dominating iron oxides in intermediate oxidation stages in the form of iron aluminates. In a system of aluminum, oxide, chromium, and silicon and in synthesis products, there are dominating mixed oxides. At surplus content of aluminum in SHS products there are present intermetallic compound products—reduced metals with aluminum.

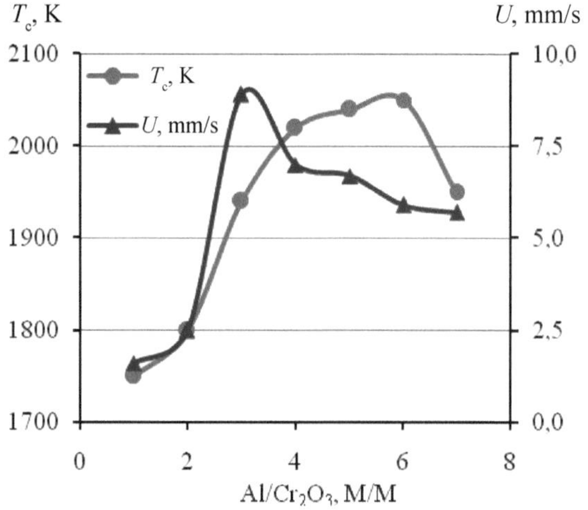

Figure 5.10 Temperatures and combustion rates of the Al-Cr_2O_3 (a) in the absence of silica sol and (b) in the presence of silica sol. Reproduced from Dilmukhambetov EE, Mansurov ZA, Fomenko SM. (2011). Aluminothermic SHS: the effect of silica sol added as a binder. *Int. J. Self-Propag. High-Temp. Synth.*, **20**(2): 82–87.

The obtained results indicate that silica hydrosols are an effective binding agent in SHS systems, allowing SHS processes and aluminothermy of metal oxides at moderate temperatures of 1000°C–1600°C, which is of important practical interest. On the basis of obtained results, carbonaceous exothermic mixtures consisting of nanodispersed silica are developed. Heat treatment of such mixtures at 850°C–900°C as the result of SHS leads to the formation of a strong refractory composite with physical and chemical characteristics and is highly competitive with graphite (Mansurov et al., 2001).

5.4 Carbonaceous SHS Refractory Materials

A general approach to the synthesis of carbonaceous refractory materials lies in conducting aluminothermic solid-phase combustion of metal oxides in a regime in the presence of carbon (Ksandopulo et al., 1996). Typically, the carbon in the form of graphite takes in amounts significantly greater than the stoichiometric ratio of the starting reagents. At high temperatures of synthesis (to the order of 1500°C) the reduced metal reacts with carbon and forms refractory carbides, for example:

$$4Al + 3TiO_2 + 3C = Al_2O_3 + 3TiC \quad (5.13)$$

$$4Al + 3SiO_2 + 3C = Al_2O_3 + 3SiC \quad (5.14)$$

Transition metal oxides, such as chromium, titanium, niobium, zirconium, and others, are used as oxidation agents, and it is possible to obtain chemically stable, high-refractory carbonaceous materials. SHS products are composite materials from refractory compounds—aluminum oxide, metal carbide, and carbon.

Previously (Vongay et al., 2004; Dilmukhambetov et al., 2006) works concerning investigations of SHS in systems containing carbon, aluminum powder, refractory clay or mortar, and magnesium sulfate as binder and oxidant were carried out. The obtained material had characteristics that allowed its use for bonding of graphite and silicon carbide products. However, despite the obvious advantage over conventional carbon-containing refractory bodies containing organic binder-binding substances, the material has negative characteristics, for example, the presence of carbides and aluminum sulfides formed in the course of secondary reactions during SHS. These compounds by reacting with water vapor are destroyed, causing distortion of the structure throughout a monolithic refractory mass.

There is a possibility to avoid the formation of carbides and aluminum sulfides during SHS by optimizing the oxygen content in the initial mixture of components and increasing the oxygen supply for the reaction of aluminothermic reduction of silicon. Simultaneously, it is desirable to completely eliminate the sulfate compounds in the composition of the initial mixture. To avoid undesirable effects as a binder, the use of colloidal silica was suggested, which was obtained by the hydrolysis of ethyl silicate (brand ES-40) and by a solution of sulfuric acid of different concentrations (Fig. 5.11). As a basic chemical process, during the development of carbonaceous SHS refractories, the process of SHS with an intermediate stage was realized. Thus the silicon carbides, titanium, zirconium, etc., are synthesized in the second step of the synthesis. An important element of the developed technology is a combination of sol-gel and SHS technology in the manufacturing process, resulting in pre-encapsulation of components of the SHS system by aggregated nanodispersed oxides, such as xerogel.

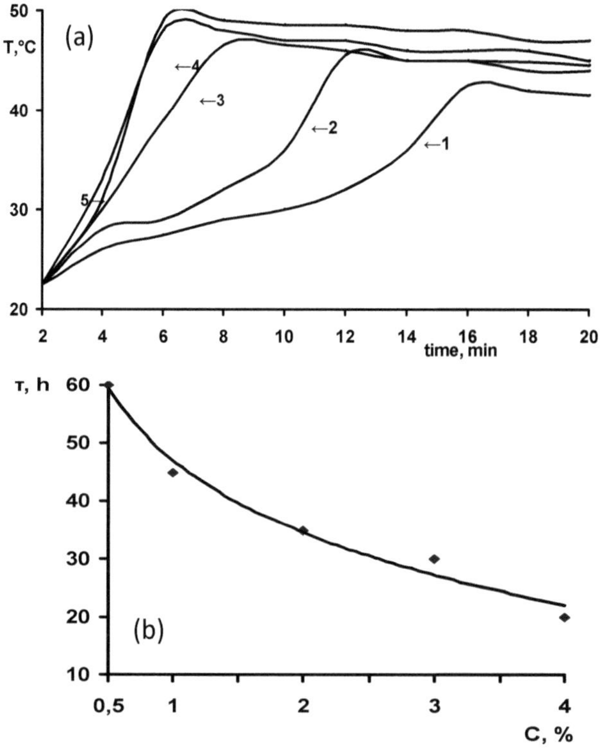

Figure 5.11 Optimal conditions for obtaining silica sol by the hydrolysis of ethyl silicate. (a) Dependence of hydrolysis temperature on the concentration of H_2SO_4: 1, 0.5%; 2, 1.0%; 3, 2.0%; 4, 3.0%; and 5, 4.0%. (b) Stable time of silica sol. Reproduced from Mansurov ZA, Fomenko SM. (2014). Carbonaceous refractory materials on SHS-technology. *Adv. Sci. Technol. Mater. Res.*, **88**: 112–128.

The use of silica sol as a binder introduced a new effect: heterocoagulation of the sol. The phenomenon of sol heterocoagulation leads to the formation of ultradispersed silica with particle sizes of 20–30 nm or less in the system. An important consequence of sol heterocoagulation is the fact that the surface of aluminum particles is blocked by nanodispersed silica and its activity is suppressed both in an alkaline medium (pH 10–11) and in an acidic environment (pH 1–2). In the absence of a sol the immediate release of hydrogen is observed. In further work silica sol was obtained by weakly acidic hydrolysis of ethyl silicate. The optimum composition of hydrolyzate, possessing adequate stability to coagulation of sol, is determined: ethyl silicate (ES), 40%–55%; 0.5% solution of sulfuric acid, 45%.

For realization of SHS to obtain the masonry mortar composition for fastening of graphite and silicon carbide products the aluminothermal system with aluminosilicate oxidants was chosen as a filler, which consists of the waste of electrode graphite in the form of grains about 1–3 mm in size and the powder with dispersibility of less than 90 microns. The exothermic mixture was shut by a hydrolyzed ethyl silicate solution with a 20%–25% content of SiO_2. The particle sizes of silica sol were 20–30 nm. The obtained initial mass demonstrated much better rheological characteristics than described in research by Ksandopulo et al. (1996). It moistened well the surface of fastened graphite samples when exposed for 4 h at 40°C–50°C, and it is completely converted to solid state due to heterocoagulation and the formation of a spatially structured matrix of silica gel, which includes the remaining components of the mixture. Thus, the material after drying is an exothermic composition containing a significant amount of ultradispersed silica oxide. After heating in a muffle furnace at 1100°C–1150°C the process of SHS is started. The velocity of the combustion wave was 2–5 mm/s at a combustion temperature of 1400°C–1700°C. After synthesis passing, a material is formed having negligible porosity and that fastens well the graphite products.

The fasten seam remains without alterations up to the heating temperature of 1800°C. In experimental destruction of fastened blocks, the failure plane of the sample always continues in the direction of destructive effects. The existence of seams of the synthesized refractory material is not evident, that is, the strength properties of synthesized SHS refractory material were similar to graphite blocks and reached 5–6 MPa. These values are highly competitive with a known high-temperature method for bonding of graphite products. It is noteworthy that the strength and adhesion properties of the dried mixture are also high before the realization of SHS.

According to XRF results the phase composition of carbonaceous refractory after SHS was comprised of silicon carbide, aluminum oxide, graphite, and elemental silicon. Aluminum carbide and metal aluminum were not detected.

Compounds that are obtained as the result of SHS and heat treatment are mostly individual refractory compounds, and their composition is a promising refractory material for use not only as masonry mortar for fastening of graphite and silicon carbide plates and blocks (Dilmukhambetov et al., 2006) but also as a ramming mixture for the production of high-resistance carbonaceous products used in the metallurgy industry (Fomenko et al., 2011).

The use of SHS carbonaceous masses for the production of crucibles for melting metals is of interest. It is known that the main wear of graphite crucible is due to its burning out at periodic thermal cycling during the operation process. In this regard, characteristics such as burning in air are very important for determination of the carbonaceous material's perspectivity. Investigation was conducted on cylindrical samples with a diameter of 2 cm and a height of 4 cm pressed by exothermic mixtures from compressed humidified silica sol. In all samples titanium oxide and aluminum silica in different ratios were used as oxidant. The carbon content in the original system was 20, 30, and 40 wt%. Then SHS samples were exposed to thermal cycling at 950°C in an air atmosphere, with 30 min exposure in each cycle. The nature of the relative change of the sample weight at thermal cycling is shown in Fig. 5.12. For comparison, the graph represents the relative variation of crucible graphite mass, which indicates its greatest burning out in air. The burning out of carbonaceous SHS refractory materials is 3–5 times lower than the burning out of crucible graphite.

The physical and mechanical properties of synthesized refractories are very important characteristics. Table 5.6 shows the strength variation of compression of carbonaceous SHS refractories as the result of thermal cycling.

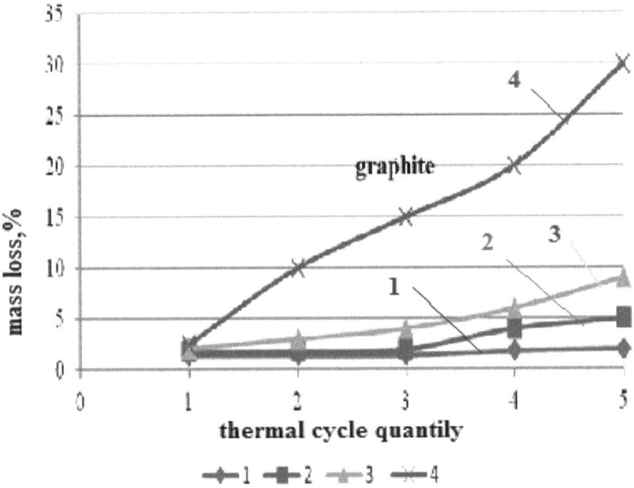

Figure 5.12 The burning degree of carbonaceous SHS refractory as the result of thermal cycling in the air. Carbon content: 1, 20%; 2, 30%; 3, 40%; and 4, graphite. Reproduced from Mansurov ZA, Fomenko SM. (2014). Carbonaceous refractory materials on SHS-technology. *Adv. Sci. Technol. Mater. Res.*, **88**: 112–128.

Table 5.6 Compression strength of carbonaceous SHS refractories before and after cyclic thermal treatment

Sample no.	1	2	3	4	5	6	Graphite	Graphite chamotte refractory
[a]σ_1, MPa	4.0	6.4	12.0	7.2	3.2	6.4	11.2	3–5
[b]σ_2, MPa	4.4	6.8	12.4	6.4	6.4	7.2	6.0	—

[a]σ_1: Compressive strength immediately after SHS
[b]σ_2: Compressive strength after 5 thermal cyclings at 900°C
Source: Reproduced from Mansurov ZA, Fomenko SM. (2014). Carbonaceous refractory materials on SHS-technology. *Adv. Sci. Technol. Mater. Res.*, **88**: 112–128.

Table 5.7 shows that as the result of heat treatments the strength of SHS refractories usually increases, while the crucible graphite shows an almost two times reduction of compressive strength. As it turns out, the multiple thermal impact hardens the material in contrast to graphite and graphite-containing traditional refractories. The method of temperature measuring for drop of standard cones was made from carbonaceous compounds. Fire resistance measurements were carried out. The temperature is measured with a thermometer pyrometric Raytek 3i1M. Measurements have shown that even at maximum heating mode of Tamman furnace to temperatures of 1850°C–1900°C softening and falling cones are not observed. These results demonstrate high refractoriness of the carbonaceous material after SHS—not lower than 1850°C.

Table 5.7 Influence of graphite content on the initial exothermal mixture for sample strength

Carbon content in mixture	Compressive strength after SHS, MPa	Compressive strength after thermal treatment at a temperature 1400°C, MPa
20	3.1	18.2
30	4.8	13.9
40	4.2	8.9
50	3.4	8.6

Source: Reproduced from Mansurov ZA, Fomenko SM. (2014). Carbonaceous refractory materials on SHS-technology. *Adv. Sci. Technol. Mater. Res.*, **88**: 112–128.

The results of x-ray phase analysis are presented in Table 5.8 and show that the high-refractory material after SHS is conditioned by a large content of high-melting compounds: aluminum oxide, titanium carbide, silicon, spinel, and carbon.

Table 5.8 Phase composition of SHS products

Silicon-containing systems		Titanium-containing systems	
Corundum	24–49	Corundum	20–45
Quartz	7–22	Quartz	0–2
Christobalite	0–3	Christobalite	0–1
Silicon	2.5–28	Titanium-disilicide	0–4
Carbon	6–44	Carbon	9–48
Silicon carbide	4–10	Titanium carbide	6–8

Source: Reproduced from Mansurov ZA, Fomenko SM. (2014). Carbonaceous refractory materials on SHS-technology. *Adv. Sci. Technol. Mater. Res.*, **88**: 112–128.

The basic refractory phases—aluminum oxide, graphite, titanium carbide, silicon, spinel—provide refractoriness of samples above 1850°C.

The experimental samples were held in molten bronze for 15 min at a temperature of 1450°C. The resistance to melting is determined by the change of sample mass by dissolving or chemical interaction with the liquid phase.

Table 5.9 shows the nature of mass variation of test specimens, at high-temperature influence of liquid metal, depending on the carbon content in the starting mixture and the synthesis conditions.

Table 5.9 Corrosion resistance of SHS refractories to molten bronze

Element	Wt%	At%
OK	30.29	43.23
CK	4.68	4.09
AlK	55.08	46.60
SiK	5.77	4.69
TiK	4.18	1.99

No.	Carbon content in the initial mixture, %	The samples synthesized in an argon atmosphere, mass variation, %	The samples synthesized in an air atmosphere, mass variation, %
1	20	4.3	4.5
2	30	3.4	3.0
3	40	0	2.3
4	50	0	4.0

Source: Reproduced from Mansurov ZA, Fomenko SM. (2014). Carbonaceous refractory materials on SHS-technology. *Adv. Sci. Technol. Mater. Res.*, **88**: 112–128.

All the samples of carbonaceous SHS refractory materials have shown high corrosion resistance to melting; the weight loss was 0%–5%.

To test the operation on corrosion resistance to melted materials from developed material by the dry-pressing method, the refractory products were fabricated in the form of cylindrical cups with a height of 80 mm, an outer diameter of 50 mm, a wall thickness of 7 mm, and a bottom thickness of 15 mm. SHS was carried out in an argon atmosphere (inert) and in an air atmosphere. The sample that was synthesized in an inert atmosphere was denser and more carbonized. Electron micrographs of samples that were synthesized in the air atmosphere show that the basis of the metal-ceramic matrix is penetrated in random ways by fibers of different thicknesses—from 30 to 1000 nm (Fig. 5.13).

Figure 5.13 Electron micrographs of SHS products synthesized in an air atmosphere. Reproduced from Mansurov ZA, Fomenko SM. (2014). Carbonaceous refractory materials on SHS-technology. *Adv. Sci. Technol. Mater. Res.*, **88**: 112–128.

Such a picture is observed during the formation of composite material carcasses from aluminum oxide, at partial oxidation of metallic aluminum in air with participation of its suboxides. The samples that were synthesized under an argon atmosphere at pressure of 10 atmospheres have entirely different structures (Fig. 5.14). Cubic and orthorhombic particles with a size of 1 micron and less are uniformly distributed throughout the volume of the composite. Furthermore, the degree of carbonation in this case is much higher because the inert atmosphere prevents the graphite oxidation and promotes the formation of carbide compounds around the carbon particles.

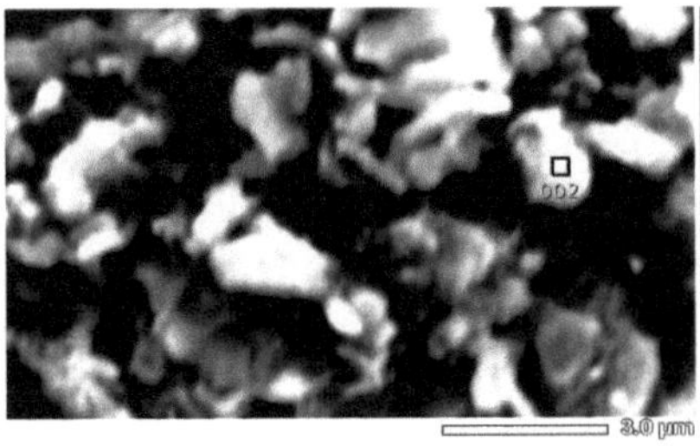

Title	:	IMG 2
Instrument voltage	:	15.00 kV
Magnification	:	x10.000
Date	:	2014.05.04
Pixel	:	1280 × 960

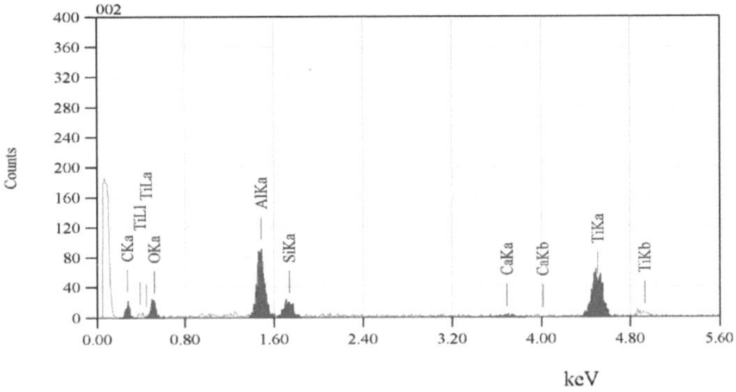

Acquisition parameters		
Instrument		6510
Acc. voltage		15.00 kV
Probe current		1.00000 nA
PHA mode		T3
Real time		30.07 s
Live time		30.00 s
Dead time		0%
Counting rate		191 cps
Energy range		0–20 keV

ZAF method standard-less quantitative analysis
fitting coefficient: 0.6699

Element	(keV)	Mass%	Error%	Atom%	Compound	Mass%	Cation	K
C K	0.277	23.80	0.96	40.26				12.2785
O K	0.525	26.14	3.12	33.21				16.7916
Al K	1.486	13.10	0.77	9.87				16.5049
Si K	1.739	3.06	0.92	2.81				3.9699
Ca K	3.690	0.86	1.65	0.43				1.5831
Ti K	4.508	33.05	2.37	14.02				48.8720
Total		100.00	—	100.00				—

Figure 5.14 Electron micrographs of SHS products synthesized in an argon atmosphere. Reproduced from Mansurov ZA, Fomenko SM. (2014). Carbonaceous refractory materials on SHS-technology. *Adv. Sci. Technol. Mater. Res.*, **88**: 112–128.

In these cylindrical crucibles experimental melting was performed (Tamman furnace): duralumin at 950°C, copper and bronze at 1400°C–1450°C, and iron at 1600°C. Figure 5.15 shows the photographs of the obtained ingot. Only on the iron ingot are metal droplets visible—the reaction product of the melt with the inner surface of the crucible. The other melts do not react with the material of the glass.

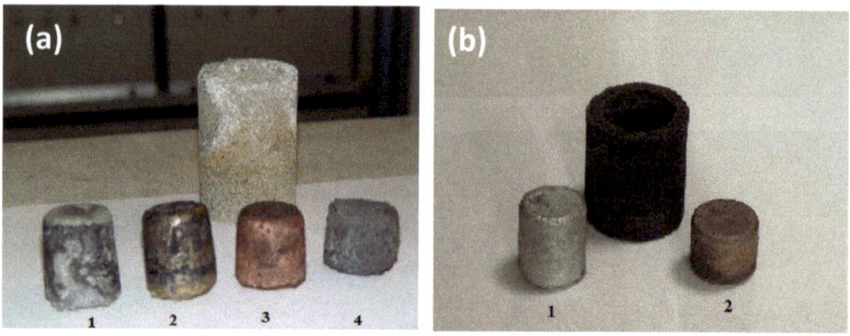

Figure 5.15 Ingot metals melted in cylindrical crucibles: (a) crucible is synthesized in an air atmosphere; (b) crucible is synthesized in an argon atmosphere; 1, duralumin; 2, bronze; 3, copper; and 4, iron. Reproduced from Mansurov ZA, Fomenko SM. (2014). Carbonaceous refractory materials on SHS-technology. *Adv. Sci. Technol. Mater. Res.*, **88**: 112–128.

So, the laboratory experimental crucibles showed unique properties and by their technical properties are considerably superior to traditional graphite. However, for the application of SHS technology at the lining of induction melting furnaces, equality of electrophysical properties between the SHS material and traditional crucible graphite is a necessary condition. The measurements showed that the lower electric conductivity of the used material compared to crucible graphite can lead to deviations of electric modes from standard furnaces. For the removal of such nonconformities the exothermic mixture's composition was optimized, with the purpose of obtaining carbonaceous composites, satisfying the operation conditions of induction furnaces. Also their electrical characteristics were determined.

SHS products of exothermic mixtures were objects of research. As reducing agents they contain the powders of aluminum, silicon, and titanium; as oxidation agents they contain silicon oxides and titanium; as fillers they contain the bout of electrode graphite and aluminosilicate mortar; and as binding material silica hydrolysis of Degussa AG production is used or a sol is obtained from ethyl silicate. Dry mixtures were etched by silica sol. The samples in cubic form were prepared by the compaction or tamping method. As a result of heterocoagulation of silica sol the samples take up the initial mechanical strength for 2–3 h. After drying, the samples were heated in an electric furnace at temperatures of 850°C–950°C, and at the same time, the process of SHS occurred. The specific electric resistance was measured in the apparatus.

Measurements of specific electric resistance showed that all synthesized materials had significant electrical conductivity. For two series of SHS products from exothermic mixtures using silicon oxides and titanium as main oxidizing agents, the values of specific electric resistance were $5 \cdot 10^{-1}$–$2 \cdot 10^{-4}$ ohm · m and $4 \cdot 10^{-3}$–$3 \cdot 10^{-5}$ ohm · m, respectively, at a temperature of 298 K. For the samples of crucible graphite the specific electric resistance is determined within $8 \cdot 10^{-6}$–$1 \cdot 10^{-5}$ ohm · m.

Measurements of specific electric resistance for silicon-containing composites and graphite evidence the increase of their conductivity with temperature increase. For titanium-containing systems, on the contrary, the temperature dependence of electrical resistance shows a decrease of conductivity at temperature increase. Such behavior is characteristic of conductivity of metallic type.

On the basis of XRF results it can be assumed that in silicon-containing systems materials, electrical conductivity is provided by agents such as silicon, carbon, and silicon carbide—substances that have a semiconducting conduction mechanism.

In the case of titanium-containing systems the basic electrical conducting phases are metal components such as Ti_5Si_3 and $TiSi_2$, having a significantly lower specific electric resistance than carbon and titanium carbide. Accordingly, their contribution to the total conductivity is predominant, which is reflected in thermal dependence of the resistivity.

The difference of electrical resistance between investigated systems and crucible graphite is quite significant, and reduction of these differences is achieved by introducing silicon powder into the titanium system. At a silicon content of 6%–8% the electrical resistances of SHS composite and crucible graphite are equalized (Fig. 5.16).

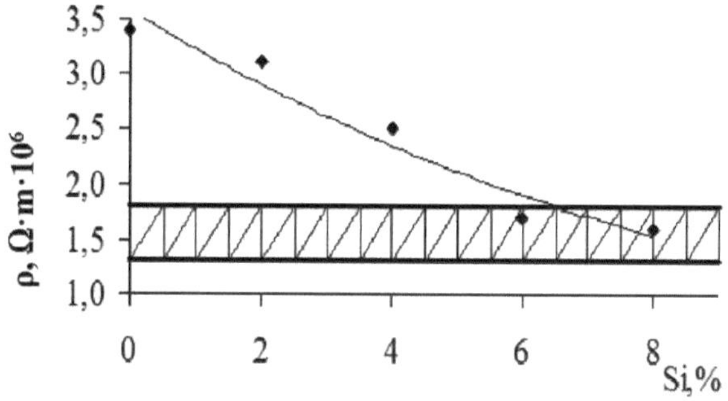

Figure 5.16 Specific electric resistance of carbonaceous titanium-silicon composite (the shaded area corresponds to the crucible graphite resistivity). Reproduced from Mansurov ZA, Fomenko SM. (2014). Carbonaceous refractory materials on SHS-technology. *Adv. Sci. Technol. Mater. Res.*, **88**: 112–128.

Thus, undertaken studies of SHS products in carbon-containing systems have shown the producibility of fire-resistant carbon-containing composite products that can replace the expensive graphite elements of lining of induction melting furnaces of ferrous and nonferrous metals without significant change of the electric mode of their operation.

5.5 Experience in Application of SHS Refractories in Thermal Generating Units

At the Institute of Combustion Problems are developed and manufactured the refractory mortars FURNON based on SHS technology (Ksandopulo and Ismailov, 1988, 1992). Refractory SHS mortars FURNON are powdered mixtures that are added to water (or covered by the water) and are used as masonry mortar or a protective coating of the lining of the thermal generating unit. Here, in an exothermic mixture aluminum is used as the reducing agent but such materials as ore concentrates, clays, natural and artificial refractory materials, and waste of mountain-processing plants are used as filling agents. Solid-phase combustion of exothermic mixtures causes refractory materials such as alumino-silicate, magnesian, and high carbon to be considered as reaction products, although the initial products are not the same. SHS mortars FURNON have the combustion temperatures 1600–2000°C and a combustion rate of 1 mm/s.

The unique feature of mortars FURNON fully appears in the welding effect of brick laying into a monolith. This effect is due to the fact that as the result of SHS not only the material synthesis of welding occurs but at the same time the chemical interaction of the welding material with brick occurs too.

The developed materials have been successfully tested in the repair of industrial burning furnaces of many enterprises: Omsk petroleum processing plant, Nicolaevsky alumina plant, Bratsky aluminum plant, Kant cement-slate plant, Semiluki refractory plant, and others. The maximum application efficiency of SHS refractories in the form of masonry mortars was achieved during repair of revolving furnaces for the cement-burning process, as well as preheat furnaces for coke and anode pastes and shaft furnaces for limestone calcining.

In the period from 1998 to the present time the Nicolaevsky alumina plant has implemented modernization and reconstruction of three-shaft limestone burning furnaces. Experts from JSC "Lipetsk steel project" (Russia) and Institute of Combustion Problems (Kazakhstan) have developed the modernization project (Dilmukhambetov et al., 2013a, 2013b). Currently, in circumstances when there is continuous growth in raw material prices, energy inputs, and transportation, as well as the development of effective calcining technologies, the creation of new and improvement of existing calcining units is an urgent task.

5.5.1 Reconstruction of Shaft Furnaces for Limestone Calcining

The existing counterflow shaft furnace for limestone calcining has a traditional arrangement—in the lined (fettle) furnace are located a heating zone, a calcination zone, and cooling and unloading zones for precalcined lime (Fig. 5.17a) (Monastyrev and

Aleksandrov, 1979). An essential feature of construction of a modernized furnace is the availability of a gas-distributing kern, which separates the calcination zone for four sectors (Fig. 5.17b). The gas-distributing kern is made from periclase-chromite bricks brand ПХСУТ.

Figure 5.18 shows the general diagram of the calcination zone. The entire construction is mounted on two air-cooled beams and one water-cooled one and is thermally insulated by heat-resistant concrete. A two-stage arrangement of eight cyclone burners on gas and the availability of the kern in the calcination zone change fundamentally the conditions of heat and mass transfer. The role of the kern is in the distribution of the multiton mass of limestone for four sectors of calcination zone, where the limestone calcining occurs relatively autonomously. This leads to a more uniform distribution of gas flow velocity, equalization of temperature field along the cross section of the furnace, and velocity increase of the limestone passage. Thus, the most responsible and loaded furnace node–gas distributing kern was demanded by the essential high requirements in terms of mechanical and thermal characteristics.

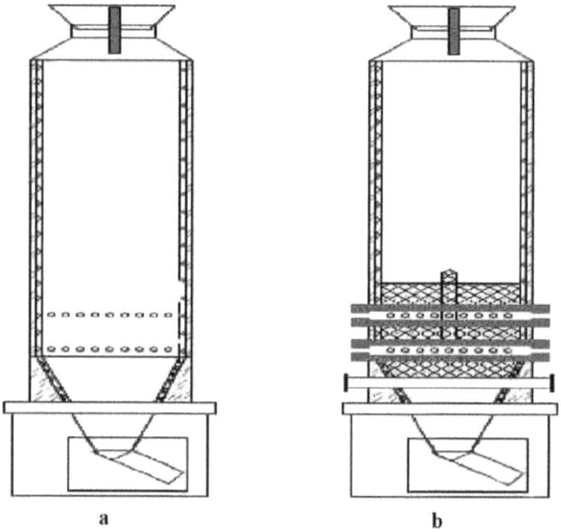

Figure 5.17 Lime-burning furnace of the Nicolaevsky alumina plant: (a) before reconstruction and (b) after reconstruction. Reproduced from Dilmukhambetov EE, Fomenko SM, Mansurov ZA, Korkembai Zh, Reshetnyak AF. (2013a). Application of SHS-refractories during limestone furnace refurbishment. *Adv. Mater. Res.*, **602–604**: 957–961.

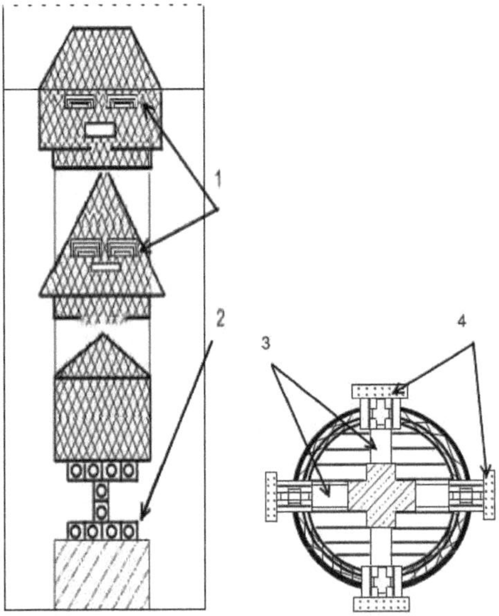

Figure 5.18 Scheme of the calcination zone of a shaft furnace: 1, air-cooled bulks of first and second sections; 2, water-cooled bulk; 3, gas duct; and 4, cyclone burners. Reproduced from Dilmukhambetov EE, Fomenko SM, Mansurov ZA, Korkembai Zh, Reshetnyak AF. (2013a). Application of SHS-refractories during limestone furnace refurbishment. *Adv. Mater. Res.*, **602–604**: 957–961.

The problem of mechanical and thermal resistance of kern has been solved with the help of SHS refractory mortars, namely their ability to weld the bricks into monolithic constructions. In the process of SHS, due to exothermic reactions in mortars, the temperature is close to the melting temperature of brick components. This provides an extremely strong bond among the bricks. After synthesis, the welding material has physicochemical and operational characteristics, similar to the material of bricks.

Stages of drying and calcining of a gas-distributing kern with the purpose of initiating the SHS in the masonry joint precede the start-up of furnaces. These operations are carried out in a closed volume of the calcination zone, where are positioned the kern, cyclone burners, the air-cooled ceiling over the kern, and thermally insulated metal constructions of the unloading area. Technical features as well as availability of metal constructions do not allow one to provide the desired temperature for synthesis in the fettle zone. Therefore, for kern laying, a special masonry mortar with a low self-ignition temperature was developed. According technical indicators for working conditions the SHS mortar fully satisfies the requirements of the project (Table 5.10), so it was used for kern laying and the shaft of the calcinating zone.

Table 5.10 Technical characteristics of SHS mortar for kern bricks, brand periclase bricks

Characteristic	Criteria
Fire resistance, °C	1650.0
Self-ignition temperature, °C	830.0
SHS temperature, °C	1800.0
Shear resistance, MPa	16.0
Compressing strength, MPa	35.0
Thermal stability (quantity of air thermal cycling)	17.0

Source: Reproduced from Dilmukhambetov EE, Fomenko SM, Mansurov ZA, Korkembai Zh, Reshetnyak AF. (2013a). Application of SHS-refractories during limestone furnace refurbishment. *Adv. Mater. Res.*, **602–604**: 957–961.

When the kern is heated by cyclone burners up to temperatures of 830°C–850°C the self-propagating high-temperature synthesis in welding seams provide the strong brick fastening and necessary mechanical strength for the entire construction. Designers, planners, and operators have used SHS refractories with maximum efficiency in solving complex engineering and technical problems. After performance of these procedures the furnace was put into service. Over a 10-year operation of calcining furnace the Nicolaevsky alumina plant has shown high technical and economic efficiency. Table 5.11 shows the comparative characteristics of furnace work of the Nicolaevsky alumina plant before reconstruction and after reconstruction, depending on the crude, unfired lime.

Key performance indicators of calcining furnaces since 1999 are shown in Table 5.12. The performance of furnaces is presented on the basis of the final product—calcined lime with a mean activity of 85%. As shown in the table, the reconstructed calcining furnaces are characterized by high utilization; the time of maintenance-free exploitation exceeds the project time by 5 years.

Table 5.11 Work characteristics of shaft furnaces of the Nicolaevsky alumina plant

Criteria	Furnace numbers 1 and 2 before reconstruction	Furnace number 1 after reconstruction	Furnace number 2 after reconstruction
Productivity, *t*/round the clock	145	195	200
Fuel consumption, kg lime	214	145	125
Lime activity, %	88	91	92
Use factor	0.8	0.95	0.95
Polluting missions content:			
NO_x, g/nm^3	0.08	0.03	0.03
SO_2, g/nm^3	0.05	Is absent	Is absent
CO, g/nm^3	18.8	0.25	0.25
Dust, g/nm^3	2.8	0.65	0.65

Source: Reproduced from Dilmukhambetov EE, Fomenko SM, Mansurov ZA, Korkembai Zh, Reshetnyak AF. (2013a). Application of SHS-refractories during limestone furnace refurbishment. *Adv. Mater. Res.*, **602–604**: 957–961.

Table 5.13 shows the comparative performance of similar shaft furnaces for limestone calcining. The table shows that reconstructed furnaces of the Nicolaevsky alumina plant according to own parameters close to the world's best samples of the firm Merz. It should be noted that the cost for reconstruction of each furnace is 600,000–800,000 USD and is compensated within 1 year of operation. Earlier, reconstruction and construction projects of shaft furnaces were not so economical or cost effective.

Table 5.12 Key performance indicators of calcining furnaces of the Nicolaevsky alumina plant

Furnace number	Start-up day	Date for complete repair	Degree of utilization	Average productivity (around the clock)	Content CaO%
1	29.03.1999	09.06.2003	0.978	173	85–92 (depending on limestone quality)
1	01.10.2003	01.02.2011	0.975	178.2	
2	30.07.2002	26.03.2007	0.971	174.1	
2	16.01.2008	25.05.2014	0.978	157.6	
3	06.08.2006	15.03.2011	0.966	161.6	
3	07.11. 2011	—	—	—	

Source: Reproduced from Dilmukhambetov EE, Fomenko SM, Mansurov ZA, Korkembai Zh, Reshetnyak AF. (2013a). Application of SHS-refractories during limestone furnace refurbishment. *Adv. Mater. Res.*, **602–604**: 957–961.

Table 5.13 Key performance indicators of calcining furnaces

Furnace type	Average productivity (around the clock)	Limestone activity, %	Specific consumption	
			Cond. fuels, kg/t	El. energy kW·h/t
Shaft transfer on coke	160–180	75–80	165–180	14–16
Shaft on masut	80–100	65–75	175–180	30–40
Shaft rimose 1.8 × 8.0 M on gas	120–150	80–87	130–165	20–25
Shaft of circular section, with estimated power 200, with beam gas burners	130–160	85–88	156–193	20–25
Two-shaft right opposite flow of "Mertz" firm	150	92–94	125–135	20–25
Modernized shaft furnaces of the Nicolaevsky alumina plant	160–175	88–92	133–140	20–25

Source: Reproduced from Dilmukhambetov EE, Fomenko SM, Mansurov ZA, Korkembai Zh, Reshetnyak AF. (2013a). Application of SHS-refractories during limestone furnace refurbishment. *Adv. Mater. Res.*, **602–604**: 957–961.

Currently the transfer shaft furnaces, working on coke, are commonly translated into liquid or gaseous fuels. The accumulated experience of reconstruction of shaft furnaces as well as new developments of the Institute of Combustion Problems may be useful for the introduction of energy-saving technologies for the production of building materials (Dilmukhambetov et al., 2013a, 2013b). Economic return after reconstruction is achieved within a year of the placement of the furnaces in service.

5.5.2 Use of Carbonaceous SHS Materials at Fettle of Melting Induction Furnaces ICT-016

The unique mechanical and electrophysical properties of carbonaceous SHS materials allow the use the SHS systems for lining of melting induction furnaces instead of graphite crucibles. We have developed a high-carbon refractory material on the basis of silicon oxides and titanium as oxidation agents in SHS systems (Fomenko et al., 2011; Dilmukhambetov et al., 2013b). A dry exothermic mixture is shut by silica sol for obtaining a pasty stuffed mass with a high content of graphite. Heterocoagulation of silica sol provides the formation of ultradispersed silicon oxide, promoting the initiation of SHS at relatively low temperatures.

The laying of high-carbon mass was carried out by stuffing by a pattern method on prepacked quartzitic basis. After finishing of lining works the mold is extracted and drying of crucible is carried out. On heating the furnace to 1000°C material SHS is initiated, as the result of which a solid product in the form of melted crucible is formed. The melting of bronze and aluminum and their alloys in the furnace IST-016 at the mechanical engineering plant have shown that the lining material has a mechanical strength that is competitive with the graphite crucible strength, low wettability by molten metal and slags, and high chemical resistance. The melting process and working regime of furnace IST-016 are shown in Fig. 5.19. After a lapse of 300 ferrous metal meltings the crucible surface remains smooth, clean, and without defects and burnouts. Under the same operating conditions graphite crucibles allow no more than 50 melts because in air graphite burns intensively and repeated thermal cycles lead to the destruction of the crucible.

Figure 5.19 Melting of bronze in an induction furnace. Reproduced from Mansurov ZA, Fomenko SM. (2014). Carbonaceous refractory materials on SHS-technology. *Adv. Sci. Technol. Mater. Res.*, **88**: 112–128.

Table 5.14 Test results of induction furnace IST-016 used at the engineering plant

Parameters	**Lining by high-carbon SHS refractory material (the development of IPG)**			
Quantity of melts	1–25	26–45	46–50	51–280
State of lining	Starting operation 90–120 min; the surface is clear and without destructions.	Starting operation 40–60 min; the surface is clear and without destructions.	Starting operation 40–60 min; the surface is clear and without destructions or slag lining	Starting operation 40–60 min; the surface is clear and without destructions; the slag lining is removed easily by power.
Parameters	**Graphite crucible according traditional technology**			
Quantity of melts	1–25	26–45	46–50	—
State of lining	Starting operation 60–90 min, surface falling, shallow honeycomb	Starting operation 60–90 min, deep honeycomb	Starting operation 60–90 min, deep burning, destruction of the furnace lining	—

Source: Reproduced from Mansurov ZA, Fomenko SM. (2014). Carbonaceous refractory materials on SHS-technology. *Adv. Sci. Technol. Mater. Res.*, **88**: 112–128.

Table 5.14 shows the test results of a high-carbon SHS material in comparison with the traditional technology of furnace lining. Expansion in the number of melts by five to six times in comparison with melting in a graphite crucible means considerable saving of expendable materials, both in natural calculation and monetarily.

5.6 Conclusion

The refractory SHS mortars represent a principally new class of refractory that is different from the traditional ones obtained by furnace technology. Their distinctive feature is the substantially comprehensive completion of the synthesis of a ceramic body during the heating stages of the lining because of the inner energy of components. Future considerations of the SHS refractories technology lie in several planes:

- The development of refractory materials with an unconventional phase composition, such as carbides, nitrides, and borides, in comparison with oxides and carbon
- Increased use of crude ore and waste of mountain-processing industry with the purpose of their utilization
- A combination of principles of sol-gel and SHS technologies when using silica sols as the bonding agent
- The development of methods and techniques for the manufacture of molded refractories and shaped products for various applications
- The lack in Kazakhstan of any significant refractory production despite the fact that in the republic there are a large number of consumers, making research and development in the field of refractory technology relevant

References

Dilmuhambetov EE, Vongai IM, Shihova LV. (2005). Aluminothermic combustion of iron oxide in SHS regime in the presence of sulfur. *Ш International Symposium "Combustion and Plasmachemistry"*, Almaty, p. 107.

Dilmukhambetov EE, Fomenko SM, Mansurov ZA, Korkembai Zh, Reshetnyak AF. (2013a). Application of SHS-refractories during limestone furnace refurbishment. *Adv. Mater. Res.*, **602–604**: 957–961.

Dilmukhambetov EE, Fomenko SM, Mansurov ZA, Reshetnyak F. (2013b). Use of SHS-refractories during limestone furnace refurbishment. *XII International Symposium of Self-propagating High-temperature Synthesis*, Texas, USA, pp. 328–329.

Dilmukhambetov EE, Mansurov ZA, Fomenko SM. (2007). Application of carbonaceous SHS-refractory materials in the induction furnace lining technology. *IX International Symposium on Self-Propagating High-Temperature Synthesis*, Dijon, France, **6**: 5.

Dilmukhambetov EE, Mansurov ZA, Fomenko SM. (2009). Synthesis of SHS- and sol-gel based carbonaceous refractory materials. *X International Symposium on Self-Propagating High-Temperature Synthesis*, Tsakhadzor, p. 370.

Dilmukhambetov EE, Mansurov ZA, Fomenko SM. (2011). Aluminothermic SHS: the effect of silica sol added as a binder. *Int. J. Self-Propag. High-Temp. Synth.*, **20**(2): 82–87.

Dilmukhambetov EE, Fomenko SM, Mansurov ZA, Vongay IM. (2006). Method of material connection, Patent 18058, Publ. 15.12.06, Bul. № 12, 4 p. with illustration.

Dubrovin AS, Pliner YuL. (1962). Temperature of non-furnace alumino-thermic process. News of SB AS SSSR, (12): 911–915.

Fomenko SM, Dilmukhambetov EE, Mansurov ZA. (2010). Manufacturing method of refractory crucible of induction furnace. Applicant and patent holder is the Institute of Combustion Problems, Republic of Kazakhstan, № 2010/0775.1, Innovative patent 24126, application form 10.06.2010, publ. 15.06.11, p. 3.

Fomenko SM, Nalibaev TA, Dilmuhambetov EE. (2003). Transmission electron microscopy of phase transitions in the system silicon oxide–aluminum. *II International Symposium "The Combustion and Plasma Chemistry,"* Almaty, pp. 107–111.

Glushko VP. (1979). *Thermal Constants of Substances.* Prod. IX, M.: Science, p. 486.

Ismailov MB, Fomenko SM. (1992). About some features of aluminothermic combustion systems. *Phys. Combust. Explos.*, **28**(1): 46–51.

Ismailov MB. (1992). *The Processes of Metallothermic Combustion and Their Application in Refractory Technology.* Dissertation: 01.04.17. Institute of Combustion Problems, Alma-Ata, p. 431.

Ksandopulo GI, Bekishev KK, Ismailov MB, others. (1986). Production process of refractory products and materials. Patent 1369211.

Ksandopulo GI, Dilmukhambetov EE, Ismailov MB, Fomenko SM, others. (1996). Fire-resistance ramming mixture for limning. Applicant and patent holder is the Institute of Combustion Problems, Republic of Kazakhstan, COUB 35/00, 35/04, Patent 4013, Publ. 16.12.96, Bul. № 4, p. 2.

Ksandopulo GI, Ismailov MB. (1992). Combustion processes involving mineral row materials and some problems of refractory synthesis. *Int. J. SHS*, **1**(3): 496–507.

Ksandopulo GI, Ismailov MB (1993). Combustion problems of gases and condensate systems. CSHS-refractories "furnon", *Alma-Ata, Prod.* **5**., p. 5.

Ksandopulo GI, Ismailov MB, Satbaev BN, others. (1987). Refractory mortar for laying of periclase-chromite products. Patent 1717586.

Ksandopulo GI, Ismailov MB. (1988). Composition for lining laying. Patent 170081.

Mansurov ZA, Dilmukhambetov EE, Ismailov MB, Fomenko SM, Vongai IM. (2001). New refractory materials on the SHS technology. *La Chimica el' Industria*, **83**: 1–6.

Mansurov ZA, Fomenko SM. (2014). Carbonaceous refractory materials on SHS-technology. *Adv. Sci. Technol. Mater. Res.*, **88**: 112–128.

Mansurov ZA. (2010). SHS refractory materials "furnon" and their practical implementations in Kazakhstan and Russia. *Adv. Sci. Technol.*, **63**: 312–321 (online available since October 27, 2010).

Merzhanov AG, Mukasyan AS. (2007). *Solid-Flame Combustion*. Torus Press, p. 336.

Merzhanov AG. (1975). *Combustion Processes in Chemical Technology and Metallurgy*. Chernogolovka, Moscow, p. 5.

Monastyrev AB, Aleksandrov AB. (1979). *Furnaces for Production of Lime*. Metallurgy, p. 232.

Nersesyan MD, Merzhanov AG, Borovinskaya IP. (1984). Production process of refractory products. Patent 1144338.

Nersesyan MD, Merzhanov AG, Borovinskaya IP. (1985). Production process of refractory products. Patent 1187410.

Novikov NP, Borovinskaya IP, Merzhanov AG. (1975). *Thermodynamic Analysis of the Reactions of Self-Propagating High-Temperature Synthesis. Sat.: The Combustion Process in Chemical Engineering and Metallurgy*. Chernogolovka, pp. 175–178.

Podergin BA, Kizimov YuA, Ivanova TA. (1974). *Temperature Measurement of Aluminothermic Reduction Reactions of Oxides by Treatment of spectral line. Theory and Technology of Metallothermic Processes.* Science, Novosibirsk, pp. 62–67.

Shabanova TA, Sarkisov PD. (2004). *The Basis of Sol-Gel Technology of Nanodispersed Silica.* IKS Academy, p. 206.

Vongay IM., Dilmukhambetov EE, Fomenko SM, Mansurov ZA. (2004). SH-synthesis of carbonaceous refractory in the presence of silica sol. *Combust. Plasmachem.*, **2**(2): 109–113.

Wolpert BA, Barzykina RA. (1988). *Theory of Combustion Waves in SHS Systems.* Preprint AS SSSR, Chernogolovka.

Chapter 6

Aliovalent Doping of Multiferroic $BiFeO_3$ Nanoparticles for Enhanced Functionality

Bhavya Bhushan[a] **and Amiya Priyam**[b]

[a]*Department of Physics, School of Applied Sciences, KIIT University, Patia, Bhubaneswar 751024, India*
[b]*Department of Chemistry, School of Physical and Chemical Sciences, Central University of South Bihar (Formerly Central University of Bihar), Vinoba Nagar, Chandauti, Gaya 823001, India*

amiya.priyam@gmail.com

$BiFeO_3$ is a smart ceramic material that possesses multiferroic properties. In this chapter we give an overview of the aliovalent doping of these multiferroic nanocrystals for the betterment of particle characteristics. We show that even by doping with nonmagnetic ions, the magnetic properties of the $BiFeO_3$ nanocrystals can be significantly enhanced. Concomitantly, morphological transformation from a faceted polygon to spherical particles (TEM analysis) and a blue shift in optical absorption spectra have also been observed. Alkaline earth metal ions, Ca^{2+}, Sr^{2+}, and Ba^{2+}, were substituted at the Bi^{3+} site (doping level 5 mol%) in the crystal lattice of $BiFeO_3$ using a sol-gel technique. All of

Smart Ceramics: Preparation, Properties, and Applications
Edited by Ajay Kumar Mishra

ISBN 978-981-4774-30-7 (Hardcover), 978-1-315-16359-8 (eBook)
www.panstanford.com

the doped and undoped $BiFeO_3$ nanocrystals show a ferromagnetic behavior. On doping with Ba^{2+}, the coercive field is enhanced by more than 700%, to a maximum value of 1.78 kOe. Similarly, Ca^{2+} doping leads to a 480% increase in the saturation magnetization, to a large value of 2.38 emu g^{-1}. Taking Sr^{2+} as a model dopant, $BiFeO_3$ nanocrystals were prepared with doping levels varying from 1% to 7% (mol%). The optimum doping level was found to be 3%, for which maximum enhancement in saturation and remnant magnetization values were obtained. Further, we also introduce a new method of equimolar codoping of Ca^{2+} and Ba^{2+} ions, which causes simultaneous enhancement of coercive field (800% increase) and saturation magnetization (600% increase). Thus, Ca-Ba equimolar codoping significantly enhances the twin aspects of magnetic properties, making them better suited for device applications.

6.1 Introduction

Ceramic materials that show coexistence of different ferroic properties in the same phase are called multiferroics (Fiebig et al., 2002; Srinivas et al., 2005; Eerenstein et al., 2006). Multiferroics have caught the imagination of the scientists around the globe primarily due to the exciting physics of these materials, that is, the coexistence and coupling of the ferroelectricity, ferromagnetism, and ferroelasticity. If these materials are also engineered to nanometer size regime, the ferroic features can be enhanced and additionally some new properties also emerge. It represents the convolution of multiferroics and nanotechnology, which can help rationally design and fabricate novel multifunctional devices such as sensors, transducers, memories, and spintronics. The present article embodies some of the interesting results and findings of the work done at the interface of multiferroics and nanoscience.

This class of materials has immense application in data storage, spin valves, spintronics, quantum electromagnets, microelectronic devices, etc. However, single-phase multiferroic materials are rare because of the contradictory requirement of d-electron occupancy in the A and B metal sites of the ABO_3-type perovskite structure (Hill, 2000). $BiFeO_3$ (BFO) is a multiferroic in which ferroelectricity is due to the presence of $6s^2$ lone pair of electrons present in Bi^{3+} ions and (anti)ferromagnetism is due to the occupancy of unpaired d-orbital electrons (Khomskii, 2006). It has a rhombohedral distorted perovskite structure with the space group R3c. BFO is ferroelectric below Curie temperature, T_C = 830°C, and a G-type antiferromagnet below Néel temperature, T_N = 370°C, with a long-range cycloidal spin arrangement of wavelength 62 nm, incommensurate with the lattice (Rivera and Schmid, 1997). It also exhibits interesting optical properties (Gao et al., 2007; Wei et al., 2008).

Despite such favorable properties, applications of BFO have been limited due to several reasons, for example, the formation of secondary phases during synthesis, weak magnetic characteristics, weak magnetoelectric coupling, and high leakage current. To overcome these problems, several researchers have attempted to modify the Bi or Fe sites of BFO by substitution with different transition metal ions or paramagnetic rare earth metal ions to get enhanced magnetic, electric, and magnetoelectric properties (Chung et al., 2006; Kumar and Yadav, 2006; Kothari et al., 2007; Yu et al., 2008). Interestingly, a new way of improving the magnetic and optical properties has also emerged in recent years. This method involves substituting BFO nanoparticles (NPs) with nonmagnetic alkaline earth metal ions, such as Ba^{2+} (1.35 Å), Sr^{2+} (1.18 Å), and Ca^{2+} (1.00 Å), having ionic radii comparable to that of Bi^{3+} (1.03 Å) ions (Li et al., 2007; Khomchenko, 2008). Through a series of works (Bhushan et al., 2009, 2010, 2012), we have attempted to address these crucial issues by synthesizing pure-phase BFO NPs doped with Ca, Sr, and Ba metal ions and by carrying out detailed studies to comprehend the thermal, optical, magnetic, and dielectric properties of these systems. The motivation for synthesis of alkali-earth-doped BFO NPs comes from the interesting magnetic behavior exhibited by $BaFeO_3$, $SrFeO_3$, and $CaFeO_3$ (Lebon et al., 2004; Dasgupta et al., 2005; Matsui et al., 2005). Clearly, iron is in +IV oxidation states in these perovskites. $CaFeO_3$ exhibits metal-to-semiconductor transition at a temperature close to room temperature. It also shows charge disproportionation transitions for iron at lower temperatures. $SrFeO_3$ shows different types of magnetoresistance effects for different oxygen stoichiometries at different temperatures, while $BaFeO_3$ is itself a multiferroic. Substitution of the divalent alkaline earth cations (Ca^{2+}, Sr^{2+}, Ba^{2+}) at the Bi^{3+} site would create a charge imbalance that can be compensated in three possible ways: (i) partial conversion of Fe^{3+} to Fe^{4+} state, (ii) charge disproportionation of the Fe^{3+} and Fe^{5+} state, and (iii) creation of oxygen vacancies. However, it is not clear which one of these factors exists predominantly in alkaline-earth-doped BFO NPs and how these factors manifest themselves into enhancement of magnetic properties.

In this chapter, we present an overview of sol-gel synthesis and characterizations of nanosized, phase-pure, single-crystalline BFO ceramics substituted with Ba^{2+}, Sr^{2+}, and Ca^{2+} ions. These samples are designated as BBFO (Ba-doped BFO), BSFO (Sr-doped BFO), and BCFO (Ca-doped BFO). It also provides an overview of the effect of the nature and concentration of alkaline earth dopants on the structural, magnetic, optical, and dielectric properties of BFO NPs. In the first part, the concentration of dopants was kept constant at 5% while the nature of ions was varied as we moved down group II of the periodic table, Ca, Sr, and Ba. One of the most interesting findings reported here is the significant improvement in magnetic properties on doping these nonmagnetic ions (Bhushan et al., 2009). Room temperature ferromagnetism has been observed in all of the alkaline-earth-doped BFO NPs. Besides, the doping also caused a shape transformation from a rhombohedral multifaceted geometry to a spherical geometry. The second part reports the effect of the concentration of the alkaline earth dopant on the properties of BFO NPs (Bhushan et al., 2010). Strontium was selected as the dopant, and its concentration was gradually varied. The highlight is the first experimental evidence of a theoretically predicted vibrational peak and the finding of an optimum dopant level for the enhanced magnetic properties. Further, simultaneous doping of two alkaline earth metals is presented in the last part. Ca^{2+} and Ba^{2+} ions were codoped into BFO NPs such that the total dopant concentration remained constant but the ratio of the two dopants was varied (Bhushan et al., 2012). The resulting variations of magnetic and optical properties are also discussed. Our results also suggest that oxygen vacancies might be responsible for the observed variations in the magnetic behavior of the alkaline-earth-doped BFO NPs.

6.2 Experimental Details

We employed a sol-gel technique, as reported previously (Park et al., 2007), with some modifications to prepare BFO and doped BFO samples. 1:1 molar ratio of $Bi(NO_3)_3 \cdot 5H_2O$ and $Fe(NO_3)_3 \cdot 9H_2O$ were added successively to ethylene glycol. In the case of substituted samples 0.95:1:0.05 molar ratio of $Bi(NO_3)_3 \cdot 5H_2O$, $Fe(NO_3)_3 \cdot 9H_2O$, and the corresponding salt—$Ba(NO_3)_2$ for BBFO, $Sr(NO_3)_2$ for BSFO, and $CaCl_2 \cdot 2H_2O$ for BCFO—were added to the solvent. The resulting mixture was stirred vigorously at 80°C for 1 h to obtain the sol. It was then left to dry overnight at 80°C in an oven. The gel was preheated at 400°C and was finally annealed at 550°C for 30 min. The structural analysis of the prepared samples was carried out by a Philips x-ray diffractometer with a cobalt target (λ = 0.178897 Å). The morphological characterization of the BFO NPs was done with a JEOL 2010 model high-resolution transmission electron microscope operating at 200 kV. All the TG-DTA and DSC data were recorded in the Seiko EXSTAR model TG6000 and DSC6300 systems while heating the sample in air. Fourier transform infrared spectroscopy (FTIR) data of all samples were recorded with a Perkin Elmer spectrum GX spectrometer in the mid-IR frequency range with a resolution of 4 cm^{-1}. UV-Vis absorption measurements were recorded on a Shimadzu UV-1601PC spectrophotometer. Room temperature Mössbauer measurements were carried out in a standard PC-based setup consisting of a 1024-channel multichannel analyzer (MCA) card, working in the constant acceleration mode. A 10 mCi ^{57}Co in a Rh matrix was used as the source. The system was calibrated with a high-purity iron foil of thickness 12 μm. A least squares fitting program LGFIT2 was used to fit the experimental data (Meerwall, 1975). All the magnetization data were taken using a Lakeshore model 7400 series vibrating sample magnetometer in the field range −10 kOe to 10 kOe at room temperature. The dielectric constant of all samples was measured on an inductance (L), capacitance (C), and resistance (R) meter (LCR-meter) (TEGAM USA, model 3550). An Ag-pellet-Ag capacitor was used for the dielectric constant measurements at room temperature. To demonstrate the magnetoelectric coupling in our samples, the dielectric constant was also measured at different applied magnetic fields at a frequency of 100 Hz at room temperature.

6.3 Results and Discussions

There are three major subsections to the discussion. The first is about the effect of the nature of alkaline earth dopants. The concentration of dopants is fixed at 5%, and the size of the ions is varied gradually, $Ca^{2+} \rightarrow Sr^{2+} \rightarrow Ba^{2+}$. The corresponding variation in structural, vibrational, optical, and magnetic properties is studied and discussed in detail.

6.3.1 Nature of the Dopant

6.3.1.1 Structural analysis

Figure 6.1 shows x-ray diffraction (XRD) patterns of all the samples confirming clean phase formation of BFO, BBFO, and BSFO NPs. Some additional peaks were observed in the XRD pattern of BCFO that correspond to some impurity phase. Observed XRD patterns could be indexed to the rhombohedral structure of BFO with the space group R3c. The average crystallite sizes of our samples as calculated from the broadening of the two most intense peaks ([104] and [110]) using the Scherrer formula were noted to be 77, 34, 29, and 28 nm for BFO, BBFO, BSFO, and BCFO, respectively. The values of the lattice strain (along [110]) were 0.00324, 0.00330, 0.00584, and 0.00505 for BFO, BBFO, BSFO, and BCFO NPs, respectively. It is also seen from Fig. 6.1a that the twin peaks such as (104) and (110), (006) and (202), and (116) and (122) merge together to form a single peak in the case of doped samples (Fig. 6.1b–d). The substantial reduction of size in the doped samples could be due to restricted crystal growth as divalent cations (Ba, Sr, and Ca) substitute trivalent cations of Bi.

A typical transmission electron microscopy (TEM) image of the BFO sample prepared by the present sol-gel method is shown in Fig. 6.2a. The average particle size estimated from TEM images for the BFO sample came out to be 95 nm. The selected area electron diffraction (SAED) pattern is presented in Fig. 6.2b, showing the presence of sharp diffraction spots that are a clear indication of well-developed single-crystalline BFO NPs. It should be noted that the electron diffraction patterns obtained from different areas of the NPs also show similar sharp diffraction spots. In the high-resolution transmission electron micrograph (HRTEM) image (Fig. 6.2c) lattice fringes can also be seen, which indicates the sample is free of defects and possesses a high degree of crystallinity. The energy dispersive spectrum of BFO nanocrystals is shown in Fig. 6.2d, which confirms the presence of Bi, Fe, and O in our prepared sample. Spectra taken at a number of selected positions of the sample show the presence of the same constituents. The Cu and C signals arise from the TEM grid.

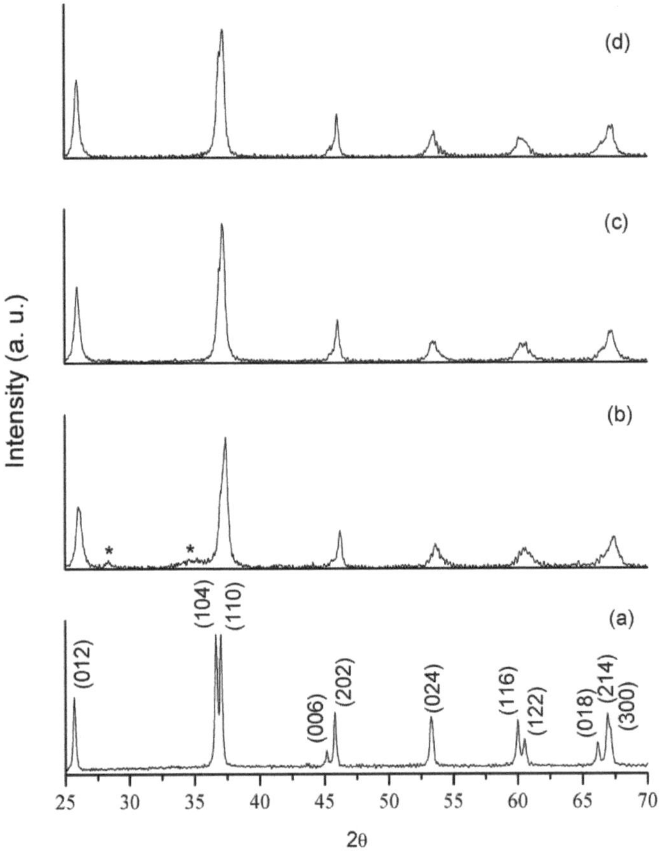

Figure 6.1 XRD patterns of (a) BFO, (b) BCFO, (c) BSFO, and (d) BBFO*: impurity phase. From Bhushan B, Basumallick A, Bandopadyay SK, Vasanthacharya NY, Das D. (2009). Effect of alkaline earth metal doping on thermal, optical, magnetic and dielectric properties of $BiFeO_3$ nanoparticles. *J. Phys. D: Appl. Phys.*, **42**: 065004. © IOP Publishing. Reproduced with permission. All rights reserved.

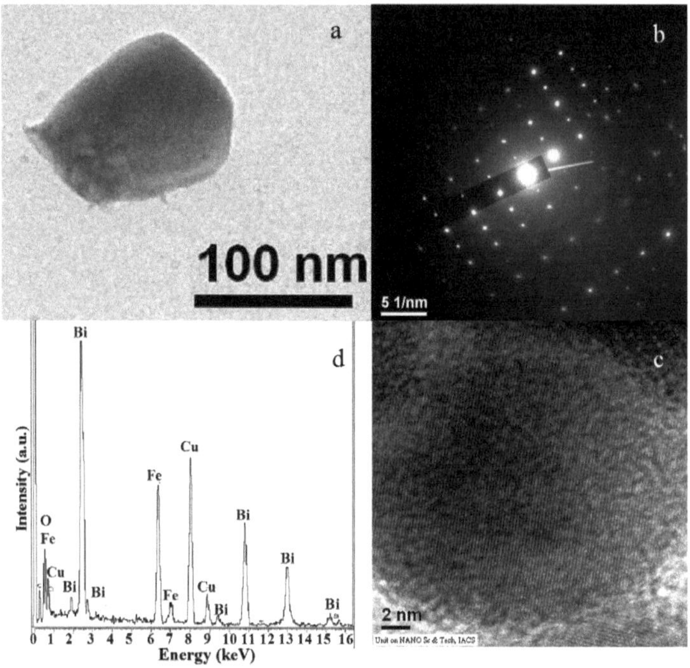

Figure 6.2 (a) TEM image of an individual BFO nanoparticle, (b) SAED pattern of BFO, (c) HRTEM image of a BFO nanoparticle, and (d) EDS of BFO nanoparticles. From Bhushan B, Basumallick A, Bandopadyay SK, Vasanthacharya NY, Das D. (2009). Effect of alkaline earth metal doping on thermal, optical, magnetic and dielectric properties of $BiFeO_3$ nanoparticles. *J. Phys. D: Appl. Phys.*, **42**: 065004. © IOP Publishing. Reproduced with permission. All rights reserved.

6.3.1.2 Thermal analysis

At room temperature BFO has rhombohedral distorted perovskite structure (α-phase), but at higher temperatures it shows changes in the crystal structure (orthorhombic β-phase and cubic γ-phase), as reported previously (Speranskaya et al., 1965). A recent work (Palai et al., 2008) also presented evidence of existence of BFO in three different phases with a modified phase diagram of the above system. They reported that α-to-β phase transition occurs at 823°C, β-to-γ phase transition takes place at 925°C, decomposition of the cubic phase into $Bi_2Fe_4O_9$ occurs between 933°C and 961°C, and above that the compound decomposes into Bi_2O_3 and Fe_2O_3.

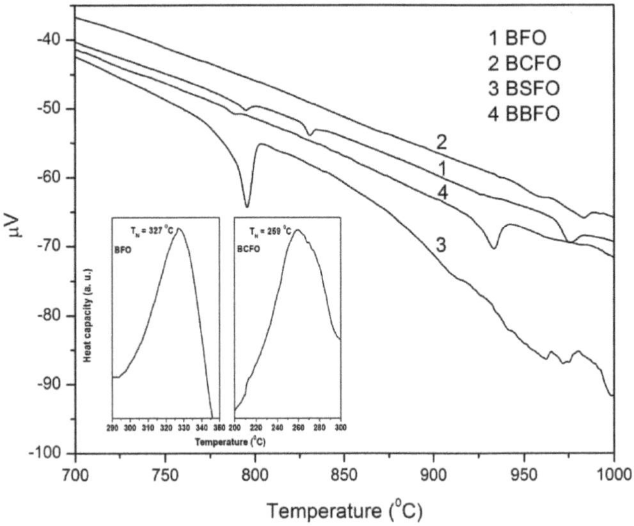

Figure 6.3 DTA spectra of all samples. Inset: DSC curves of BFO and BCFO samples. From Bhushan B, Basumallick A, Bandopadyay SK, Vasanthacharya NY, Das D. (2009). Effect of alkaline earth metal doping on thermal, optical, magnetic and dielectric properties of $BiFeO_3$ nanoparticles. *J. Phys. D: Appl. Phys.*, **42**: 065004. © IOP Publishing. Reproduced with permission. All rights reserved.

We carried out differential thermal analysis (DTA) measurement to observe the changes, if any, in the first-order structural transition temperatures of BFO NPs with respect to bulk BFO and the subsequent effect after doping. A DTA graph of the samples that shows various endothermic peaks is shown in Fig. 6.3. It should be noted that no changes in weight are observed in the thermogravimetric (TG) analysis (graph not shown) of all the samples in the temperature range of 700°C–1000°C. It is evident from Fig. 6.3 that for the pristine sample, α-to-β phase transition occurs at a relatively lower temperature of 795°C, while β-to-γ phase transition occurs at 831°C. Decomposition of the cubic phase into $Bi_2Fe_4O_9$ occurs between 927°C and 975°C, and above that the compound decomposes into Bi_2O_3 and Fe_2O_3. However, all the four endothermic peaks are not visible in the doped samples and the trends are also different. This may happen in doped BFO NPs as these transformations may be influenced by the replacement of ions and the creation of oxygen vacancies. Differential scanning calorimetry (DSC) data, as shown in the inset of Fig. 6.3, gives the second-order phase change occurring in the samples. As evident from the figure, T_N in the case of BFO NPs is 327°C, which is lower in comparison to that of the BFO bulk sample (T_N = 370°C). In the case of doped NPs, T_N = 259°C, 315°C, and 295°C for BCFO, BSFO, and BBFO, respectively. It is quite evident that T_N has shifted toward room temperature in the case of doped samples with a maximum shift in the Ca-doped BFO NPs, which is quiet desirable for obtaining room-temperature magnetoelectric coupling (Yang et al., 2005).

6.3.1.3 UV-Vis absorption and bandgap

Figure 6.4 shows the UV-Vis spectra of all samples. As seen from the spectra, pristine BFO NPs exhibit an absorption peak at 530 nm, which shifts to 405, 401, and 410 nm in the case of BBFO, BSFO, and BCFO, respectively. The bandgap was calculated from the point of inflection in the first derivative plot of the

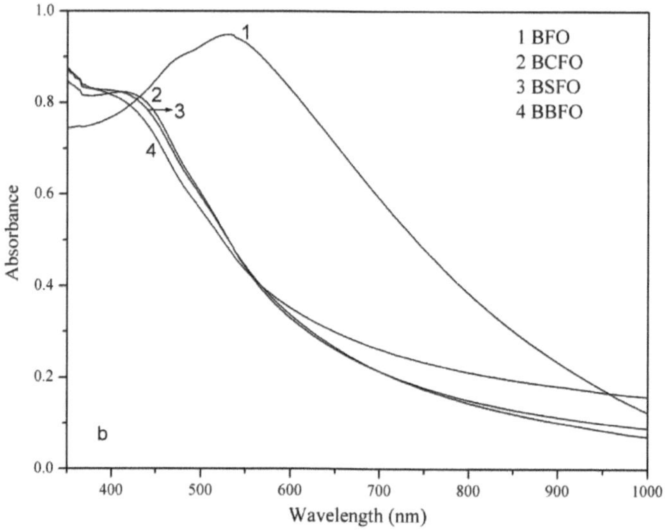

Figure 6.4 UV-Vis absorption spectra of pristine and doped BFO nanoparticles. From Bhushan B, Basumallick A, Bandopadyay SK, Vasanthacharya NY, Das D. (2009). Effect of alkaline earth metal doping on thermal, optical, magnetic and dielectric properties of $BiFeO_3$ nanoparticles. *J. Phys. D: Appl. Phys.*, **42**: 065004. © IOP Publishing. Reproduced with permission. All rights reserved.

UV-Vis absorption spectra of NP systems (Vishwanatha and Sarma, 2006). The bandgap of BFO NP is ~1.8 eV, which is in agreement with the earlier reported values (Gao et al., 2007; Higuchi et al., 2008; Palai et al., 2008). However, for all the doped samples the calculated bandgap is ~2.3 eV. A similar increase in the bandgap was also observed on doping of CdSe and ZnO nanocrystals with Mg^{2+} ions (Cho et al., 2007; Kwak et al., 2007), which also belongs to the same class of alkaline earth metals. The prominent shift in the absorption spectra of doped samples in comparison to pristine BFO is because of the rearrangement of molecular orbitals due to the presence of foreign elements, which results in a change in the band structure of the doped NPs. As the bandgap of alkaline earth ferrates (~3 eV) (Kozhevnikov et al., 2000; Dasgupta et al., 2005; Feng et al., 2007) is larger than that of BFO, incorporation of these alkaline earth metals in the BFO lattice results in intermediate bandgap values.

6.3.1.4 Hyperfine analysis

The hyperfine properties of BFO and doped BFO NPs were studied by ^{57}Fe Mössbauer spectroscopy. Charge compensation due to the substitution of divalent ions in BFO may cause either the formation of oxygen vacancies or the conversion of Fe^{3+} ions into Fe^{4+} ions. Earlier it was reported that room-temperature ferromagnetism in Ba-doped BFO might be due to conversion of Fe^{3+} ions to Fe^{4+} ions (Li et al., 2007). The presence of Fe^{2+} ions was also proposed as one of the reasons for the weak ferromagnetic nature of doped BFO samples (Gao et al., 2006).

Using Mössbauer spectroscopic analysis we can precisely determine the oxidation state of Fe in doped BFO samples. Figure 6.5 shows the room-temperature Mössbauer spectra of all samples. Apart from BSFO all samples show a combination of a sextet and a minor doublet. This minor doublet may be due to superparamagnetic relaxation of finer particles, as reported previously (Park et al., 2007). Hyperfine parameters of all the samples extracted from Mössbauer data are tabulated in Table 6.1. The obtained value of the hyperfine magnetic field is consistent with previous reports (Lebeugle et al., 2007; Park et al., 2007). The large linewidths obtained in our samples were

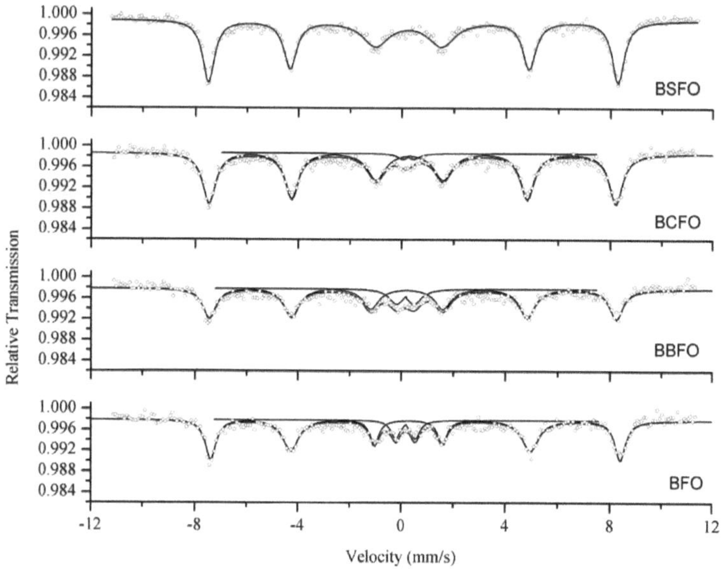

Figure 6.5 Mössbauer spectra of all samples taken at room temperature. From Bhushan B, Basumallick A, Bandopadyay SK, Vasanthacharya NY, Das D. (2009). Effect of alkaline earth metal doping on thermal, optical, magnetic and dielectric properties of $BiFeO_3$ nanoparticles. *J. Phys. D: Appl. Phys.*, **42**: 065004. © IOP Publishing. Reproduced with permission. All rights reserved.

Table 6.1 Mössbauer parameters of all samples

Sample name		IS[a] (mm/s) (±0.01 mm/s)	QS[b] (mm/s) (±0.01 mm/s)	H_{int}[c] (Tesla) (±0.3 T)	Line width (mm/s)	% Intensity
BFO	S[d]	0.38	0.33	49.05	0.56	86.42
	D[e]	0.18	0.75	—	0.44	13.58
BBFO	S	0.30	0.23	48.7	0.69	83.25
	D	0.16	0.66	—	0.70	16.75
BCFO	S	0.32	0.19	48.7	0.69	96.28
	D	0.17	0.44	—	0.53	3.72
BSFO	S	0.30	0.24	48.9	0.80	100

[a]IS = Isomer shift
[b]QS = Quadrupole splitting
[c]H_{int} = Internal magnetic field
[d]S = Sextet
[e]D = Doublet

assigned to the presence of different nonequivalent Fe sites in the crystal structure (Blaauw and van der Woude, 1973). However, later it was suggested that line broadening arose from the slight modulation of the hyperfine energies as the magnetic moments rotate with respect to the *c* axis of the crystal (Lebeugle et al., 2007). The isomer shift data of all samples confirm that Fe is present only in the Fe^{3+} state. No signature of Fe^{4+} or Fe^{2+} was found in our samples by Mössbauer spectroscopy.

6.3.1.5 Magnetization analysis

In Fig. 6.6 hysteresis loops of BFO and doped BFO NPs at room temperature are shown. For all the samples we can observe sizable hysteresis with finite values of the coercive field, remnant magnetization, and saturation magnetization, which are tabulated in Table 6.2. It should be noted that saturation in all **M–H** loops is easily achieved at an applied field of 10 kOe. The weakly ferromagnetic property in our substrate-free BFO NPs as obtained at room temperature has originated due to two main reasons: the canting of the spin arrangement and the suppression of the spiral spin arrangement. The presence of a spiral spin arrangement of wavelength 62 nm is responsible for the suppression of magnetization in bulk BFO. But BFO NPs having a particle size of the order of this spiral spin structure exhibit room-temperature ferromagnetism, as observed in our NP samples (Sosnowska et al., 2002). Also, numerous magnetization studies of antiferromagnetic NPs have shown that the magnetization in large applied fields is considerably larger than that in the corresponding bulk materials. It was suggested by Néel (1961) that this might be due to the finite number of magnetic moments in NPs, which may lead to a difference in the numbers of spins in the two sublattices because of random occupancy of lattice sites. This results in an uncompensated magnetic moment that leads to enhanced magnetic properties. The largest value of coercive field, H_C, is obtained for the BBFO sample, which is ~78 kOe. Such a high value of H_C at room temperature is quite significant. For the BCFO sample, saturation magnetization, M_S, at 10 kOe is ~2.38 emu/g, which is substantially higher as compared to our other samples. This value of M_S for Ca-doped BFO is even larger when compared to some of the ferro- or paramagnetic metal ions–doped BFO samples, as reported earlier (Palkar et al., 2004;

Das et al., 2007; Khomchenko et al., 2008). The obtained high values of H_C for BBFO and M_S for BCFO are attributed to the presence of a dopant. As these divalent metal ions (Ca^{2+}, Sr^{2+}, Ba^{2+}) get substituted at trivalent Bi sites in the BFO nanocrystal, a sort of charge imbalance could arise, which may destabilize the system. To maintain the electroneutrality, one oxygen vacancy would be created for every two alkali earth ions introduced into the BFO lattice. Thus created oxygen vacancies may contribute to the enhanced magnetic properties of the doped BFO samples, as established by Rao and coworkers (Sundaresan et al., 2006). These vacancies also reduce the oxygen content to some extent and give rise to nonstoichiometry, which may be expressed as $Bi_{0.95}X_{0.05}FeO_{3-\delta}$, where $\delta \leq 0.025$.

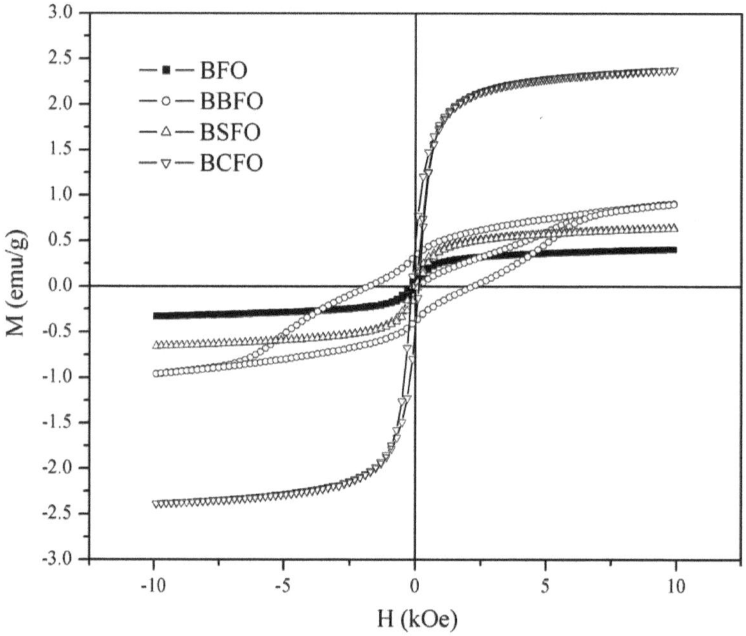

Figure 6.6 M–H curve of BFO and doped BFO nanoparticles recorded at room temperature. From Bhushan B, Basumallick A, Bandopadyay SK, Vasanthacharya NY, Das D. (2009). Effect of alkaline earth metal doping on thermal, optical, magnetic and dielectric properties of $BiFeO_3$ nanoparticles. *J. Phys. D: Appl. Phys.*, **42**: 065004. © IOP Publishing. Reproduced with permission. All rights reserved.

Table 6.2 Magnetic parameters at room temperature

Sample	Particle size[a] (nm)	Dopant	Ionic radius of dopant (Å)	Coercive field (kOe)	Remnant magnetization (emu/g)	Saturation magnetization (emu/g)
BFO	77	—	—	0.22	0.09	0.41
BCFO	28	Ca^{2+}	1.00	0.14	0.46	2.38
BSFO	30	Sr^{2+}	1.18	0.09	0.073	0.64
BBFO	34	Ba^{2+}	1.35	1.78	0.33	0.91

[a]As obtained from XRD data

The effect of a reduced particle size in the case of doped samples with respect to pristine is not so prominent as the values of $\mathbf{H}_C$ and $\mathbf{M}_S$ for BFO NPs of a similar particle size are quite low (Park et al., 2007) as compared to that in our doped NPs. The observation of higher coercivity and saturation magnetization in some of the doped NPs may help in finding a way for improving magnetization in BFO, which, in turn, will lead to many technological applications.

6.3.1.6 Dielectric analysis

The dielectric constant (ε_r) of all samples as a function of frequency at room temperature is illustrated in Fig. 6.7. It can be seen that the dielectric constant decreases with increasing frequency. However, all the samples show similar behavior. They all have a Debye-like relaxation. At frequency f = 1 MHz, ε_r was found to be 45.5, 31.4, 18.8, and 4.5 for BFO, BCFO, BSFO, and BBFO, respectively. Some authors have reported very high values of ε_r in BFO ceramics and thin films (Palkar et al., 2004; Kumar and Yadav, 2006; Li et al., 2007), but that might be due to Maxwell–Wagner contribution or by the substrate-induced strain in the case of thin films (Kamba et al., 2007). As compared to the pristine BFO, all the doped samples show a lower value of dielectric constant. This is consistent with the increase in bandgap (Section 6.3) as wide-bandgap materials are expected to have lower dielectric constants following the inverse proportionality relation (Pan et al., 004; Peacock and Robertson,

2007). Further, an interesting trend emerges if we look at the ε_r values of the doped BFO samples. It can be noticed that the dielectric constant decreases as the ionic radius of the dopant increases. It is in agreement with earlier studies where a similar correlation between the ionic radii and the dielectric constant was obtained (Raju et al., 1993; Chou et al., 2007).

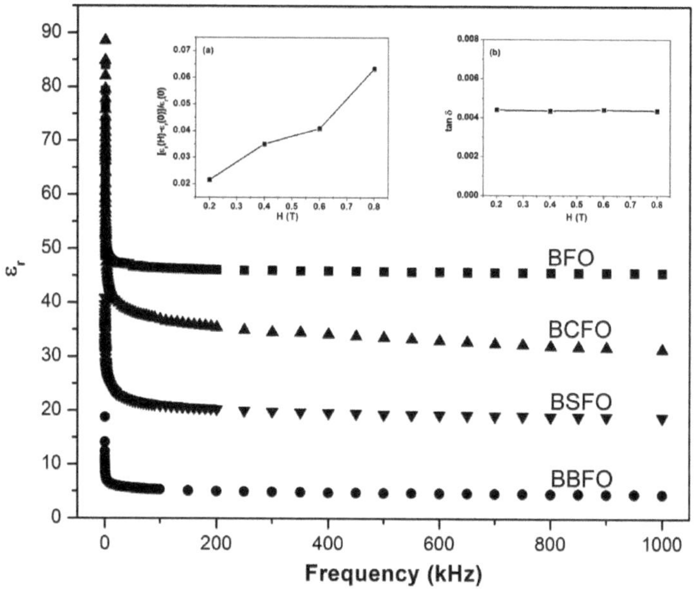

Figure 6.7 Dielectric constant of BFO and doped BFO nanoparticles recorded at room temperature as a function of frequency. Insets: (a) Variation of dielectric constant and (b) dielectric loss with applied magnetic field for BCFO nanoparticles. From Bhushan B, Basumallick A, Bandopadyay SK, Vasanthacharya NY, Das D. (2009). Effect of alkaline earth metal doping on thermal, optical, magnetic and dielectric properties of $BiFeO_3$ nanoparticles. *J. Phys. D: Appl. Phys.*, **42**: 065004.

One of the simplest and widely used methods to show multiferroic coupling is the examination of the dielectric constant with the applied magnetic field. In a multiferroic material, the dielectric constant is perturbed by the onset of magnetic ordering, which itself is affected by the applied magnetic field. We have shown here the variation of dielectric constant as a function of applied magnetic field, that is, $[\varepsilon_r(\mathbf{H}) - \varepsilon_r(0)/\varepsilon_r(0)]$

to describe the magnetoelectric coupling. As shown in inset (a) of Fig. 6.7, ε_r increases with an increasing magnetic field, giving a positive coupling of 0.063 at an applied magnetic field of 0.8 T for the BCFO sample. This change is quite significant at room temperature. The corresponding dielectric loss (tan δ), as presented in inset (b) of Fig. 6.7, is minimal at room temperature.

6.3.2 Concentration of the Dopant

To further understand the effect of concentration of the dopant, we synthesized a series of Sr-doped BFO NPs and investigated their structural, optical, and magnetic properties. A sol-gel route was employed to prepare $Bi_{1-x}Sr_xFeO_3$ NPs, with x = 0, 0.01, 0.03, 0.05, and 0.07, hence designated as BFO, BSFO1, BSFO3, BSFO5, and BSFO7 respectively. The dependence of the structural, magnetic, and optical properties on the concentration of the Sr dopant is discussed next.

6.3.2.1 Structural and morphological analysis

Figure 6.8a shows XRD patterns of all the samples, and it confirms the pure phase formation of BFO and BSFO NPs. The XRD patterns indicate the rhombohedral structure of BFO with the space group R3c. However, some additional peaks were observed in BSFO7 that might be due to some impurity phase. The average crystallite size of all samples as calculated using the Scherrer formula and the value of the lattice strain along the [110] axis. The variation of the average crystallite size and lattice strain along the [110] direction with the dopant concentration is depicted in Fig. 6.8b. We could observe an exponential reduction in crystallite sizes and a linearly increasing trend in the value of the lattice strain as the concentration of Sr in BFO is increased. With an increase in the dopant concentration the crystal growth of the NPs might be somewhat arrested, which results in the formation of smaller particles. As can be seen in Fig. 6.8c there are several twin peaks in the XRD pattern of BFO, viz., [104] and [110], [006] and [202], [116] and [122], [018] and [214], etc., and with an increase in the concentration of Sr the relative intensities of these peaks change. The ratio of the relative intensities of such twin peaks is shown in Fig. 6.1c, which are a linear function of the dopant concentration. The ratio of the intensities of the peaks reflected by the crystal planes [110] and [104] is ~1 in the pristine sample, but this ratio increases to 1.39 in the 3% Sr-doped BFO and finally becomes 1.85 in the 7% Sr-doped BFO sample. In the case of the peaks reflected by [116] and [122] crystal planes, we can see (Fig. 6.8a) that the intensity of the stronger peak, that is, the peak corresponding to [116] plane continuously decreases as the dopant concentration increases and finally it becomes even weaker than the intensity of the peak reflected by the [122] plane.

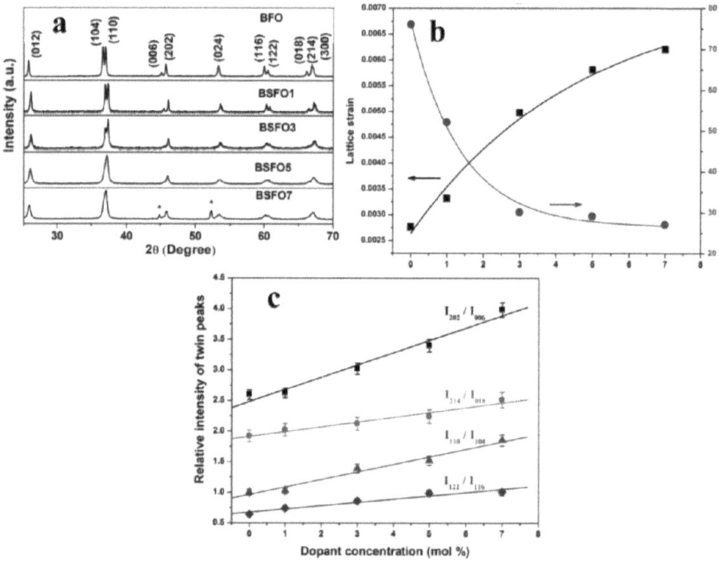

Figure 6.8 (a) X-ray diffraction patterns of all the samples*: the impurity phase. (b) Effect of dopant concentration on lattice strain and crystallite size and (c) change in the relative intensities of indicated twin peaks of BFO and Sr-doped BFO nanoparticles as a function of dopant concentration.

This quantitative analysis shows that an increase in the Sr concentration in BFO leads to a gradual increase in lattice deformation. Clearly some changes occurred in the crystal structure of BFO as the foreign element (Sr) was introduced in the pure sample.

Figure 6.9a shows a TEM image of the BSFO5 sample. The particle size distribution is presented in Fig. 6.9b. The average particle size calculated from the histogram is found to be 33.3 ± 2.2 nm, which is in good agreement with the results obtained from XRD data. The relative percentage size distribution is found to be 6.6%, which shows that the synthesized NPs have high monodispersity. An HRTEM image from a portion of the BSFO1 NP is shown in Fig. 6.9c. Clear lattice fringes are visible, indicating the defect-free nature of the sample with a high degree of crystallinity. SAED data taken from individual particles of the pristine sample (Fig. 6.9d) exhibit sharp diffraction spots that confirm the well-developed single-crystalline nature of BFO NPs.

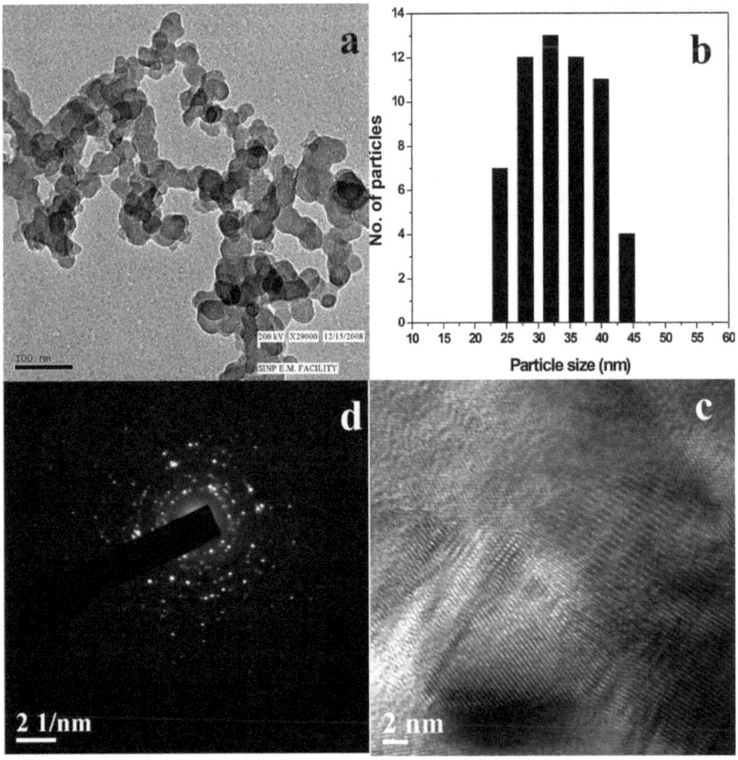

Figure 6.9 (a) TEM image of BSFO5 nanocrystals, (b) particle size distribution of BSFO5 nanocrystals, (c) HRTEM image of BSFO1 sample, and (d) SAED pattern of a BFO nanoparticle.

HRTEM of an individual NP (see Fig. 6.2) of the pristine sample and the 5% Sr-doped BFO samples revealed a clear shape transformation from an asymmetric rhombohedral shape faceted with sharp edges and corners for BFO to a perfectly spherical shape for BSFO5. This remarkable shape change occurred after doping the pristine BFO NPs with Sr. Such shape transformation from nearly rhombohedral to spherical geometry could result in the loss of structural anisotropy in the nanocrystalline systems. The changes that occurred in the intensities of the twin peaks in the XRD pattern might be a consequence of such loss of anisotropy in the nanocrystals that were clearly evidenced in our system (Bhushan et al., 2009).

6.3.2.2 FTIR and UV-Vis absorption analysis

Figure 6.10a shows the FTIR spectra of BFO and BSFO5 recorded at room temperature. One of the fundamental absorptions is observed in the wavelength region 440–460 cm^{-1}. This band is attributed to the bending vibration of the Fe–O bond (nearly at 443 cm^{-1}) in the FeO_6 octahedral unit (Rao and Rao, 1970) and also to the BiO_6 octahedral structure unit that exhibits absorption at 450 cm^{-1} (Som et al., 1992). However, the most important as well as complex part of the spectra is the region from 480 to 660 cm^{-1}. To extract more information the spectra were deconvoluted with the Gaussian-type function. The deconvolution of the complex spectral region in the component bands was done in accordance with the second derivative method, which can provide more details and highlight small differences in an IR spectrum (Priyam et al., 2010). The deconvoluted spectra with the respective second derivative plot are shown in Fig. 6.10b and Fig. 6.10c for BFO and BSFO5 NPs, respectively. In both cases the spectral region from 480 to 660 cm^{-1} is resolved into three bands: the first one, at 526 cm^{-1}, corresponds to the Bi–O bond in the BiO_6 octahedral structure unit (Som et al., 1992); the second one, at 549 cm^{-1}, corresponds to the Fe–O stretching vibration present in the FeO_6 octahedral unit (Rao and Rao, 1970); and the third one, at 593 cm^{-1}, corresponds to the out-of-phase vibrations of basis atoms of oxygen normal to the [111] plane of the rhombohedral BFO. It should be noted that the peak at 593 cm^{-1} was predicted theoretically earlier (Tütüncü et al., 2008) and has been experimentally evidenced for the first time in this work to the best of our knowledge. The percentage intensity of the peaks at 526 cm^{-1} and 593 cm^{-1} decreased in Sr-doped BFO NPs. This is consistent with our findings that Sr substitutes Bi in the lattice, which leads to structural changes, as discussed in the previous section.

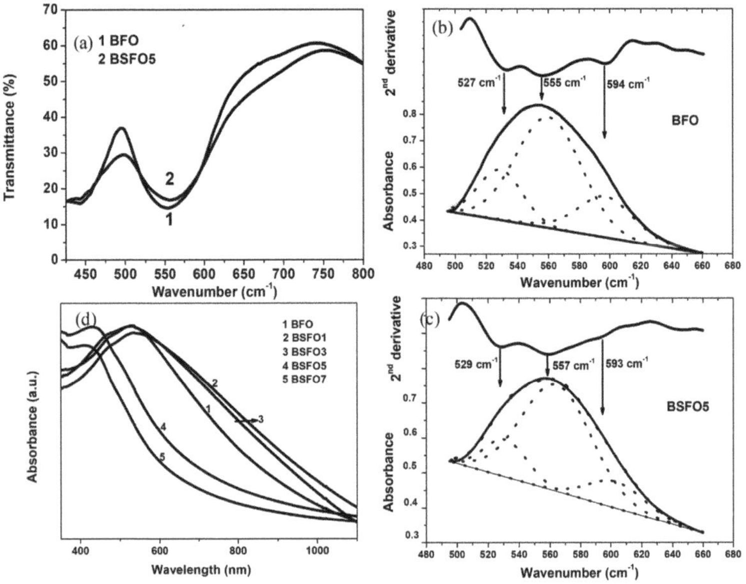

Figure 6.10 (a) FTIR spectra of BFO and BSFO5 nanocrystals. Second derivative FTIR spectrum and deconvolutions of (b) BFO and (c) BSFO5 in the region 490–660 cm^{-1}. (d) UV-Vis spectra of pristine and Sr-doped BFO nanoparticles.

The UV-Vis absorption spectra of BFO and BSFO NPs are shown in Fig. 6.10d. The absorption peaks for BFO, BSFO1, and BSFO3 are seen at 529 nm, 534 nm, and 530 nm, respectively, which are very close to each other. This indicates negligible change in the electronic structure of BFO with an increase in the dopant concentration up to 3%. However, as we move to a higher dopant concentration of 5% (BSFO5) and 7% (BSFO7), a significant shift of ~100 nm toward the shorter wavelength is observed. The bandgaps as calculated from the point of inflection in the first derivative plot of the UV-Vis absorption spectra (Vishwanatha et al., 2006; Priyam et al., 2010) of all the NP samples are ~1.8 eV for BFO, BSFO1, and BSFO3 and ~2.3 eV for BSFO5 and BSFO7. The widening of the bandgap in the BSFO5 and BSFO7 samples could be due to increased lattice strain as supported by XRD results. Though lattice strain increases linearly with dopant concentration (Fig. 6.9b), there may be a threshold value of strain for a particular dopant concentration beyond which rearrangements in the molecular orbitals take place that get reflected as a change in bandgap. It can also be noted that in the case of BFO, the bandgap is determined by the energy separation of the top of O 2p–Fe 3d mixed valence band and the bottom of the Fe 3d conduction band (Higuchi et al., 2008). The doping of divalent Sr ions at trivalent Bi sites in BFO nanocrystals gives rise to a sort of charge imbalance that is compensated by nonstoichiometry of oxygen, which may be expressed as $Bi_{1-x}Sr_xFeO_{3-\delta}$, where δ denotes the fraction of the created oxygen vacancies and depends on the dopant concentration. The creation of such oxygen vacancies can also contribute to the observed blue shift in the samples having higher doping concentrations (Roy et al., 2007). It is to be noted that the span of bandgap values for the prepared BFO and BSFO NPs is 1.8–2.3 eV, which is well within the visible portion of sunlight. BFO with a bandgap of 2.18 eV has been shown to possess good photocatalytic ability in the visible light range to decompose organic contaminants (Gao et al., 2007). Our results clearly show that the bandgap of BFO can be tuned within the visible spectrum domain by Sr doping and, thus, it increases the operational range of these photocatalytic materials. Moreover, the increase in the bandgap to higher energies at higher dopant concentrations may lead to the enhancement of photocatalytic efficiency as deduced from the correlation of bandgap and photocatalytic efficiency (Datta et al., 2008).

6.3.2.3 Magnetization analysis

DC magnetization measurements at room temperature were performed on the pristine sample and all Sr-substituted NPs in order to investigate their magnetic properties and the effect of

Sr doping in BFO. From the hysteresis curves shown in Fig. 6.11, it is to be noted that saturation is achieved in all samples within an applied field of 10 kOe. In contrast to earlier reports (Jacobson et al., 1975; Palkar et al., 2004; Das et al., 2007) on bulk BFO, NPs of BFO show ferromagnetic ordering, and similar ordering has been observed in all Sr-substituted samples. It is clear that Sr substitution has enhanced the magnetic properties of BFO. The ferromagnetic ordering as obtained in our samples could be for three reasons. First, in the case of antiferromagnetic systems, as the particle size decreases, a net magnetic moment is produced due to nonexact compensation of the two magnetic sublattices (Néel, 1961). Such uncompensated spin moments at the surface of the BSFO NPs may be a primary cause for the observed ferromagnetism. Second, the presence of spiral spin arrangement is a well-known reason for the suppression of magnetization in bulk BFO (Ederer et al., 2005; Park et al., 2007). This cycloidal spin structure is modified when the particle size of BFO is comparable to or less than its wavelength (62 nm), which may lead to the observed ferromagnetic ordering at room temperature as indicated in earlier reports (Park et al., 2007; Bhushan et al., 2009). Third, substitution of the divalent Sr ion at the trivalent Bi site can lead to a charge imbalance, which can be compensated by conversion of Fe^{3+} ions to Fe^{4+} ions or a charge disproportionation of Fe^{3+} and Fe^{5+} (Bocquet et al., 1992) or by the creation of oxygen vacancies. The possibility of existence of Fe^{4+} ions is ruled out by Mössbauer spectroscopic data. So the only way to maintain the electroneutrality is the formation of oxygen vacancies, and these created oxygen vacancies may be a possible reason for the enhancement in magnetic properties of the doped NPs (Gan et al., 1998).

Among all BSFO samples, BSFO3 exhibits the largest value of saturation magnetization, $\mathbf{M}_S$ = 1.37 emu/s, and remnant magnetization, $\mathbf{M}_R$ = 0.32 emu/g, which are quite substantial values at room temperature. The dependence of saturation magnetization on Sr concentration is shown in inset (a) of Fig. 6.11. With an increase in Sr concentration an increase in saturation magnetization is observed, which is attributed to increased concentration of oxygen vacancies. However, a noticeable decrease in $\mathbf{M}_S$ value is observed for the BFO NPs having

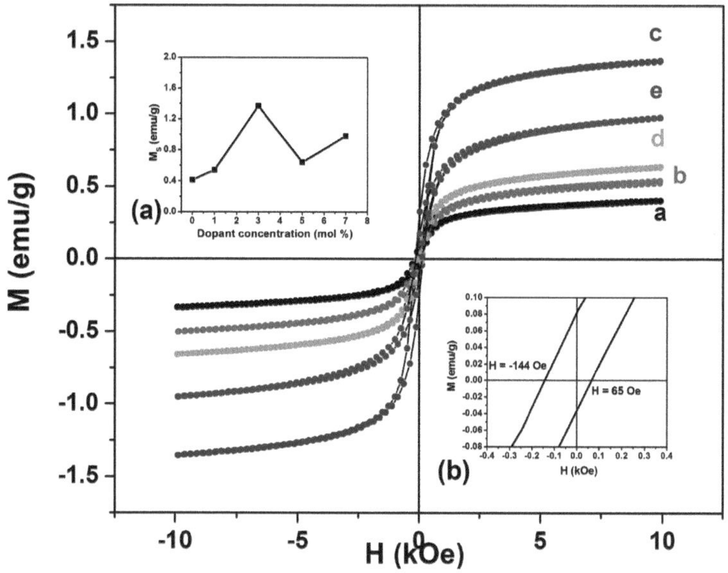

Figure 6.11 **M–H** curve of (a) BFO, (b) BSFO1, (c) BSFO3, (d) BSFO5, and (e) BSFO7 nanoparticles recorded at room temperature. Insets: (a) Variation of saturation magnetization with varying Sr concentrations and (b) shifting in the hysteresis loop for the BSFO1 nanoparticles.

a higher Sr concentration in comparison to BSFO3. One of the possible reasons for the observed reduction in the values of saturation magnetization with an increase in the dopant concentration is traced to be the increased lattice strain (see Fig. 6.8b). Increased epitaxial strain in thin films is a known cause for reduction in $\mathbf{M}_S$ values (Gan et al., 1998). Similarly, an enhanced value of lattice strain caused by the substitution of foreign elements may cause a reduction in the value of saturation magnetization in BSFO NPs having a higher Sr concentration. Ederer et al. (2005) had shown by first principles density functional theory calculations that magnetization is weakly dependent on oxygen vacancies and strain in BFO. But our results indicate that oxygen vacancies and lattice strain play a significant role in deciding the value of saturation magnetization in a Sr-doped BFO NP system. At a certain limit of doping, the increased lattice strain may become dominant over the effect of increased oxygen vacancies, resulting in a reduction in the value of saturation magnetization. From this observation we can conclude that the limit of Sr doping in BFO nanocrystals for enhancement of magnetic properties is close to 3 mol%.

A shift of the hysteresis loops toward the negative magnetic field axis was seen in all the samples that support Néel's model of magnetization. A typical shift obtained for BSFO1 NPs is shown in inset (b) of Fig. 6.11. Such shift in hysteresis indicates the presence of exchange coupling between the antiferromagnetic core and ferromagnetic surface in the case of the BSFO NP system (Park et al., 2007; Tian et al., 2009). For a BFO nanocrystal the H_{EB} is ~92.5 Oe, which is highest among all the samples. The value of HEB deceases as the concentration of Sr increases in the BSFO nanocrystalline samples. However, a 7% Sr-doped BFO sample shows discontinuation of this trend with an increment in the HEB value, which might be due to the presence of the impurities as indicated by the XRD data.

6.3.3 Equimolar Codoping

From the abovementioned results, it is quite evident that the doping A-site of BFO by alkaline earth ions has been proved to be an efficient way to enhance the various properties of BFO NPs. If we restrict ourselves to magnetic properties only, we find that maximal enhancement in saturation magnetization and coercive field was obtained in Ca-doped BFO and Ba-doped BFO NPs, respectively. To get maximum functionality, we need a single material to exhibit the best magnetic characteristics. So, we evolved a strategy for the simultaneous incorporation of two alkaline earth metals, Ca and Ba, into the A-site of BFO in order to get simultaneous enhancement in the twin aspects of magnetic properties, coercive field and saturation magnetization. We replaced 5 mol% of Bi by Ca and Ba ions, and their relative concentrations were also varied. The best result was obtained at the equimolar codoping of Ca-Ba (i.e., 2.5 mol% each), in which simultaneous enhancement of saturation magnetization and coercive field was observed (Bhushan et al., 2012). The details are discussed below.

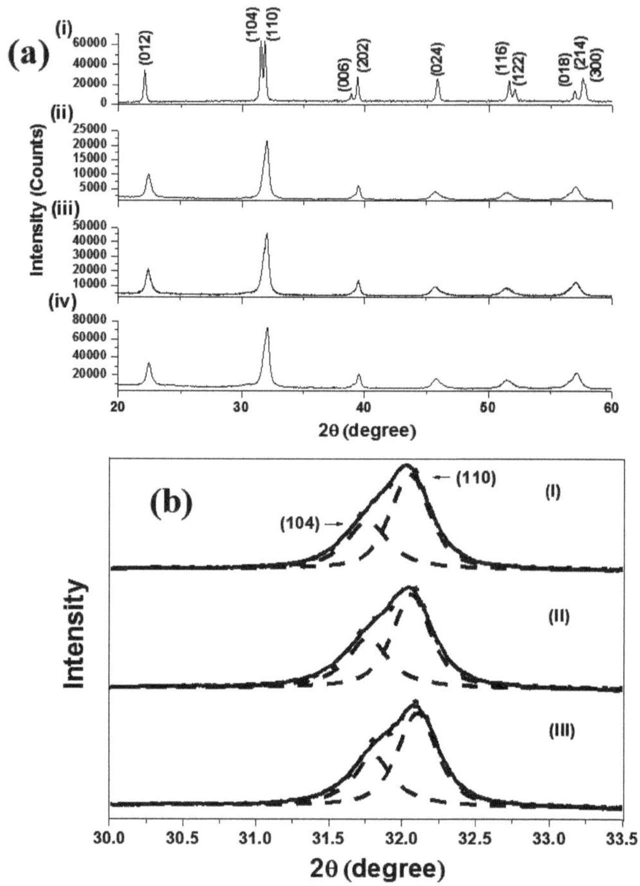

Figure 6.12 (a) X-ray diffraction patterns of all the samples and (b) change in the relative intensities of indicated twin peaks of BFO and Ca-Ba-codoped BFO nanoparticles. Reprinted from Bhushan B, Das D, Priyam A, Vasanthacharya NY, Kumar S. (2012). Enhancing the magnetic characteristics of $BiFeO_3$ nanoparticles by Ca, Ba co-doping. *Mater. Chem. Phys.*, **135**: 144–149, Copyright (2012), with permission from Elsevier.

6.3.3.1 Structural and morphological analysis

Figure 6.12a shows the XRD patterns of BFO and all the Ca-Ba-codoped BFO NPs, confirming the clean phase formation of the codoped samples. Observed XRD patterns could be indexed to a rhombohedral structure of BFO with the space group R3c. The average crystallite sizes of our samples as calculated using the

Scherrer formula were noted to be 77 nm for BFO while for all the three codoped samples it reduced to 22 nm. The substantial reduction of size in the codoped samples could be due to a restricted crystal growth as divalent cations (Ba and Ca) substitute trivalent cations of Bi. There is a remarkable change in the XRD patterns of the pristine and codoped samples. The twin peaks, such as (104) and (110), (006) and (202), and (116) and (122), were well separated in BFO NPs but seem to merge together in the case of Ca-Ba-codoped BFO NPs. To get a clear picture in the case of codoped samples, the peak comprising (104) and (110) was deconvoluted. As clear from Fig. 6.12b, the twin peaks' structure is retained in Ca-Ba-codoped BFO NPs. However, the ratio of intensities of the twin peaks changed dramatically in the codoped BFO NPs in comparison to the pristine one. The ratio of the intensities of the peaks reflected by the crystal planes [110] and [104] is ~1 in the pristine sample, which changed to ~1.6 in all the codoped samples. We note that the intensity of twin peaks depends only on the total number of dopant atoms present in the lattice and the relative concentrations of Ca and Ba in the codoped samples do not have any effect. In the case of single dopants, however, the ratio of twin peaks varied linearly with a change in the concentration of the dopant, as observed in the case of Sr-doped BFO NPs (Bhushan et al., 2010). The quantitative analysis shows that the introduction of Ca and Ba ions in BFO leads to some lattice deformation and change in the crystal structure.

6.3.3.2 Magnetization analysis

The prime objective of this work was to enhance the magnetic properties of BFO after Ca-Ba codoping. In the previous section, we observed that after 5 mol% of Ca doping in BFO the saturation magnetization increased by six times but the coercive field decreased significantly. On the other hand when we doped BFO with 5 mol% of Ba ion the coercive field increased by approximately eight times with respect to the pristine sample while there was minimal increment in the saturation magnetization. So, to get enhancement in both coercive field and saturation magnetization we prepared Ca-Ba-codoped BFO NPs. The idea was that when both Ca and Ba would be present at the Bi site of BFO, the overall magnetic properties, that is, saturation magnetization, remnant magnetization, and coercive field should increase. Figure 6.13 shows the **M–H** loops of all the Ca-Ba-codoped and pristine BFO NPs. The magnetization data were extracted from hysteresis loops, which is shown in Fig. 6.14. Among all the codoped samples, C1B4, that is, BFO NPs doped with 4 mol% Ba and 1 mol% Ca exhibited the maximum coercive field (4058 Oe). The enhancement is astonishingly high, more than 1700%, with respect to the pristine sample. In contrast, much lower enhancement was observed for BFO NPs singly doped with Ca and Ba at the same total dopant concentration of 5 mol%. As discussed in previous sections, 5 mol% Ba–doped BFO NPs showed a 700% increase while 5 mol% Ca–doped BFO NPs showed a 36% decrease. At 5 mol% of total dopant concentration, Ca-Ba codoping yields a significantly improved result. Looking at saturation magnetization, we find that the maximum enhancement, of 600%, is attained for C4B1 NPs. In comparison, the M_S was enhanced by 480% and 125%, respectively, for 5 mol% Ca–doped and 5 mol% Ba–doped BFO NPs. For technological applications, it is desirable that the two aspects of enhanced magnetic characteristics should coexist in one material. This was achieved for Ca-Ba-equidoped (2.5 mol% each) BFO NPs, in which the coercive field shows an increase of about 800% with a concomitant increase in saturation magnetization by 600%. It can be readily inferred from these results that codoping is better than single doping of alkaline earth metal ions.

NPs often exhibit unusual magnetic behaviors different from that of their bulk counterparts owing to finite size effects. In antiferromagnetic NPs, due to the large surface-to-volume ratio, the contribution of uncompensated spins at the surface becomes higher. It is a well-studied fact that the finite magnetic moment of NPs of antiferromagnetic materials is due to the presence of these uncompensated spins at the surface of the particles (Park et al., 2007). The enhanced magnetic properties of Ca-Ba-codoped BFO NPs in comparison to pristine BFO may be due to the presence of more uncompensated spins at the surface of codoped BFO NPs because of their smaller size. Another reason for the observed ferromagnetism in BFO and Ca-Ba-codoped BFO NPs is the modulation of the spiral spin structure characteristic of BFO (Park et al., 2007). When the particle size is of the order of or less than 62 nm, this spiral spin structure changes so as to result in enhancement in magnetic properties. The substitution of divalent Ca and Ba ions at the trivalent Bi site can lead to a charge imbalance, which can be compensated by the conversion of Fe^{3+} ions to Fe^{4+} ions or by the creation of oxygen vacancies. Mössbauer spectroscopic data ruled out the possibility of the existence of Fe^{4+} ions. So the only way to maintain the electroneutrality is the formation of oxygen vacancies, and these created oxygen vacancies may be a possible reason for the enhancement in magnetic properties of the doped NPs (Sundaresan et al., 2006). However, the most important point is that it is the nature of the dopant and the doping amount that play a major role in determining the magnetic properties of BFO NPs.

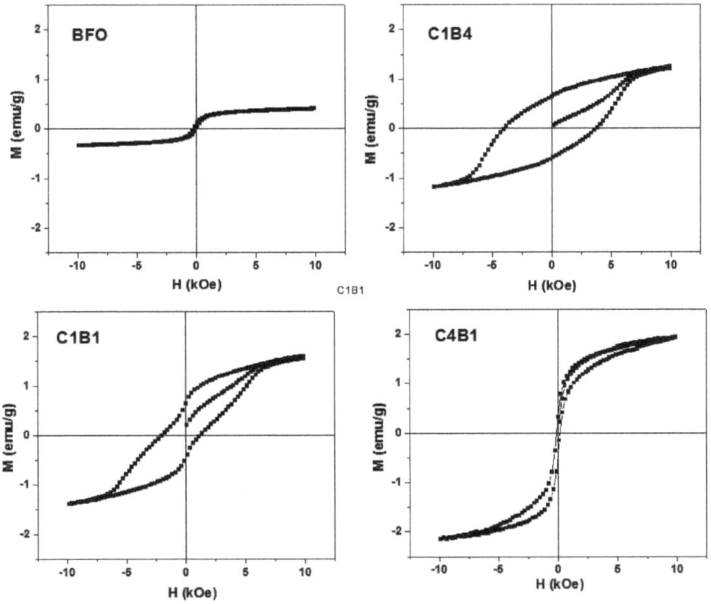

Figure 6.13 **M–H** curve of all the samples recorded at room temperature. Reprinted from Bhushan B, Das D, Priyam A, Vasanthacharya NY, Kumar S. (2012). Enhancing the magnetic characteristics of $BiFeO_3$ nanoparticles by Ca, Ba co-doping. *Mater. Chem. Phys.*, **135**: 144–149, Copyright (2012), with permission from Elsevier.

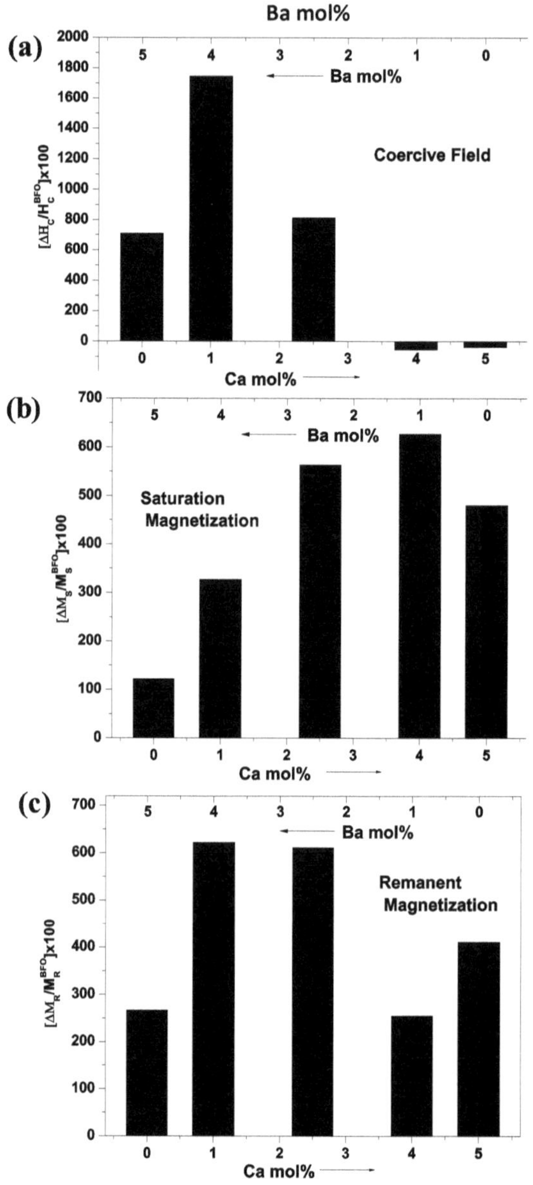

Figure 6.14 Percentage change in (a) coercive field, (b) saturation magnetization, and (c) remnant magnetization with varying concentrations of Ca and Ba ions. Reprinted from Bhushan B, Das D, Priyam A, Vasanthacharya NY, Kumar S. (2012). Enhancing the magnetic characteristics of $BiFeO_3$ nanoparticles by Ca, Ba co-doping. *Mater. Chem. Phys.*, **135**: 144–149, Copyright (2012), with permission from Elsevier.

6.4 Conclusions

The A-site doping of BFO with alkaline earth dopants has been found to be an important tool to tailor the structural, magnetic, and optical properties of BFO NPs. A facile sol-gel route has been developed to prepare pristine and doped BFO NPs having good phase purity. In the first part of the study, the dopant concentration was kept constant while the radii of the dopant ions (Ca^{2+}, Sr^{2+}, Ba^{2+}) were varied as we moved down group II (of the periodic table of elements). The characteristic structural phase transitions of BFO are found to occur at much lower temperatures in the pristine NPs. A reduction of Néel temperature has been observed in the doped samples in comparison with the pristine one. In the doped samples, no transformation of Fe^{3+} ions into Fe^{4+}/Fe^{2+} ions could be detected by Mössbauer spectroscopy. We also carried out dielectric constant measurement of all the samples. Ferromagnetism is observed in all the doped and pristine BFO NPs, while it should be noted that bulk BFO is antiferromagnetic at room temperature.

Next, we particularly focused on comprehending the concentration dependence of the alkaline earth dopants with regard to improving functionalities of BFO NPs. Sr^{2+} was chosen as the model dopant ion for this investigation. As the Sr concentration was increased, a reduction in particle size was noticed, which was also accompanied by a shape transformation from a multifaceted polygon (pristine) to a spherical one (5 mol% Sr–doped BFO), as seen in TEM images, which indicates the loss of structural anisotropy. The pristine sample and the sample having a lower dopant concentration (up to 3 mol%) exhibited absorption peaks ~530 nm while a significant shift, of ~100 nm, toward the shorter wavelength is observed for a higher dopant concentration. We carried out second derivative FTIR analysis in order to locate and identify each band in the congested spectral region as obtained in the raw FTIR spectra of the BFO and doped BFO NPs. All the samples exhibited ferromagnetic ordering of spins, and saturation of magnetization was also achieved at room temperature within an applied field of 10 kOe and the best results were obtained for 3 mol% Sr–doped samples.

In the studies, we found the maximum increase in coercive field and saturation magnetization while doping BFO with Ba ions and Ca ions, respectively. So, to get the maximum functionality with a nonmagnetic dopant we further codoped BFO with Ba and Ca ions. The ratio of Ba and Ca ions were varied, while keeping the substitution percentage (5%) constant. We obtained the best magnetic parameters with enhancement in both coercive field and saturation magnetization in a single BFO sample having an equimolar concentration (2.5%) of Ba and Ca ions. An increment of more than 18 times in the value of coercive field in the C1B4 sample and an increment of approximately 7 times in the saturation magnetization in C1B1 and C4B1 samples in comparison to the pristine BFO NPs are the most important findings of this work. Although modification of the spiral spin structure, uncompensated spin residing at the surface of the NPs, and the presence of oxygen vacancies are the prime causes of this enhancement, it is finally the choice of the dopant ions and the doping concentration that played a key role in achieving such incredible results. Low magnetic characteristics were one of the main hurdles in the way of BFO's technological applications. These new findings will certainly enhance the prospects of this potential functional material to be used for novel device fabrications.

Acknowledgments

The authors thank the Department of Science and Technology, Government of India, for research funding under various schemes (grant ref. nos. SB/FT/PS-31/2013, SB/FT/CS-84/2011, and SR/NM/NS-1047/2012).

References

Bhushan B, Basumallick A, Bandopadyay SK, Vasanthacharya NY, Das D. (2009). Effect of alkaline earth metal doping on thermal, optical, magnetic and dielectric properties of $BiFeO_3$ nanoparticles. *J. Phys. D: Appl. Phys.*, **42**: 065004.

Bhushan B, Basumallick A, Vasanthacharya NY, Kumar S, Das D. (2010). Sr-induced modification of structural, optical and magnetic properties in $Bi_{1-x}Sr_xFeO_3$ (x = 0, 0.01, 0.03, 0.05 and 0.07) multiferroic nanoparticles. *Solid State Sci.*, **12**: 1063–1069.

Bhushan B, Das D, Priyam A, Vasanthacharya NY, Kumar S. (2012). Enhancing the magnetic characteristics of $BiFeO_3$ nanoparticles by Ca, Ba co-doping. *Mater. Chem. Phys.*, **135**: 144–149.

Blaauw C, van der Woude F. (1973). Magnetic and structural properties of $BiFeO_3$. *J. Phys. C: Solid State Phys.*, **6**: 1422.

Bocquet AE, Fugimori A, Mizokawa T, Saitoh T, Namatame H, Suga S, Kimizuka N, Takeda Y, Takano M. (1992). Electronic structure of $SrFe^{4+}O_3$ and related Fe perovskite oxides. *Phys. Rev. B*, **45**: 1561.

Cho JY, Kim IK, Jung IO, Moon JH, Kim JH. (2009). Effects of Mg doping concentration on the band gap of $ZnO/Mg_xZn_{1-x}O$ multilayer thin films prepared using pulsed laser deposition method. *J. Electroceram.*, **23**: 442.

Chou X, Zhai J, Jiang H, Yao X. (2007). Dielectric properties and relaxor behavior of rare-earth (La, Sm, Eu, Dy, Y) substituted barium zirconium titanate ceramics. *J. Appl. Phys.*, **102**: 084106.

Chung CF, Lin JP, Wu JM. (2006). Influence of Mn and Nb dopants on electric properties of chemical-solution-deposited $BiFeO_3$ films. *Appl. Phys. Lett.*, **88**: 242909.

Das SR, Choudhary RNP, Bhattacharya P, Katiyar RS, Dutta P, Manivannan A, Seehra MS. (2007). Structural and multiferroic properties of La-modified $BiFeO_3$ ceramics. *J. Appl. Phys.*, **101**: 034104.

Dasgupta TS, Popović ZS, Satpathy S. (2005). Density functional study of he insulating ground states in $CaFeO_3$ and $La_{1/3}Sr_{2/3}FeO_3$ compounds. *Phys. Rev. B*, **72**: 045143.

Datta A, Priyam A, Bhattacharyya SN, Mukherjea KK, Saha A. (2008). Temperature tunability of size in CdS nanoparticles and size dependent photocatalytic degradation of nitroaromatics. *J. Colloid Interface Sci.*, **322**: 128.

Ederer C, Spaldin NA. (2005). Influence of strain and oxygen vacancies on the magnetoelectric properties of multiferroic bismuth ferrite. *Phys. Rev. B*, **71**: 224103.

Ederer C, Spaldin NA. (2005). Weak ferromagnetism and magnetoelectric coupling in bismuth ferrite. *Phys. Rev. B*, **71**: 060401(R).

Eerenstein W, Mathur ND, Scott JF. (2006). Multiferroic and magnetoelectric materials. *Nature*, **442**: 759–765.

Feng HJ, Liu FM. (2007). Electronic structure of $BaFeO_3$: an abinitio DFT study. eprint arXiv: 0704.2985, 1–6.

Fiebig M, Lottermoser T, Fröhlich D, Goltsev AV, Pisarev RV. (2002). Observation of coupled magnetic and electric domains. *Nature*, **419**: 818–820.

Gan Q, Rao RA, Eom CB, Garrett JL, Lee M. (1998). Direct measurement of strain effects on magnetic and electrical properties of epitaxial $SrRuO_3$ thin films. *Appl. Phys. Lett.*, **72**: 978.

Gao F, Cai C, Wang Y, Dong S, Qiu XY, Yuan GL, Liu ZG, Liu JM. (2006). Preparation of La-doped $BiFeO_3$ thin films with Fe^{2+} ions on Si substrates. *J. Appl. Phys.*, **99**: 094105.

Gao F, Chen X, Yin K, Dong S, Ren Z, Yuan F, Yu T, Zou Z, Liu JM. (2007). Visible-light photocatalytic properties of weak magnetic $BiFeO_3$ nanoparticles. *Adv. Mater.*, **19**: 2889–2892.

Higuchi T, Liu YS, Yao P, Glans PA, Guo J, Chang C, Wu Z, Sakamoto W, Itoh N, Shimura T, Yogo T, Hattori T. (2008). Electronic structure of multiferroic $BiFeO_3$ by resonant soft x-ray emission spectroscopy. *Phys. Rev. B*, **78**: 085106.

Hill NA. (2000). Why are there so few magnetic ferroelectrics? *J. Phys. Chem. B*, **104**: 6694–6709.

Jacobson AJ, Fender BEF. (1975). A neutron diffraction study of the nuclear and magnetic structure of $BiFeO_3$. *J. Phys. C*, **8**: 844.

Kamba S, Nuzhnyy D, Savinov M, Šebek J, Petzelt J, Prokleška J, Haumont R, Kreisel J. (2007). Infrared and terahertz studies of polar phonons and magnetodielectric effect in multiferroic $BiFeO_3$ ceramics. *Phys. Rev. B*, **75**: 024403.

Khomchenko VA, Kiselev DA, Vieira JM, Jian L, Kholkin AL, Lopes AML, Pogorelov YG, Araujo JP, Maglione M. (2008). Effect of diamagnetic Ca, Sr, Pb, and Ba substitution on the crystal structure and multiferroic properties of the $BiFeO_3$ perovskite. *J. Appl. Phys.*, **103**: 024105.

Khomskii DI. (2006). Multiferroics: different ways to combine magnetism and ferroelectricity. *J. Magn. Magn. Mater.*, **306**: 1–8.

Kothari D, Reddy VR, Gupta A, Phase DM, Lakshmi N, Deshpande SK, Awasthi AM. (2007). Study of the effect of Mn doping on the $BiFeO_3$ system. *J. Phys.: Condens. Matter*, **19**: 136202.

Kozhevnikov VL, Leonidov IA, Patrakeev MV, Mitberg EB. (2000). Electrical properties of the ferrite $SrFeO_y$ at high temperatures. *J. Solid State Chem.*, **158**: 320–326.

Kumar M, Yadav KL. (2006). Study of room temperature magnetoelectric coupling in Ti substituted bismuth ferrite system. *J. Appl. Phys.*, **100**: 074111.

Kwak WC, Kim TG, Chae WS, Sung YM. (2007). Tuning the energy bandgap of CdSe nanocrystals via Mg doping. *Nanotechnology*, **18**: 205702.

Lebeugle D, Colson D, Forget A, Viret M, Bonville P, Marucco JF, Fusil S. (2007). Room-temperature coexistence of large electric polarization and magnetic order in $BiFeO_3$ single crystals. *Phys. Rev. B*, **76**:024116.

Lebon A, Adler P, Bernhard C, Boris AV, Pimenov AV, Maljuk A, Lin CT, Ulrich C, Keimer B. (2004). Magnetism, charge order, and giant magnetoresistance in $SrFeO_{3-\delta}$ single crystals. *Phys. Rev. Lett.*, **92**: 037202.

Li M, Ning M, Ma Y, Wu Q, Ong CK. (2007). Room temperature ferroelectric, ferromagnetic and magnetoelectric properties of Ba-doped $BiFeO_3$ thin films. *J. Phys. D: Appl. Phys.*, **40**: 1603.

Matsui T, Taketani E, Tsuda H, Fujimura N, Morii K. (2005). Improvement of magnetization and leakage current properties of magneto-electric $BaFeO_3$ thin films by Zr substitution. *Appl. Phys. Lett.*, **86**: 082902.

Meerwall EV. (1975). A least-squares spectral curve fitting routine for strongly overlapping Lorentzians or Gaussians. *Comput. Phys. Commun.*, **9**(2): 117.

Néel L. (1961a). Superparamagnétisme des grains très fins antiferro-magnétiques. *Compt. Rend.*, **252**: 9, 4075.

Palai R, Katiyar RS, Schmid H, Tissot P, Clark SJ, Robertson J, Redfern SAT, Catalan G, Scott JF. (2008). β phase and γ–β metal-insulator transition in multiferroic $BiFeO_3$. *Phys. Rev. B*, **77**: 014110.

Palkar VR, Kundaliya DC, Malik SK, Bhattacharya S. (2004). Magneto-electricity at room temperature in the $Bi_{0.9-x}Tb_xLa_{0.1}FeO_3$ system. *Phys. Rev. B*, **69**: 212102.

Pan LK, Ee YK, Sun CQ, Yu GQ, Zhang QY, Tay BK. (2004). Band-gap expansion, core-level shift, and dielectric suppression of porous silicon passivated by plasma fluorination. *J. Vac. Sci. Technol. B*, **22**: 583.

Park TJ, Papaefthymiou GC, Viescas AJ, Moodenbaugh AR, Wong SS. (2007). Size-dependent magnetic properties of single-crystalline multiferroic $BiFeO_3$ nanoparticles. *Nano Lett.*, **7**: 766–772.

Peacock PW, Robertson J. (2002). Band offsets and Schottky barrier heights of high dielectric constant oxides. *J. Appl. Phys.*, **92**: 4712.

Priyam A, Blumling DA, Knappenberger Jr. (2010). Synthesis, characterization, and self-organization of dendrimer-encapsulated HgTe quantum dots. *Langmuir*, **26**: 10636.

Raju KCJ, Sivasubramanian V, Pragasam R, Viswanathan B, Murthy VRK. (1993). Contributions to the dielectric constant of the system $BaLn_2Ti_4O_{12}$ from packing fraction and nephelauxetic ratio. *J. Appl. Phys.*, **74**: 1968.

Rao GVS, Rao CNR. (1970). Infrared and electronic spectra of rare earth perovskites: ortho-chromites, manganites and ferrites. *Appl. Spectrosc.*, **24**: 436–445.

Rivera JV, Schmid H. (1997). On the birefringence of magnetoelectric $BiFeO_3$. *Ferroelectrics*, **204**: 23–33.

Roy SC, Sharma GL, Bhatnagar MC. (2007). Large blue shift in the optical band-gap of sol–gel derived $Ba_{0.5}Sr_{0.5}TiO_3$ thin films. *Solid State Commun.*, **141**: 243.

Som KK, Molla S, Bose K, Chaudhury BK. (1992). Nonlinear physical properties of amorphous $Bi_4Sr_3Ca_3Cu_yO_x$ semiconducting oxides with y between 0 and 5. *Phys. Rev. B*, **45**: 1655.

Sosnowska I, Schäfer W, Kockelmann W, Andersen KH, Troyanchuk IO. (2002). Crystal structure and spiral magnetic ordering of $BiFeO_3$ doped with manganes. *Appl. Phys. A*, **74**: S1040–S1042.

Speranskaya EI, Skorikov VM, Rode EY, Terekhova, VA. (1965). Phase diagram of the system of bismuth oxide-iron oxide. *Bull. Acad. Sci. USSR*, **5**: 874.

Srinivas S, Li JY. (2005). The effective magnetoelectric coefficients of polycrystalline multiferroic composites. *Acta Mater.*, **53**: 4135–4142.

Sundaresan A, Bhargavi R, Rangarajan N, Siddesh U, Rao CNR. (2006). Ferromagnetism as a universal feature of nanoparticles of the otherwise nonmagnetic oxides. *Phys. Rev. B*, **74**: 161306(R).

Tian ZM, Yuan SL, Liu L, Yin SY, Jia LC, Duan HN, Li P, Wang CH, Li JQ. (2009). Synthesis and exchange bias effect of $CuFe_2O_4/NiO$ nanocomposites. *Smart Mater Struct.*, **18**: 015018.

Tütüncü HM, Srivastava GP. (2008). Electronic structure and lattice dynamical properties of different tetragonal phases of $BiFeO_3$. *Phys. Rev. B*, **78**: 235209.

Tütüncü HM, Srivastava GP. (2008). Electronic structure and zone-center phonon modes in multiferroic bulk $BiFeO_3$. *J. Appl. Phys.*, **103**: 083712.

Vishwanatha R, Sarma DD. (2006). Study of the growth of capped ZnO nanocrystals: a route to rational synthesis. *Chem. Eur. J.*, **12**: 180–186.

Wei J, Xue D, Xu Y. (2008). Photoabsorption characterization and magnetic property of multiferroic $BiFeO_3$ nanotubes synthesized by a facile sol–gel template process. *Scr. Mater.*, **58**: 45–48.

Yang CH, Koo TY, Jeong YH. (2005). How to obtain magnetocapacitance effects at room temperature: the case of Mn-doped $BiFeO_3$. *Solid State Commun.*, **134**: 299–301.

Yu B, Li M, Liu J, Guo D, Pei L, Zhao X. (2008). Effects of ion doping at different sites on electrical properties of multiferroic $BiFeO_3$ ceramics. *J. Phys. D: Appl. Phys.*, **41**: 065003.

Chapter 7

Ferroelectric Nanoceramic Materials

K. Sivasakthi,[a] S.Varun,[b] and S. C. G. Kiruba Daniel[b]

[a]*Department of Theoretical Physics, University of Madras, Chennai, Tamil Nadu 600025, India*
[b]*Department of Instrumentation and Applied Physics, Indian Institute of Science (IISc), Bengaluru, Karnataka 560012, India*

crossguevara@gmail.com

Ferroelectric nanoceramic materials have a huge potential in integrated electronics as passive components, nonvolatile ferroelectric memories, and optoelectronic devices. They are mainly utilized in the form of thin films in the recent past more than as powders. Special attention is given to the piezoelectric properties of ferroelectric nanoceramic powders as well as thin films. In this chapter, the basics of nanocrystalline ferroelectrics and their synthesis routes will be discussed, which will be followed by a special emphasis on the sol-gel processing of prominent ferroelectric nanoceramics and their sintering methods. The important applications currently pursued by means of these advanced materials are compiled for further research developments.

Smart Ceramics: Preparation, Properties, and Applications
Edited by Ajay Kumar Mishra

ISBN 978-981-4774-30-7 (Hardcover), 978-1-315-16359-8 (eBook)
www.panstanford.com

7.1 Introduction

Interest in nanocrystalline ceramic materials has increased considerably because of their size-sensitive interesting new properties. Some of the prominent properties include mechanical, electrical, optical, and magnetic, which are characteristic of dense ceramics. Thus, the nanoscale structure leading to high-density grain boundaries and interfaces attributes to the new properties (Shukla and Choudhary, 2010). Coprecipitation, sol-gel, inert gas condensation, chemical vapor deposition, high-energy milling, and hydrothermal reactions are some of the common synthetic routes adopted for the past 20 years for the synthesis of high-purity nanocrystalline oxide and nonoxide powders. Hence, there is considerable progress in the synthesis of nanocrystalline ceramics.

In the field of electroceramics, this progress is additionally driven by the technology factor that is imperative to miniaturization because of the small device sizes and the increasing desire to impart multifunctionality to a single component. This trend is expected to continue into the foreseeable future, hence the increasing importance of the fabrication processes, such as sintering, that may enable nanotechnology-based devices (Shukla and Choudhary, 2010).

While fabrication of nanoceramics logically starts from nanosized powders, the passage from nanosized powders to dense ceramics maintaining the nanometric microstructure (100 nm) has proven to be challenging for the following reasons. First, agglomeration of powders increases with decreasing particle size and the size of the agglomerates rather than the size of the nanopowders define the characteristic length scales in sintering and final microstructure. Second, the final stage of sintering is always accompanied by rapid grain growth, which is an increasing driving force with decreasing grain size, making it difficult to maintain the microstructure while improving density. Third, even though there are conventional ceramic methods to enhance densification kinetics and suppress grain growth, they either depend on additives, which might adversely alter the physical properties, or resort to brute force, such as large applied pressure, which increases the cost and limits the process flexibility. Therefore, further research is required to overcome or circumvent these problems in order to achieve efficient fabrication of nanoceramics.

We will discuss an outline of different synthesis methods, in particular sol-gel processes, and the major applications of ferroelectric nanoceramics materials. We will also discuss the size effects in ferroelectric systems that have attracted a great deal of attention due to their potential application in the field of information technology.

7.2 Types of Nanoceramic Materials

Perovskite-type oxides (ABO_3) have drawn considerable attention recently and have extended their applicability to numerous fields, including microelectronics. This is mainly due to their excellent physical properties, such as ferroelectric, magnetoelectric, piezoelectric, and electrooptic effects. Titanate ceramics have gained much attention among the major perovskite-type oxides. Due to their high dielectric properties, they serve as better materials for room-temperature applications (Shukla and Choudhary, 2010).

Barium titanate ($BaTiO_3$) has a perovskite structure and is one of the most important electroceramics. It is extensively used as high-dielectric-constant capacitors, positive temperature coefficient (PTC) resistors, transducers, and ferroelectric memories. Barium titanate with different dopants is considered to be the most useful ferroelectric materials, which has a lot of potential applications as multilayer capacitors, thermistors, varistors, energy converting systems, etc. (Choudary and Thakur, 2009).

Lead titanate ($PbTiO_3$), which exhibits a perovskites structure and has a very high Curie temperature of 490°C, belongs to an important ferroelectric and piezoelectric family (Shukla and Choudhary, 2010). The other materials are $SrTiO_3$, $PbZrO_3$, and $BiFeO_3$.

7.3 Synthesis of Nanoceramic Materials (Other Than Sol-Gel)

In the past, most synthetic procedures for the preparation of perovskite crystals included high-temperature (1000°C) sintering followed by annealing. Recently, extensive efforts have been focused on the establishment of moderate reaction conditions, especially lowering the synthesis temperature for high-quality nanocrystals. In the past, ceramic methodology was usually utilized to produce perovskite oxides. The individual metal oxide components or their precursors are stoichiometrically mixed and heated together to form perovskite oxides, but it requires long diffusion times and high temperatures. The major drawbacks of this type of methodology are poorly controlled crystal morphology, grain size, and grain shape of the final material. Therefore, various other synthetic methods for perovskite ferroelectric nanomaterials have been developed (Nuraje and Su, 2013). Barium titanate has many advantages over other ceramic materials. For example, it is mechanically and chemically very stable; therefore, it exhibits ferroelectric properties much above room temperature, and it can be easily prepared and used in the ceramic polycrystalline form (Choudhary et al., 2009).

The various synthesis techniques are high-temperature sintering, sol-gel technique, coprecipitation, hydrothermal synthesis, solvothermal synthesis, electrospinning method, emulsion precipitation method, reverse microemulsion, pyrolysis and decomposition of bimetallic alkoxide precursors in the presence of coordinating ligands, liquid-solid-solution (LSS) phase transfer, composite-hydroxide-mediated approach, bioinspired synthesis methods, etc.

7.3.1 Two-Photon Lithography

A laser technique called two-photon lithography is adopted to etch out a polymer into a 3D structure. The laser hardens the spots that it exposes and leaves the rest unhardened. The unhardened material is then dissolved to produce a shell. The shell is then coated with ceramic, metals, metallic glass, etc. In the finished state, the nanotruss of the ceramic can be flattened and reverted to its original state (Shukla and Choudhary, 2010).

7.3.2 Hydrothermal Synthesis

Hydrothermal synthesis is carried out in a closed chamber, which is followed by heating of the solution. The temperature ranges from 200°C to 250°C. The desirable morphology of nanoceramics can be attained by tuning the temperature and giving appropriate pressure. This method is also known as the autoclave method. The precursor solution for the medium will be normally an aqueous solution. This is considered to be the best approach to synthesizing nanoceramics since it gives the user many advantages in terms of tuning the various parameters. They offer very stable products without the incorporation of unwanted products. Hydrothermal synthesis is also used to prepare $BaTiO_3$, $BiFeO_3$, $BaSrTiO_3$, $SrTiO_3$, $PbTiO_3$, and $PbZrTiO_3$ nanomaterials.

Using the hydrothermal method, various $BaTiO_3$ structures can be fabricated. $BaTiO_3$ nanoparticles with various morphologies, including nanowires, nanosheets, nanocubes, hexagonal nanoparticles, nanorods, starfish-like nanostructures, and sword-like nanostructures were successfully synthesized.

Solvothermal synthesis is another common way of fabricating nanoceramic structures. It is almost similar to hydrothermal synthesis, wherein organic solvents are used instead of aqueous media.

7.3.3 Template Synthesis

Template synthesis involves fabrication of nanostructures using a suitable template. This serves as a filling material for suitable polymeric precursors. The solution containing a nanoceramic material with the template is heated to a particular temperature, followed by an etching process to remove that template. The advantages of using template-mediated synthesis are structures of nanoarrays can be fabricated, synthesis can be size controlled, and a large number of nanowires and tubes can be fabricated.

7.4 Sol-Gel-Based Synthesis of Ferroelectric Nanoceramic Materials

The sol-gel process, also known as chemical solution deposition, is a widely adapted method for the synthesis of nanoceramics. This involves a chemical solution, or the sol, and precursors including metal oxides and metal chlorides form a colloid system by undergoing hydrolysis and polycondensation reactions, which can be modulated by an acid or a base. The sol and gel are mixed to produce an oxide material that is generally a type of ceramic. In the recent past, sol-gel-method-based ferroelectrics have fueled research in areas of catalysis and sensors (Lu et al., 1997), membranes (Caruso et al., 2001), coatings, nanoreactors (Liu et al., 1991), electronic applications (Fendler, 2001), etc. Sol-gel-processed nanoceramics may lead to different shapes based on the reactions conditions (Table 7.1).

Table 7.1 Different shapes of sol-gel-processed ferroelectric nanoceramics

Experimental methods	Compounds	Morphology	References
Sol-gel	$BaTiO_3$	Tube	(Chen, 2009)
Sol-gel	$BaTiO_3$	Fiber	(Li, 2009)
Sol-gel	$Bi_4Ti_3O_{12}$	Tube	(Anokhin, 2011)
Sol-gel	$K_{0.5}Bi_{0.5}TiO_3$	Wire	(Hou, 2006)
Sol-gel	$Na_{0.5}Bi_{0.5}TiO_3$	Whisker	(JingBing, 2004)
Sol-gel	$Bi_2T_2O_7$	Tube	(Zhou, 2006)

With the increase in temperature, the particles precipitate and the solvent evaporates, which results in particle grain sizes ranging from tens of nanometers to a few microns. The particles are heated by a process called sintering to produce a solid product, which results in an increase in grain size. This method could also be applied to produce a nanocomposite by heating the gel on a thin film to form a nanoceramic layer on top of the film.

Table 7.2 describes the various synthesis methods followed by sintering used for the synthesis of ferroelectric nanoceramic materials. Initially C. Mazdiyasni et al. established that well-sintered, dense ferroelectric ceramics can be obtained at temperatures as low as 900°C, when sol-gel powders prepared from solutions of metal alkoxides are employed for sintering (Sakka, 2005a).

A typical sol-gel method in the preparation of ceramics starts from a solution that consists of metal compounds, such as metal alkoxides and acetylacetonates, as the source of oxides, alcohols as the solvent, water as a hydrolysis agent, and an acid or base catalyst (Sakka, 2005b). Important considerations include the stability of the solution with time and a low tendency of the

Table 7.2 Common ferroelectric nanoceramic materials and their synthesis routes

Nanoceramic material	General formula	Grain size	Synthesis route	Reference
Barium strontium zirconate	$Ba_{1-x}Sr_xZrO_3$	17–52 nm	Citrate precursor route	(Al-Hartomy et al., 2013)
Barium lead zirconate	$Ba_{1-x}Pb_xZrO_3$ $(0 \leq x \leq 1)$	20–60 nm	Reverse micellar route (avoiding alkoxides)	(Ahmad et al., 2006)
	$Ba_4(Sm_{1-x}Nd_x)$ 9.3-Ti_8O_{54}	—	Citrate sol-gel synthesis (modified Pechini method)	(Yue et al., 2004)
	$(Na_{0.5}Bi_{0.5})0.94$ $Ba_{0.06}TiO_3$	—	Sol-gel auto combustion method	(Huo et al., 2007)
Barium lead titanate	$Ba_{0.8}Pb_{0.2}TiO_3$	60 nm	Ball milling, spark plasma sintering	(Mudinepalli et al., 2014)
Lead-free niobate	$(Na_{0.5}K_{0.5})NbO_3$	260 nm	Spark plasma sintering	(Mudinepalli et al., 2014)
Barium lead titanate	$Ba_{0.8}Pb_{0.2}TiO_3$	200 nm	Microwave sintering	(Mudinepalli et al., 2013)
Bismuth ferrite	$BiFeO_3$	300 nm	Conventional sintering	(Song et al., 2015)
Bismuth ferrite	$BiFeO_3$	200 nm	Spark plasma sintering	(Song et al., 2015)
Sodium bismuth titanate	$Na_{0.5}Bi_{0.5}TiO_3$	200 nm	Spark plasma sintering	(Mudinepalli et al., 2016)
Lead-free multiferroic composites of NCC-NBT	$xNi_{0.83}Co_{0.15}$ $Cu_{0.02}Fe_{1.9}O_{4-\delta}$- $(1-x)Na_{0.5}Bi_{0.5}TiO_3$	—	Spark plasma sintering	(Mudinepalli et al., 2014)
Lead-free multiferroic composites of NCMF-NBT	$xNi_{0.93}Co_{0.02}$ $Mn_{0.05}Fe_{1.95}O_4$- $(1-x)$ $Na_{0.5}Bi_{0.5}TiO_3$	—	Spark plasma sintering	(Mudinepalli et al., 2015)

(Continued)

Table 7.2 (*Continued*)

Nanoceramic material	General formula	Grain size	Synthesis route	Reference
Lead-free multiferroic composites of MFO-NBT	$Na_{0.5}Bi_{0.5}TiO_3$-$MnFe_2O_4$	—	Spark plasma sintering	(Praveena et al., 2014)
Lead-free multiferroic composites of CoFO-NBT	(1–*x*)NBT-(*x*)CFO	—	Solid-state double sintering	(Babu et al., 2011)
Lead-free multiferroic composites of NFO-NBT	(1–*x*)NBT-(*x*)NFO	—	Sol-gel	(Babu et al., 2011)

solution to precipitate. Starting materials are usually inorganic metal salts or metal organic compounds such as metal alkoxides (Brinker and Scherer, 1990), which are popular precursors because they react readily with water (Schwartz et al., 1997). That means in the hydrolysis reaction a hydroxyl ion becomes attached to the metal atom.

Some of the most commonly used organic precursors for sol-gel film formation are metal alkoxides $M(OR)_z$, where R stands for an alkyl group (C_xH_{2x+1}). Metal alkoxides $M(OR)_n$ are derivatives of alcohols, R-OH, which are easily removable via hydrolysis and thermal treatment, heaving high-purity unhydrated oxides (Hubert-pfulzgraf, 2003; Kessler et al., 2006). The advancements in this field during the last 25 years (a few decades), including both the studies of their molecular and crystal structure and the reactivity of these compounds, have considerably changed the image of these compounds in the eyes of chemists and material scientists (Yanovskaya et al., 1992; Sakka, 2005c; Carreon et al., 2006). Chemical precursors always play a key role in sol-gel coating by directly affecting the porosity, refractive index, hardness, and other performance characteristics of the resultant coatings and thin films. The precursor is frequently a decisive factor for production in terms of cost and usability (Hubert-pfulzgraf, 1987; Sanchez and Livage, 1990; Hubert-pfulzgraf, 1995; Yaturova et al., 2002).

Conventionally, the alkoxide is dissolved in alcohol and hydrolyzed by the addition of water under acidic, neutral, or basic conditions, although film formation is also possible by the deposition of alkoxides followed by exposure to moisture. Chemical reactivity of metal alkoxides toward hydrolysis and condensation mainly depends on the positive charge of the metal atom, d_M, and its ability to increase its coordination number N. As a general rule, the electronegativity of metal atoms decreases and their size increases when going down the periodic table, from the top right to the bottom left. The chemical reactivity of the corresponding alkoxides toward hydrolysis and condensation then increases (Lee and Pope, 1994). Metal compounds undergo hydrolysis and polycondensation at near-room temperature, giving rise to a sol (a substance that contains a continuous solid skeleton enclosing a continuous liquid phase) in which polymers or fine particles are dispersed. Hydrolysis induces polycondensation, forming dimers, trimers, and oligomers, and these particles are arranged in a 3D network to form a solid skeleton (by processes like Brownian motion of charged particles and their clusters). Important parameters such as chemical conditions, pH value, water content, temperature, and type of alcohol determine if a nucleophilic or electrophilic attack occurs, which allows to vary the structure from an open fragile to a compact network. Also, the pore and grain sizes can be adjusted in a controlled manner by varying the pH value, the aging time and medium, the drying conditions, and the sintering temperature (Lisinski et al., 2010). Further reaction connects the particles, solidifying the sol into a wet gel that contains water and solvents. Usually, various shapes are formed during the transformation from sol to gel.

Gelation can be produced (i) by the rapid evaporation of the solvent, as occurs during the preparation of films or fibers, and (ii) after a sol is cast into a mold, in which case it is possible to make objects of desired shape. After gelation, the gel is typically aged and then desiccated. Syneresys, dehydroxylation, and structural changes can occur during this stage. The removal of the entrapped solvent during desiccation results in the formation of micropores, and consequently, structures with high internal surface areas. Structural rearrangements and consolidation processes that occur in this stage also have significant effects on the gel structure. Aging of a gel may involve further condensation, dissolution, and reprecipitation of monomers or phase transformation within the solid or liquid phases. Vaporization of water and solvents produces a dry gel, one of the final products. Drying by evaporation under normal conditions gives rise to capillary pressure, which causes shrinkage of the gel network (Fig. 7.1). Most gels are amorphous, even after drying, but many crystallize when heated. Once a gel has been densified, it is equivalent to a ceramic made by a conventional method. The final stages close to being dense essentially depend on thermophysical properties of the materials. A well-known problem in sintering of ceramic powders (several hundred degrees and higher temperatures) is that the full dense materials are difficult to achieve. Generally, the thermal energy to arrange the molecules into crystal lattice positions during polycondensation is not sufficient and the chemical conditions of polycondensation might work against it (there is a crossover from open fractal structures in acidic conditions to compact structures in base conditions). In the conversion of the dried gel to the amorphous oxide phase residual alkoxy and hydroxy groups are removed exothermically from the gel through further condensation reactions, with an overall weightlessness. Further structural rearrangements also occur, resulting in changes in the bulk density, skeletal density, microstructure, and pore structure of the gel. Next, the preparation of crystalline ceramics from metal alkoxides, densification, and crystallization may proceed. The ability to prepare dense crystalline ceramics is expected to be dependent on the initial hydrolysis conditions because hydrolysis controls the structure of the polymeric gels and thus influences the rearrangement processes leading to densification and crystallization. The crystallization stage is responsible for the behavior of the conductivity under high electric fields.

The sol-gel method or precipitation, spray pyrolysis, or emulsion techniques are used to form ultrafine and uniform ceramic powders. An advantage of the sol-gel technique is the opportunity to design new structures of materials by varying different parameters. For instance, xerogels, solid materials with a 3D network of colloidal particles, are derived from a dried gel body and exhibit a high porosity of around 50% and very small pores (1–10 nm) (Lisinski et al., 2010). The microstructure and orientation degree of sol-gel-derived ferroelectric thin films are highly dependent on the chemical precursors and employed conditions of solution preparation (Schwartz, 1989; Schwartz et al., 1997). Scanning electron microscopy (SEM) images of a sol-gel-processed ferroelectric material subjected to different sintering methods, leading to different morphologies, are shown in Fig. 7.2.

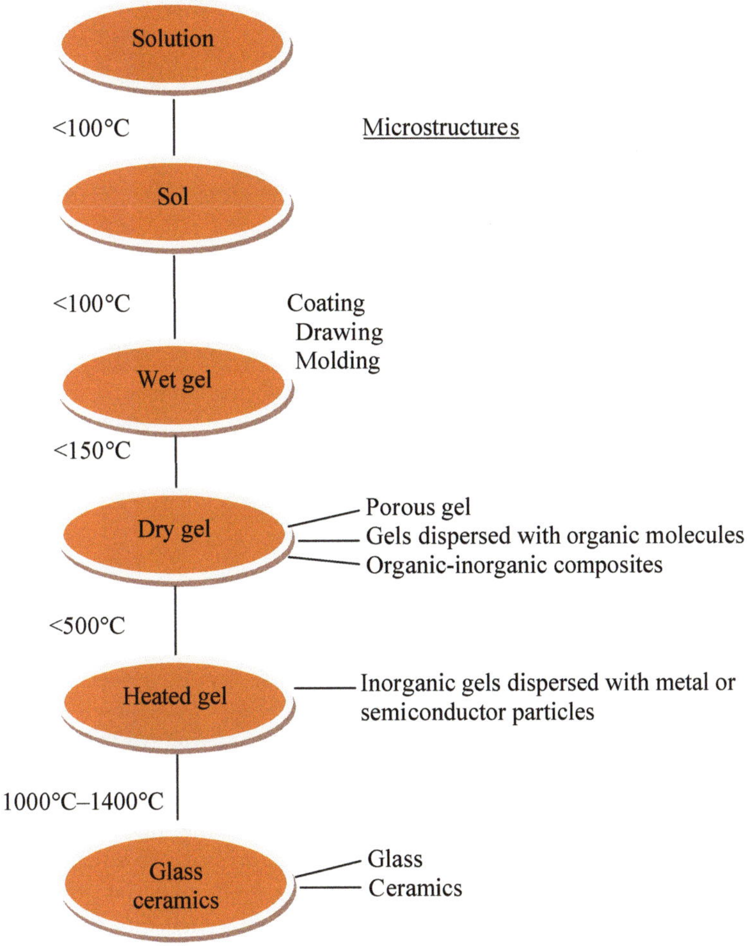

Figure 7.1 Typical products obtained during different stages of the sol-gel process.

Figure 7.2 Scanning electron microscopy images of ferroelectric $BaTiO_3$ nanoceramics: (a) three-step high-pressure sintering method at 6 GPa and 1000°C for 5 min, (b) three-step high-pressure sintering method at 6 GPa and 1100°C for 5 min, and (c) direct sintering method at 6 GPa and 1000°C for 5 min, as reported by Xiao et al. (2007). Reproduced from Xiao, CJ, Zhang, WW, Chi, ZH, Li, FY, Feng, SM, Jin, CQ, Wang, XH, Li, LT, Chen, RZ. (2007). Ferroelectric $BaTiO_3$ nanoceramics prepared by a three-step high-pressure sintering method. *Physica Status Solidi* (a), **204**(3): 874–880, with permission from Wiley.

7.4.1 Different Ferroelectric Ceramics Obtained by Sol-Gel Methods

Sol-gel synthesis has been a very successful and efficient route for obtaining high-quality ferroelectric/piezoelectric materials (Bradley et al., 1978; Schwartz, 1989; Schwartz et al., 1997; Brunckova et al., 2004; Marat et al., 2004; Chilibon et al., 2005; Cheng and Ho, 2010), etc. A few of the ferroelectric materials achieved by sol-gel processing are presented in subsequent sections.

7.4.1.1 Lead titanate

Lead titanate ($PbTiO_3$) is the parent material of a number of well-known doped derivates and solid solutions, for example, lanthanum-doped $PbTiO_3$, Pb(Zr, Ti)O_3, and Pb(Mg, Nb)O_3–$PbTiO_3$ (Setter et al., 2006). Schwartz (1989) described the lead titanate sol-gel processing starting from metal alkoxides to yield ceramics: solution preparation (stock solution); mixing of equal volumes of solvent (methoxiethanol-additive water) and solution; gelation of the alkoxide solution through the hydrolytic polycondensation reactions, which occur following the addition of water, to obtain wet gel and then an aging process; desiccation and removal of volatiles (drying); heat treatment of desiccated gel to obtain an amorphous oxide; and crystallization by heat treatment to obtain perovskite $PbTiO_3$. $PbTiO_3$ ceramics are prepared utilizing as precursors titanium (IV) isopropoxide, $Ti(OCH[CH_3]_2)_4$, of 99.99% purity and lead acetate trihydrate, $Pb(CH_3COO)_2 \cdot 3H_2O$, of 99% purity and 2-methoxyethanol, $CH_3OCH_2CH_2OH$, as the solvent. Preparation of $PbTiO_3$ gel was realized according to Schwartz's studies (1989), with some modifications, such as the reaction time and hydrolyses conditions. Also, by utilizing nitrogen atmosphere flux the distillation process was faster (Marat et al., 2004; Chilibon et al., 2005). The gel was prepared by addition of equal volumes of precursor solution (sol) and a solution containing water (R_w = 2.5) and 2-methoxyethanol as the solvent. Hereby, the gelation of solutions was controlled through a hydrolytic polycondensation by addition of water (Fig. 7.3). In the drying process, the densification and removal of volatile compounds induce some modifications in the gel structure and solvent removal leads to microporosity and, therefore, the area surface increases. During the gel conversion into amorphous powder through a thermal process, hydroxyl groups and some organic residuals are exothermically removed and induce appropriate mass decrease. After the water removal and burning of residual compounds, the network collapses, while the structure relaxes, and viscous sintering continues until the porous phase is eliminated. Densification of the amorphous gels by a thermal process implies structure relaxation, polymer condensation, and viscous sintering. In the preliminary drying gel process, the gel is heated at 200°C for 12 h in a glass bottle, which is put into a silicon oil bath, and all this time it is rotated by a spinner. After that, the amorphous powder is heated in an oven at 300°C for 24 h. Next, the drying process is done in an oven at 500°C for 2 h, and we obtain an amorphous powder. After that, the amorphous powder is crystallized in an oven at 800°C for 2 h and cooled in air (Chilibon et al., 2005).

For $PbTiO_3$, PTCa (calcium-modified lead titanate), lead acetate trihydrate, calcium acetate hydrate, $Ca(CH_3COO)_2 \cdot nH_2O$, and titanium tetraisopropoxide are used as precursor materials and 1,2-propanediol $CH_3CH(OH)\text{-}CH_2OH$ is used as the solvent (Lgreja et al., 1999; Dias et al., 2004). The acetates and solvent are added into a sealed trineck flask and filled with dry nitrogen and the dissolution is performed at 80°C for 1 h. Distillation of water is performed for 12 h at the solvent boiling point. The solution is cooled to 50°C and a solution of titanium tetraisopropoxide with a solvent is added dropwise to the solution through an isobaric funnel. The resulting solution is distilled for 3 h at the solvent boiling point to remove reaction by-products. After the distillation the solution is refluxed for 5 h and adjusted to 1 M. A solution of water and solvent (with the same volume as that of the previous solution) is added to start hydrolysis and polycondensation. The amount of water added is R_w = 2.5 $(H_2O)/(Ti)$ = 2.5. A transparent gel is obtained after 30 min at 70°C. The resulting gel is dried at 120°C for 6 h in air. The dried gel is heat-treated at 300°C for 12 h in an oven to obtain the green powder.

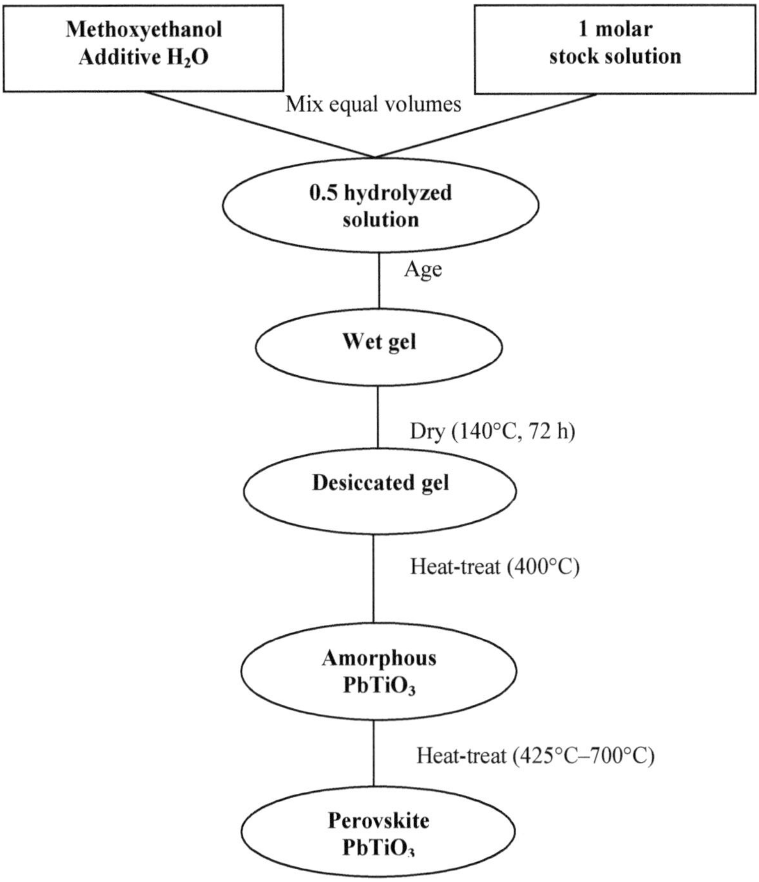

Figure 7.3 Schematic representation of overall synthesis of $PbTiO_3$ ceramics (adapted from Schwartz, 1989).

7.4.1.2 BST synthesis

An original sol-gel process to prepare the $(Ba,Sr)TiO_3$ (BST) is described by Yang et al. (2002). The reagent-grade barium acetate, $Ba(CH_3COO)_2$; strontium acetate, $Sr(CH_3COO)_2$; and tetrabutyl titanate, $Ti(OC_4H_9)_4$, are raw materials. Solvents (acetic acid and alcohol), small amounts of acetylacetone ($CH_3COCH_2COCH_3$), and acetic acid are added. The sol is relatively stable and becomes a gel in a few days. A BST amorphous gel is prepared by the sol-gel process and calcined in the 2.45 GHz multimode microwave furnace to synthesize BST nanopowder at 900°C. The fine BST ceramics sintered by the microwave sintering technique display a lower dielectric loss than that of conventional samples, indicating a reduction in the influence of defects with the microwave process. The BST gel is dried at 100°C for 24 h and then annealed in the microwave furnace to obtain the BST nanopowders. The nanopowders are granulated and isostatically pressed. As a comparison, the conventional process for the preparation of the BST ceramics is as follows: the reagent grade $BaCO_3$, $SrCO_3$, and TiO_2 powders are raw materials, and these powders are mixed in the desired stoichiometry and milled for 10 h. The powders are calcined in a muffle furnace at 1100°C for 2 h (Yang et al., 2002).

7.4.1.3 Lead zirconate titanate

In a ceramic preparation of lead zirconate titanate (PZT) by sol-gel, there are many precursors available, but a typical process employs titanium isopropoxide, zirconium propoxide, and lead acetate. Other starting raw materials can be Ti isopropoxide, Zr *n*-butoxide, and Pb acetate, with 2-methoxiethanol as solvent (Chilibon et al., 2005). It is usually to make up stock solutions. The first stock solutions are of the stoichiometrically correct mix of Pb, Zr, and Ti precursors, and the second solution is the hydrolysis solution. The solution is aged overnight. Methoxyethanol is then added. Another PZT synthesis route by the sol-gel method is described previously (Suarez et al., 2009). The starting reagents for the sol-gel PZT solution were (i) lead (II) acetate trihydrate, $Pb(OAc)_2 \cdot 3H_2O$, 99.8% pure; (ii) glacial acetic acid (HOAc), 99.7% pure; (iii) zirconium (IV) propoxide, $Zr(OPr)_4$, 70 wt.% in 1-propanol; (iv) titanium (IV) propoxide, $Ti(OiPr)_4$, 97% pure; (v) acetylacetone (AcacOH), 99% pure; and (vi) 2-methoxyethanol (2-MOE), 100% pure. Lead acetate was dissolved in acetic acid (1:3 molar ratio) while stirring and refluxed at 115°C for 3 h, and a thick transparent solution (solution A) was obtained. Zirconium and titanium alkoxides were mixed with acetylacetone (1:3.06 molar ratio), and this mixture was stirred and refluxed at 90°C for 4 h to obtain solution B. Afterward, Solutions A and B were mixed together as appropriate amounts of solvent (2-MOE) were slowly added for controlling pH, viscosity, and liquid surface tension. After stirring for 24 h a light-yellow

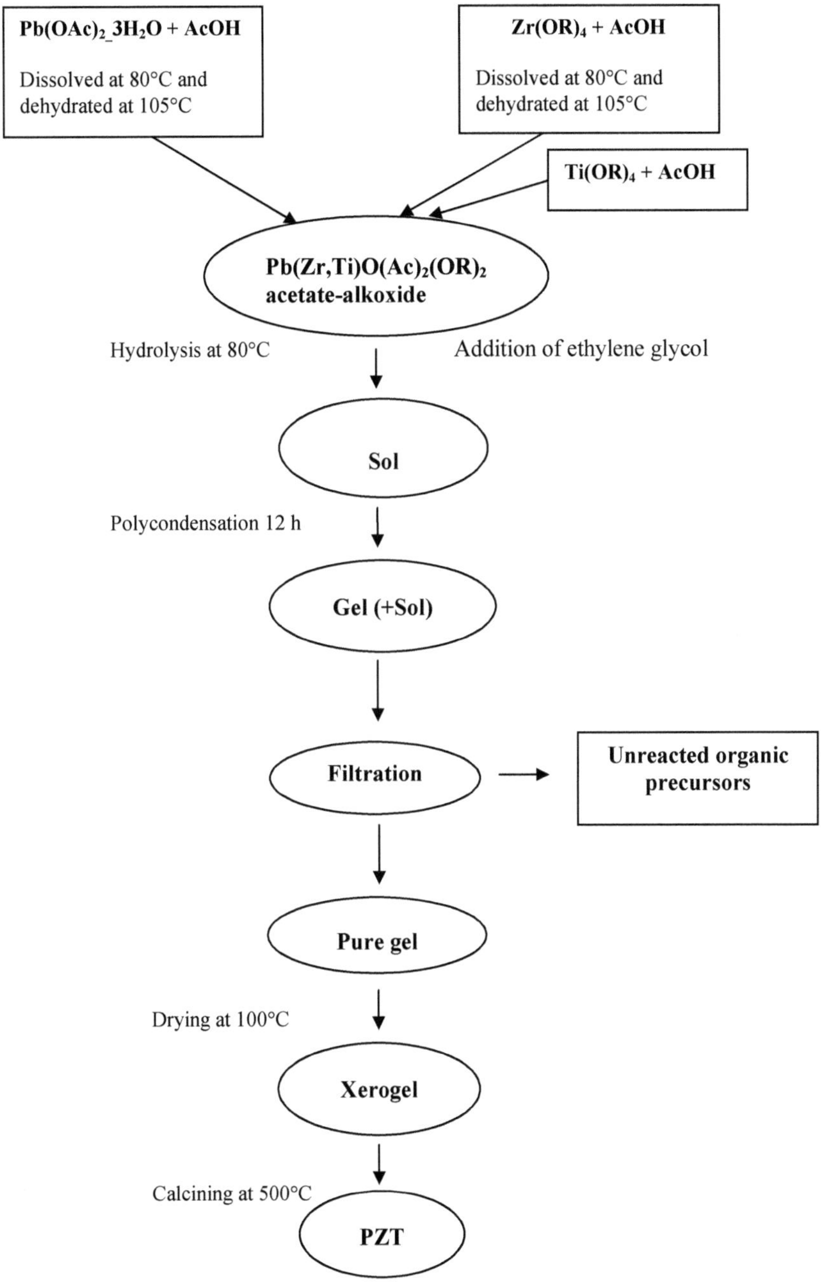

Figure 7.4 Synthesis of lead zirconium titanate by the sol-gel process.

solution was obtained at room temperature. Resulting solutions were heated at 100°C until dried and thermally treated to analyze the phase evolution from the amorphous PZT sol-gel network to the expected final perovskite structure. The other preparation route of PZT powders by sol-gel synthesis used as basic precursors lead acetate trihydrate, $Pb(OAc)_2 \cdot 3H_2O$, and zirconium acetylacetonate, $Zr(Ac\text{-}CHAc)_4$, (Brunckova et al., 2004). The solutions were dehydrated at 105°C for 2 h and after cooling to 80°C they were mixed with the acetate solution of titanium orthotetrabuthylate, $Ti(OC_4H_9)_4$, in the required ratio Pb:Zr:Ti = 1:0.53:0.47 (Fig. 7.4). A pure perovskite phase $Pb(Zr_{0.53}Ti_{0.47})O_3$ was formed by calcination at 500°C.

In other cases, PZT and Nb-doped PZT (PZTN) ceramics were fabricated from triol sol-gel derived powders (Sangsubun et al., 2009). The pellets were sintered at temperatures in the range of 1050°C–1200°C for 2 h. Coexistence of tetragonal and rhombohedral phases was clearly observed in PZT ceramics sintered at 1150°C, while PZTN ceramics showed coexistence of these phases at 1100°C. Densities of undoped PZT ceramics were rather low but could be improved by Nb addition. Grain growth inhibition was also observed in all PZTN samples. The dielectric constant was found to depend mainly on the density of the ceramics. An addition of Nb dopant to PZT improved dielectric constant values regardless of the sintering temperature.

7.4.1.4 PMZT (Mn-doped PZT)

In the case of PMZT (Mn-doped PZT) powder's achievement by sol-gel, precursor solutions of $Pb(OAc)_2 \cdot 3H_2O$, and $Mn(OAc)_2$ for PMZT, were dissolved in a methanol/methanol amine (CH_3OH/MEA) mixture with gentle heating of $Zr(OnPr)_4PrOH$ and $Ti(OnBu)_4$ and then mixed separately with CH_3CH_2OH. A stoichiometric amount of acetic acid was added to this mixed solution and continuously stirred at room temperature for 1 h. The Pb solution was added to the Zr/Ti solution, and the mixture was stirred for 2 h. The resulting yellow solution was filtered through a 0.2 um filter and ethylene glycol was added.

7.4.1.5 Lead magnesium niobium titanate

The preparation of the precursor solution for lead magnesium niobium titanate, $Pb(Mg_{1/3}Nb_{2/3})0.7Ti_{0.3}O_3$ (PMNT) was broadly analogous with the PZT process. The precursors are prepared by reacting magnesium ethoxide, $Mg(OC_2H_5)_2$, 98%; niobium ethoxide, $Nb(OC_2H_5)_5$, 99.95%; and titanium isopropoxide, $Ti([CH_3]_2CHO)_4$, with lead acetate trihydrate, $Pb(CH_3COO)_2 \cdot 3H_2O$, 99.9% in 2-methoxyethanol ($CH_3OCH_2CH_2OH$) as the solvent according to the formula $Pb(Mg_{1/3}Nb_{2/3})_{0.7}Ti_{0.3}O_3$. All the starting precursors are dissolved in 2-methoxyethanol separately, and the B-site precursors are mixed together and refluxed for 1 h. A lead acetate solution is added last and stirred for 1 h. An appropriate amount of ethyl acetoacetate ($CH_3COCH_2COOC_2H_5$) 99% is added dropwise to the solution as a chelating agent. The concentration of the final solution is adjusted to 0.8 M after partial hydrolysis by adding the appropriate amount of water: propanol H_2O (1:4) mixture. The whole process for the precursor solution preparation is performed in a nitrogen gas atmosphere. After that partially hydrolyzed stock solution is aged for 24 h. Also, the precursor sol solution for magnesium oxide thin-film fabrication can be prepared by the reaction of magnesium acetate tetrahydrate, $Mg(CH_3COO)_2 \cdot 4H_2O$, in ethanol solvent using nitric acid as a catalyst (Choi and Hwang, 2000).

7.4.1.6 Strontium titanate ceramics

Strontium acetate, $Sr(CH_3COO)_2$, and titanium (IV) isopropoxide, $Ti(OC_3H_7)_4$, were used as the precursors for the fabrication of the ferroelectric strontium titanate ($SrTiO_3$) ceramics by the sol-gel processing technique (Devi et al., 2004). The method of stock solution preparation for strontium titanate powder is dissolution of strontium acetate in ethanol and acetic acid solvents and mixing up of the resultant solution with titanium (IV) isopropoxide in equimolecular amounts. The fired powder was amorphous, but the annealing at 700°C for 1 h in air yielded cubic phase crystalline $SrTiO_3$ powder. Ceramic samples were prepared by pressing and sintering the crystalline powder.

7.4.1.7 Lithium tantalate and lithium niobate

Lithium niobate ($LiNbO_3$) and lithium tantalate ($LiTaO_3$) present large nonlinear optical coefficients (Lisinski et al., 2010) and, therefore, are suitable materials for the challenge to combine the potential of random systems with nonlinear optical properties. Porous lithium tantalate ($LiTaO_3$) and lithium niobate ($LiNbO_3$) xerogels were prepared using the sol-gel method (Nashimoto and Cima, 1991; Ono and Hirano, 2002; Takahashi et al., 2004; Lisinski et al., 2010). For instance, as starting materials for stoichiometric $LiTaO_3$ gels, Lisinski et al. (2010) used lithium ethoxide powder and liquid tantalum ethoxide. A complex alcoxide solution Li:Ta = 1:1 was prepared by ethanol (VWR) and acetic acid and stirred for 1 h in an argon glove box. After 24 h the liquid formed a wet gel at 50°C and then was dried for 2 weeks. The resulting xerogels were sintered at different temperatures for 48 h. The $LiNbO_3$ xerogels were prepared in the same manner as the $LiTaO_3$ xerogels, but lithium ethoxide and niobium ethoxide were used as precursors.

7.4.1.8 Nanosized alloy-metal oxide composite and ceramics

Nanosized alloy-metal oxide composite and ceramics were prepared using single-source alkoxide precursors (Lecerf et al., 2001). The molecule $Ni_2Sn_2(OBut)_8$ was used in a low-pressure chemical vapor deposition (CVD) process to grow composite thin films where the binary alloy Ni_3Sn_4 was dispersed in a SnO_2 matrix. The mixed-metal alkoxides $[BaZr(OH)(OPri)_5(PriOH)_3]_2$, $[BaTi(OH)(OPri)_5(PriOH)_3]_2$, and $Ba_2TiZr(OPri)_{12}$ were used as molecular precursors for size-controlled synthesis (sol-gel) of the ceramics $BaTiO_3$, $BaZrO_3$, and $BaZr_{0.5}Ti_{0.5}O_3$, respectively. The complete crystallization of the pervoskite structure was at a relatively lower temperature of 600°C for the binary oxides and at 400°C for the ternary oxide system.

7.5 Applications of Ferroelectric Nanoceramics

Ferroelectric nanoceramics provide a new platform for wide range of applications (Fig. 7.5). It includes solar energy conversion and memory devices, and it can be extended to healthcare materials. Due to remarkable properties of ferroelectric nanoceramics they can be used in fabrication of cheaper memory devices, dynamic random access memory (DRAM), and other materials. One of the most promising applications of sol-gel-processed ferroelectric nanoceramics is ferroelectric bypass capacitors for modern telecommunication engineering. Ferroelectric random access memories (FeRAMs) are one of the main applications of perovskite materials to replace magnetic core memory, magnetic bubble memory systems, and electrically erasable read-only memory for many applications (Chilibon et al., 2012). The other major applications of ferroelectric nanoceramics are microsensors, solar electricity, photocatalyst devices, and other semiconductor devices.

Figure 7.5 Schematic representation of different applications of ferroelectric nanoceramics.

7.6 Conclusions

Ferroelectric ceramics are one of the most promising ceramic materials with a wide range of applications. Hence, low-cost synthesis methods are crucial for mass production. Sol-gel-based synthesis provides a cost-effective synthesis process for most of the ferroelectric nanoceramic materials. In this chapter we have gone through different synthesis methods, with an emphasis on sol-gel synthesis and various sol-gel processes used for the synthesis of prominent ferroelectric nanoceramic materials. This will lead to further research on ferroelectric nanoceramic materials for varied applications.

References

Al-Hartomy OA, Ubaidullah M, Kumar D, Madani JH, Ahmad T. (2013). Dielectric properties of $Ba_{1-x}Sr_xZrO_3$ ($0 \leq x \leq 1$) nanoceramics developed by citrate precursor route. *J. Mater. Res.*, **28**(08): 1070–1077.

Ashby MF. (2005). *Materials Selection in Mechanical Design*, 3rd ed. Pergamon Press, Tarrytown.

Babu SN, Hsu JH, Chen YS, Lin JG. (2011). Magnetoelectric response in lead-free multiferroic $NiFe_2O_4$–$Na_{0.5}Bi_{0.5}TiO_3$ composites. *J. Appl. Phys.*, **109**: 07D904.

Babu SN, Min SG, Malkinski L. (2011). Electric field tunable magnetic properties of lead-free $Na_{0.5}Bi_{0.5}TiO_3/CoFe_2O_4$ multiferroic composites. *J. Appl. Phys.*, **109**: 07D911.

Bradley DC, Mahsotra RC, Gaur GP. (1978). *Metal Alkoxides*. Academic Press, New York.

Brinker CJ, Scherer GW. (1990). *Sol-Gel Science: The Physics and Chemistry of Sol-Gel Processing*. Academic Press, New York.

Brunckova H, Medvecky L, Brianc J, Saksl K. (2004). Influence of hydrolysis conditions of the acetate sol-gel process on the stoichiometry of PZT powders. *Ceram. Int.*, **30**: 453–460.

Carreon MA, Guliants VV, Yuan L, Hughett AR, Dozier A, Seisenbaeva GA, Kessler VG. (2006). Mesoporous nanocrystalline mixed metal oxides from heterometallic alkoxide precursors: cobalt–nickel oxide spinels for propane oxidation. *Eur. J. Inorg. Chem.*, **24**: 4983–4988.

Caruso RA, Antonietti M. (2001). Sol-gel nanocoating: an approach to the preparation of structured materials. *Chem. Mater.*, **13**(10): 3272–3282.

Cheng CC, Ho TT. (2010). Dielectric characteristics and orientation of piezoelectric ($PbZr_xTi_{1-x}O_3$, PZT) films fabricated by sol–gel spray and spin techniques. *J. Supercond. Novel Magn.*, **23**(6): 1215–1218.

Chilibon I, Marat-Mendes J, Igreja R, Dias CJ. (2005). Ultrasound role in sol–gel processing of $PbTiO_3$ ceramics. *J. Optoelectron. Adv. Mater.*, **7**(5): 2727–2735.

Choi H, Hwang S. (2000). Sol–gel derived magnesium oxide precursor for thin-film fabrication. *J. Mater. Res.*, **15**(4): 842–845.

Choudhary RNP, Thakur AK. (2009). Thermal, structural and complex impedance analysis of Mn^{4+} modified $BaTiO_3$ electroceramic. *J. Phys. Chem. Solids*, **70**: 1401–1407.

Clapsaddle BJ, Gash AE, Satcher JH Jr, Simpson RL. (2003). Silicon oxide in an iron(III) oxide matrix: the sol–gel synthesis and characterization of Fe–Si mixed oxide nanocomposites that contain iron oxide as the major phase. *J. Non-Cryst. Solids*, **331**: 190–201.

Clapsaddle BJ, Sprehn DW, Gash AE, Satcher JH Jr, Simpson RL. (2004). A versatile sol–gel synthesis route to metal–silicon mixed oxide nanocomposites that contain metal oxides as the major phase. *J. Non-Cryst. Solids*, **350**: 173–181.

Devi A, Sharma H, Sarma H. (2004). Sol–gel processed strontium titanate ceramics. *Ferroelectr. Lett.*, **31**(3–4): 73–78.

Dias CJ, Marat-Mendes RM, Inacio P, Marat-Mendes JN, Duas Gupta DK. (2004). Recent advances in ceramic-polymer film for detection of proteins. *IEEE Trans. Dielectr. Electr. Insul.*, **11**(1): 35–40.

Dong W, Zhu C. (2002). Use of ethylene oxide in the sol-gel synthesis of α-Fe_2O_3 nanoparticles from Fe (III) salts. *J. Mater. Chem.*, **12**(6): 1676–1683.

Fabrizioli P, Burgi T, Baiker A. (2002). Environmental catalysis on iron oxide–silica aerogels: selective oxidation of NH_3 and reduction of NO by NH_3. *J. Catal.*, **206**(1), 143–154.

Fendler JH. (2001). Chemical self-assembly for electronic applications. *Chem. Mater.*, **13**(10), 3196–3210.

Gash AE, Satcher Jr JH, Simpson RL. (2003). Strong akaganeite aerogel monoliths using epoxides: synthesis and characterization. *Chem. Mater.*, **15**(17), 3268–3275.

Gash AE, Tillotson TM, Satcher Jr JH, Hrubesh LW, Simpson RL. (2001). New sol-gel synthetic route to transition and main-group metal oxide aerogels using inorganic salt precursors. *J. Non-Cryst. Solids*, **285**(1), 22–28.

Gash AE, Tillotson TM, Satcher Jr JH, Poco JF, Hrubesh LW, Simpson RL. (2001). Use of epoxides in the sol-gel synthesis of porous iron (III) oxide monoliths from Fe (III) salts. *Chem. Mater.*, **13**(3), 999–1007.

Hubert-Pfalzgraf LG. (2003). Some trends in the design of homo- and heterometallic molecular precursors of high-tech oxides. *Inorg. Chem. Commun.*, **6**: 102–120.

Hubert-Pfalzgraf LG. (1987). Alkoxides as molecular precursors for oxide-based inorganic materials: opportunities for new materials. *New J. Chem.*, **11**(10), 663–675.

Hubert-Pfalzgraf LG. (1992). Metal alkoxides and β-diketonates as precursors for oxide and non-oxide thin films. *Appl. Organomet. Chem.*, **6**(8), 627–643.

Hubert-Pfalzgraf LG. (1995). Chapter 2. In *Processing of Ceramics.* Lee BI, Pope EJA, Dekker M (eds.). Marcel Dekker, New York.

Igreja R, Dias CJ, Marat-Mendes JN. (1999). Processing and characterization of sol–gel derived modified $PbTiO_3$ for ferroelectric composites. *Ferroelectrics*, **228**: 99–109.

Igreja R, Dias CJ, Marat-Mendes JN. (1999). The role of crystallization process of ferroelectric ceramic gels on the electroactive properties of sol–gel ceramic/polymer composites. *10th International Symposium on Electrets*, pp. 631–634.

Itoh H, Tabata T, Kokitsu M, Okazaki N, Imizu Y, Tada A. (1993). Preparation of SiO_2-Al_2O_3 gels from tetraethoxysilane and aluminum chloride. *J. Ceram. Soc. Jpn.*, **101**(1177), 1081–1083.

Kessler VG, Spijksma GI, Seisenbaeva GA, Hakansson S, Blank DHA, Bouwmeester HJM. (2006). New insight in the role of modifying ligands in the sol–gel processing of metal alkoxide precursors: a possibility to approach new classes of materials. *J. Sol-Gel Sci. Technol.*, **40**(2–3): 163–179.

Lecerf N, Mathur S, Shen H, Veith M, Hüfner S. (2001). Chemical vapour and sol–gel syntheses of nano-composites and -ceramics using metal-organic precursors. *Scr. Mater.*, **44**(8–9): 2157–2160.

Lee BI, Pope EJA. (1994). In *Chemical Processing of Ceramics.* Lee BI (ed.). Marcel Dekker, New York.

Leite ER, Carreno NLV, Longo E, Pontes FM, Barison A, Ferreira AG, Maniette YM, Varela JA. (2002). *Chem. Mater.*, **14**: 3722.

Lisinski S, Ratke L, Schaniel D, Jungk T, Soergel E, Boysen H, Woike T. (2010). Second-harmonic generation in nano-structured $LiTaO_3$- and $LiNbO_3$-xerogels with randomly oriented ferroelectric grains. *Opt. Mater.*, **32**(4): 504–509.

Liu C, Martin CR. (1991). Composite membranes from photochemical synthesis of ultrathin polymer films. *Nature*, **325**(6330): 50.

Livage J, Sanchez C, Henry M, Doeuff S. (1989). The chemistry of the sol-gel process. *Solid State Ionics*, **32–33**(Part 2): 633–638.

Lu YF, Ganguli R, Drewien CA, Anderson MT, Brinker CJ, Gong WL, Guo YX, Soyez H, Dunn B, Huang MH, Zink JI. (1997). *Nature*, **389**: 364.

Mansuy C, Mahiou R, Nedelec JM. (2003). A new sol-gel route to Lu_2SiO_5 (LSO) scintillator: powders and thin films. *Chem. Mater.*, **15**(17), 3242–3244.

Marat-Mendes J, Chilibon I, Igreja R, do Carmo M, Dias C, Inácio P. (2004). Influence of ultrasounds in the preparation and properties of $PbTiO_3$ ceramics by sol–gel processing, European micro and nano systems 2004. *EMN04 Conference, Advances & Applications for Micro & Nano Systems, ESIEE*, Noisy le Grand, Paris, France, pp. 149–153.

Maxim N, Overweg A, Kooyman PJ, van Wolput JHMC, Hanssen RWJM, van Santen RA, Abbenhuis HCL. (2002). Synthesis and characterization of microporous Fe–Si–O materials with tailored iron content from silsesquioxane precursors. *J. Phys. Chem. B*, **106**(9): 2203–2209.

Mudinepalli VR, Leng F, Reddy MP, Lin WC, Murty BS. (2016). Structural, dielectric and ferroelectric properties of lead-free $Na_{0.5}Bi_{0.5}TiO_3$ ceramics prepared by spark plasma sintering technique. *Indian J. Phys.*, **90**: 131–138.

Mudinepalli VR, Song S, Li J, Murty BS. (2013). A comparative study of structural and electrical properties of $Ba_{0.8}Pb_{0.2}TiO_3$ nanocrystalline ceramics prepared by microwave and spark plasma sintering. *Mater. Chem. Phys.*, **142**(2): 686–691.

Mudinepalli VR, Song SH, Murty BS. (2014). Enhanced magnetoelectric properties in lead-free $Ni_{0.83}Co_{0.15}Cu_{0.02}Fe_{1.9}O_{4-\delta}$–$Na_{0.5}Bi_{0.5}TiO_3$ composites by spark plasma sintering. *Scr. Mater.*, **82**: 9–12.

Mudinepalli VR, Song SH, Li JQ, Murty BS. (2015). Magnetoelectric properties of lead-free $Ni_{0.93}Co_{0.02}Mn_{0.05}Fe_{1.95}O_4$–$Na_{0.5}Bi_{0.5}TiO_3$ multiferroic composites synthesized by spark plasma sintering. *J. Magn. Magn. Mater.*, **386**: 44–49.

Nashimoto K, Cima MJ. (1991). Epitaxial $LiNbO_3$ thin-films prepared by a sol–gel process. *Mater. Lett.*, **10**(7–8): 348–354.

Nuraje N, Su K. (2013). Perovskite ferroelectric nanomaterials: review. *Nanoscale*, **5**: 8752–8780.

Ono S, Hirano S. (2002). Processing of highly oriented lithium tantalate films by chemical solution deposition. *J. Mater. Res.*, **17**(10): 2532–2539.

Praveena K, Varma KBR. (2014). Improved magneto-electric response in $Na_{0.5}Bi_{0.5}TiO_3$–$MnFe_2O_4$ composites. *J. Mater. Sci.: Mater. Electron.*, **25**(1): 111–116.

Sakka S. (2005). *Handbook of Sol–Gel Science and Technology: Processing, Characterization and Applications*, Vol. I Sol–Gel Processing. Kozuka K (vol. ed.).

Sakka S. (2005). *Handbook of Sol–Gel Science and Technology: Processing, Characterization and Applications*, Vol. III Application of Sol–Gel Technology. Sakka S (vol. ed.).

Sakka S. (2005). *Handbook of Sol–Gel Science and Technology: Processing, Characterization and Applications*, Vol. II Characterization and Properties of Sol–Gel Materials and Products. Almeida RM (vol. ed.).

Sanchez C, Livage J. (1990). Sol-gel chemistry from metal alkoxide precursors. *New J. Chem.*, **14**: 513.

Sangsubun C, Watcharapasorn A, Jiansirisomboon S. (2009). Sol–gel derived PZT and PZTN ceramics: a role of sintering temperature. *Ferroelectrics*, **382**: 1–152.

Schwartz RW, Voigt JA, Tuttle BA, Payne DA, Reichert TL, DaSalla RS. (1997). Comments of the effects of solution precursor characteristics and thermal processing conditions on the crystallization behavior of sol–gel derived lead titanate thin films. *J. Mater. Res.*, **12**(2): 444–456.

Schwartz RW. (1989). *Chemical Processing of $PbTiO_3$ by Coprecipitation and Sol–Gel Methods. The Role of Powders and Gel Characteristics on Crystallization Behavior.* Thesis, University of Illinois at Urbana–Champaign, USA.

Setter N, Damjanovic D, Eng L, Fox G, Gevorgian S, Hong S, Kohlstedt H, Kingon A, Park N, Stephenson G, Stolitchnov I, Tagantsev A, Taylor D, Yamada T, Streiffer S. (2006). Ferroelectric thin films: review of materials, properties and applications. *J. Appl. Phys.*, **100**(5): 051606-1-46 (a review paper).

Shukla A, Choudhary RNP. (2010). Ferroelectric phase-transition and conductivity analysis of La^{3+}/Mn^{4+} modified $PbTiO_3$ nanoceramics. *Physica B*, **405**: 2508–2515.

Song SH, Zhu QS, Weng LQ, Mudinepalli VR. (2015). A comparative study of dielectric, ferroelectric and magnetic properties of $BiFeO_3$ multiferroic ceramics synthesized by conventional and spark plasma sintering techniques. *J. Eur. Ceram. Soc.*, **35**(1): 131–138.

Suárez-Gómez A, Saniger-Blesa JM, Calderón-Piñar F. (2009). The effects of aging and concentration on some interesting sol–gel parameters: a feasibility study for PZT nanoparticles insertion on in-house prepared PAA matrices via electrophoresis. *J. Electroceram.*, **22**: 136–144.

Takahashi M, Yamauchi K, Yagi T, Nishiwaki A, Wakita K, Ohnishi N, Hotta K, Sahashi I. (2004). Preparation and characterization of high-quality stoichiometric $LiNbO_3$ thick films prepared by the sol-gel method. *Thin Solid Films*, **458**(1): 108–113.

Tillotson TM, Gash AE, Simpson RL, Hrubesh LW, Satcher JH Jr, Poco JF. (2001). Nanostructured energetic materials using sol–gel methodologies. *J. Non-Cryst. Solids*, **285**: 338–345.

Tillotson TM, Sunderland WE, Thomas IM, Hrubesh LW. (1994). Synthesis of lanthanide and lanthanide-silicate aerogels. *J. Sol–Gel Sci. Technol.*, **1**(3): 241–249.

Wang TT, Herbert JM, Glass AM. (eds.) (1988). *The Applications of Ferroelectric Polymers*. Chapman and Hall, New York.

Xiao, CJ, Zhang, WW, Chi, ZH, Li, FY, Feng, SM, Jin, CQ, Wang, XH, Li, LT, Chen, RZ. (2007). Ferroelectric $BaTiO_3$ nanoceramics prepared by a three-step high-pressure sintering method. *Physica Status Solidi* (a), **204**(3): 874–880.

Ya Turova N, Turevskaya EP, Kessler VG, Yanovskaya MA. (2002). *The Chemistry of Metal Alkoxides*. Kluwer, Dordrecht.

Yang W, Chang A, Yang B. (2002). Preparation of barium strontium titanate ceramic by sol-gel method and microwave sintering. *J. Mater. Synth. Process.*, **10**: 303–309.

Yanovskaya MI, Turevskaya EP, Kessler VG, Obvintseva IE, Ya Turova N. (1992). Application of metal alkoxides in the synthesis of oxides. *Integr. Ferroelectr. Int. J.*, **1**(2–4): 343–352.

Zhang Q, Whatmore RW. (2001). Sol–gel PZT and Mn-doped PZT thin films for pyroelectric applications. *J. Phys. D: Appl. Phys.*, **34**: 2296–2301.

Chapter 8

Chemistry behind the Performance of Ceramic Membranes and Their Future in Membrane Technology

Derrick S. Dlamini, Nomcebo P. Khumalo, Simphiwe Zwane, Ajay K. Mishra, and Bhekie B. Mamba

University of South Africa, Nanotechnology and Water Sustainability Research Unit, College of Science Engineering and Technology, Science Campus, Florida, Johannesburg, South Africa

dlamids@unisa.ac.za

Incorporation of nanoparticles within a thin-film layer to develop a nanoarchitectured membrane for separations is a promising research route to promote chemical, thermal, and mechanical stability of the thin layer. The pioneering work for the thin-film nanocomposite (TFN) membrane for RO applications was reported by Hoek and his coworkers in 2007. The authors introduced a novel concept in preparing a composite RO membrane by adding zeolite nanoparticles into the matrix of a thin PA selective layer. The TFN membrane was tested and compared to the control TFC membrane. The results showed that compared with the control TFC membrane, the TFN membranes could exhibit better water permeability without compromising NaCl rejection when both

Smart Ceramics: Preparation, Properties, and Applications
Edited by Ajay Kumar Mishra

ISBN 978-981-4774-30-7 (Hardcover), 978-1-315-16359-8 (eBook)
www.panstanford.com

types of membranes were tested under the same conditions. It is widely believed that the nano-/mesopores of nanoparticles could provide more water channels and/or high porosity but reject ions.

8.1 Introduction

The lack of safe drinking water is one of the major challenges faced by the world, especially in developing countries. Over 783 million people have no access to safe drinking water worldwide (United Nations, 2012). The availability of clean water is necessary for all essential designated uses (e.g., domestic, industrial, and agricultural use). Therefore, the quality of the water supply is a basic concern for human health (Soticha et al., 2014).

The formation of disinfection by-products (DBPs) during water treatment processes (such as chlorination) and contamination of water by agricultural and industrial activities have remained the predominant challenges of providing safe water supply to communities in developing countries (Ulucan et al., 2013). Agricultural activities release residuals that may degrade drinking water quality, especially in rural areas where groundwater and surface water are the primary sources of drinking water. The water from these sources is collected for use without any further treatment. However, the extent and magnitude of water quality degradation imposed by the agricultural sector are difficult to assess and control because of its nonpoint nature (Osteen et al., 2012).

DBPs occur in drinking water as a result of reactions between natural organic matter (NOM) and inorganic disinfectants or oxidants, such as chlorine, chloramine, and ozone (Kitis et al., 2004). Trihalomethanes (THMs), nitrosamines, and haloacids (HAAs) are the most carcinogenic and predominantly abundant DBPs formed by disinfecting with chlorine during the water treatment process. For this reason, the US Environmental Protection Agency (EPA) (2001) has established the maximum allowable contaminant level of 0.08 mg/L for total THMs and of 0.06 mg/L for HAAs. This is because the occurrence of contaminants such as DBPs in drinking water is a major concern due to their carcinogenic nature and association with reproductive failure (Graves et al., 2001).

Currently, chlorination is the most prevalent and cost-effective disinfection process used in small treatment systems, especially in developing countries (Richardson, 2005). However, this technology creates the problem of DBPs such as polychlorinated biphenyls (PCBs).

Another study aimed to create awareness of the presence and quantity of DBP levels in regularly used public tap water in South Africa was undertaken by Nothnagel et al. (2008). Preliminary results obtained during concentration analysis of four THMs (trichloromethane, dichlorobromomethane, dibromochloromethane, and tribromomethane) indicated that the concentration of THMs in tap water in South Africa was higher than the target range established by the US EPA (0–0.08 μg/mL). It was recommended that urgent attention be given to the regulation and monitoring of DBPs in drinking water in South Africa (Nothnagel et al., 2008).

In another study, Mhlongo et al. (2009) monitored the prevalence of nitrosamines, specifically *N*-nitrosodimethylamine (NDMA) in water supplies in South Africa. Water samples were collected from three different water treatment plants, namely Sedibeng (in Free State), Rand Water (in Gauteng), and Magaliesburg (in the northwest) water treatment plants. Solid-phase microextraction (SPME) and gas chromatography–mass spectrometry (GC-MS) were used for the quantification of NDMA concentrations in the water samples. Trace amounts of NDMA were detected at the Sedibeng water treatment plant and its entire distribution network. The study also focused on the removal of nitroamines using water-insoluble cyclodextrin (CD) polymers from the water samples. From the results, it was concluded that the application of these nanosponges can be useful for the mitigation of trace DBPs such as NDMA (Mhlongo et al., 2009).

Membrane filtration technologies have shown to be a promising alternative for consumption water production and water recycling. The concept of filtration is based on size exclusion effects, and therefore, contaminants that are larger than the membrane pore size can be effectively retained. In the last few decades, the interest in ceramic membranes has increased concurrently with the development of new processes and new fields of applications like water treatment. In water treatment, the ceramic membrane should turn contaminated feed water into clean water flow (permeate) and a concentrated waste stream (Belibi et al., 2015).

Several studies have shown that virus removal using ceramic filters made of raw materials (e.g., diatomaceous earth) (Michen et al., 2011) or chemically synthesized ceramics (e.g., zirconia) is possible (Meyn et al., 2012). In studies where pretreatment by coagulation/flocculation was performed, high virus removal was reported even for pore sizes larger than the viruses (Meyn et al., 2012; Meder et al., 2013). Other than size exclusion methods of separation, functionalized ceramic particles (e.g., amino-silanized) and ceramic components (e.g., MgO-doped colloidal zirconia) can be applied for virus adsorption (Hong and He, 2012).

8.2 Polymeric versus Ceramic Membranes

It is also possible to use ceramic materials as additives in preparation of polymeric membranes to improve the hydrophilicity of polysulfone (PSf) membranes. Ceramic membranes have been widely used in industries for application in high temperatures and aggressive chemicals because of their high chemical and thermal stability. These membranes are specifically insensitive to swelling in any solvent, and they are highly resistant to a range of cleaning agents in order to minimize fouling (Hong and He, 2012).

Polymeric membranes as well as ceramic membranes can be applied for the removal of dyes. Dyes are recalcitrant organic molecules that have a complex aromatic structure. This makes it difficult for them to be removed from wastewaters using other filtration techniques such as sand and slow filtration. They are usually biologically nondegradable, resistant to aerobic digestion, and stable to light heat and oxidizing agents. For the past three decades dyes have been removed from wastewaters by treatment and decolorization processes. Membrane separation and adsorption have been considered the best methods for dye removal. However, current membranes applied in dye removal are susceptible to fouling, which degrades the performance of the membrane, resulting in maintenance cost.

8.2.1 Polymeric Membranes

Polymeric membranes have unique properties that make them ideal materials for use in water treatment. These properties include high contamination removal capacity, flexibility of operation, and availability of polymeric membrane materials (Jhaveri and Murthy, 2016). However, one of the major problems faced by water treatment industries in applying membranes for wastewater and water treatment is the costs associated with polymer membrane application. These costs include operational costs and maintenance costs, such as cleaning of fouled membranes. Moreover, high costs are also incurred while setting up the membrane filtration plant.

Fouling decreases permeate flux, increasing the maintenance costs. Fouling is the accumulation of rejected macromolecules on the membrane surface or membrane pores (Aryanti et al., 2015; Jhaveri and Murthy, 2016). This changes the membrane surface properties, resulting in depreciation in membrane efficiency to remove target pollutants in addition to the reduction of permeate flux. Fouling, therefore, reduces membrane life span (Sun, 2007). Potential membrane foulants include NOM, trace organic compounds, and microorganisms (Sukhishvili and Granick, 2002; Kochkodan and Hilal, 2015; Jhaveri and Murthy, 2016).

The use of nanoparticles in modifying polymeric membranes in an attempt to reduce biofouling and enhance permeate flux has received much attention during the last few years (Ng et al., 2013; Bet-moushoul et al., 2015). Integration of nanomaterials into polymeric membranes for their application in drinking water and wastewater treatment processes is rapidly growing in an attempt to increase hydrophilicity and permeate flux, thereby overcoming the inherent problem of membrane fouling (Amin et al., 2014). Incorporating nanomaterials in membranes results in the formation of *enhanced membranes*, which are reactive rather than just being a physical barrier for the passage of solutes (Ranade and Bhandari, 2014). The hybrid low-fouling inorganic/organic composite polymeric membranes can be prepared by directly coating or depositing inorganic/organic particles onto membrane surfaces (Wang et al., 2013).

According to Ng et al. (2013), the incorporation of nanoparticles into polymeric matrices results in a nanocomposite membrane with a synergistic effect (Ng et al., 2013). This is because advanced nanocomposite membranes could be designed to meet specific water treatment applications by adjusting their structure and physicochemical properties, such as pore size, permeability, selectivity, hydrophilicity, conductivity, mechanical strength, and thermal stability and also the antiviral and antibacterial properties of polymeric membranes (Jeong et al., 2007; Li et al., 2008).

The synergistic effects resulting from incorporating nanoparticles in membranes include photocatalysis, adsorption, and disinfection (antimicrobial, antiviral, or antifungal) (Li et al., 2008). Adsorption is mainly applied for color and odor removal and removal of acids, organic metals, and refractory contaminants from the chemical industry (Ranade and Bhandari, 2014). The adsorption process is recognized as the most promising and effective fundamental approach in wastewater and water treatment due to its flexibility in terms of design and operation (Ranade and Bhandari, 2014).

As a result, multiple water treatment goals are achieved with the use of nanoengineered membranes. In addition, due to increased hydrophilicity of nanocomposite membranes, the fouling tendencies of the composite membranes are minimized. As an example, polymeric and ceramic membranes enhanced with TiO_2 were found to be fouling resistant and highly efficient in removing a number of organic contaminants and pathogenic microorganisms in the presence of UV-A irradiation (Li et al., 2008). Therefore, developing nanocomposite membranes with various functions (e.g., catalytic, photocatalytic, antiviral, and antimicrobial) can be pursued to increase membrane efficiency in water treatment and, therefore, prevent potential human health hazards and environmental impacts as a result of polluted water.

Defects are also an issue in nanoenabled polymeric membranes. Incorporation of nanoparticles within thin-film layers of membranes to develop nanoarchitectured membranes for separations is a promising research route to promote chemical, thermal, and mechanical stability of the thin layers. The pioneering work on the thin-film nanocomposite (TFN) membranes for reverse osmosis (RO) applications was reported by Jeong et al. (2007). Due to the significance of the subject, a lot of publications have followed since 2007. Jeong et al. (2007) introduced a novel concept in preparing a composite RO membrane by adding zeolite nanoparticles into the matrix of a thin polyamide (PA) selective layer.

8.2.2 Ceramic Membranes

Ceramic-based membranes occupy an important position in the water and wastewater treatment industry because of their high chemical, thermal, and mechanical stability, which makes them suitable for use in harsh conditions such as corrosive and high-temperature environments. However, because of the limitation of their pore sizes, directly employing ceramic membranes for the separation of emulsified oil/water mixtures would cause a severe fouling problem and greatly compromise fluxes (Hua et al., 2007; Zhou et al., 2010; Zhu et al., 2014).

According to Zhu et al. (2014), among the various ceramic filtration membranes, zirconia ceramic membranes exhibit a better performance than others for oily wastewater treatment. Zhou and coworkers (2010) successfully prepared a nanosized zirconia-particle-modified Al_2O_3 microfiltration (MF) membrane using $ZrCl_4$ as a precursor. It was reported that the modified membrane exhibited improved and stable flux for oil/water emulsion separation compared with the Al_2O_3 MF membrane without modification because of the high hydrophilicity of ZrO_2.

The general synthesis routes for ceramic filtration membranes involve a sol-gel process followed by sintering of the initial ceramic into ceramic microcrystals. The preparation processes usually cause the resulting membranes to exhibit relatively rough surfaces. It is well known that surface roughness is an important factor that affects membrane performance. The increase of surface roughness will correspondingly increase the surface area, which consequently causes a more serious fouling problem than smoother surfaces (Weir et al., 2001).

The development of low-cost ceramic membranes based on cheap natural materials such as clays and apatite has recently appeared as an efficient solution to treat wastewater (Weir et al., 2001; Palacio et al., 2009; Harabi et al., 2014). Efforts in this direction include the addition of natural clay to alumina to decrease the production costs of composite ceramic membranes by lowering the firing temperature required to achieve high mechanical resistance and porosity. The main role of the clay is to act as an internal binder after its phase transformation upon firing.

To reduce the costs of ceramic membrane fabrication, recent research works are focused on the use of cheaper raw materials such as natural raw clay (Palacio et al., 2009), dolomite, kaolin (Harabi et al., 2014), and waste materials such as fly ash (Tewari et al., 2010), to mention just a few. Regarding mesoporous and microporous ceramics, these materials have pore sizes that range from 50 nm to below 1 µm. This interval of pore size is commonly beyond the pore size range for which the powder processing routes are mostly suitable. Other approaches include the sol-gel process or the preparation of synthetic zeolites and pillared clays. To obtain the composite ceramic membrane the active layer has to be deposited on a symmetric macroporous substrate known as the support that provides mechanical strength for an overlying thin active layer. In the main process to make ceramic membranes, the first step is to obtain a good dispersion of small particles and the second is to deposit such dispersion on a support by a slip-casting method. The support provides mechanical strength to a membrane active layer to withstand the stress induced during water filtration. Simultaneously, the support enables the membrane to have a low resistance to the filtrate flow. The support has to be chemical stable and has heat shock stability and high porosity for reasons such as stability during steam cleaning in applications that require sterility. The high hardness of ceramics also gives the membranes better wear resistance.

According to Almandoz et al. (2004) the selective layer used in liquid permeation applications consists of a porous layer, usually symmetric, with a pore density of 109 pores/cm^2 approximately and a porosity (or void volume fraction) between 30% and 70%. The pore size is in a range of 50–1000 nm in the case of MF and 2–50 nm for ultrafiltration (UF) membranes (Almandoz et al., 2014).

In comparison to polymeric membranes, the production costs of ceramic membranes are three to five times higher, but this can be compensated by their longer lifetime of up to 10 years instead of 1 year for polymeric membranes (Lee et al., 2013). Additional disadvantages of ceramic membranes are their brittle behavior and higher material weight compared to polymeric membranes. To decrease the membrane cost, regional raw materials such as natural ball clay, bentonite, feldspar, quartz, alumina, and chemical additives can be used to prepare composite ceramic membranes.

The interest in low-cost ceramic membranes has recently increased since they combine high thermal and mechanical stability, long life, and good chemical stability with economy. The properties of the ceramic membranes are mainly determined by their composition, the sintering temperature, and the pore former content. The general compositions of low-cost ceramic membranes are very wide, depending on the nature of their raw materials. These materials include kaolin, sepiolite, apatite, perlite, and phosphate or a mixture of some of them, among many others (Loi-Brügger et al., 2006; Lorente-Ayza et al., 2015). The pore former is necessary to reach the optimum permeability level. One of the widely used pore formers in the ceramic membranes' compositions is starch. Several studies reveal that starch has been used in a proportion between 2 and 20 wt% (Ranade and Bhandari, 2014). Starch and other organic pore formers generate pores during their burning out during sintering. Lorente-Ayza et al. (2015) state that the addition of starch granules to a mixture of inorganic raw materials yields ceramic membranes of greater porosity, tailored pore size, and higher permeability (Lorente-Ayza et al., 2015).

Recently, even though the use of organic-material-based membranes is highly developed, ceramic membranes present a number of advantages (Fakhfakh et al., 2010). Compared to polymeric membranes, ceramic membranes show outstanding chemical, thermal, and mechanical stability. The importance of porous ceramic membranes is increasing in water purification because monomodal and narrow pore size distributions in the meso- and macroporous range can be tailored depending on applied processing parameters such as initial ceramic particle size, shaping technique, and drying and sintering conditions. Because of their high mechanical strength ceramic membranes can withstand high pressure loads, and therefore, they are able to bear high water permeate flux rates. Another advantage of ceramic membranes is that they can be easily cleaned by back flushing or thermal, acidic, or basic treatment without corroding the surface and affecting the pore morphology. Another advantage of ceramic membranes is that they do not show any swelling behavior during water filtration, maintaining their structural compactness. Swelling (plastification) of the membrane by the more soluble component (i.e., water) should result in an increase of (nonselective) diffusibility in the membrane. On the flipside, ceramic membrane fabrication still remains highly expensive from a technical and economic point of view due to the use of expensive powders such as alumina, zirconia, titania, and silica. These powders are an advantage in the environmental field where great volumes of wastewater are generally treated and hence membranes are required to show a great surface area.

Compared to their polymeric counterparts, ceramic membranes are physically and chemically more durable; hence they have been widely used in highly specialized applications under extreme operating conditions such as high temperature or with aggressive solvents, beyond the capacity of typical polymeric membranes (Fakhfakh et al., 2010). These specialized filtration applications can be found in many industries, such as food processing, beverage production, chemical manufacturing, and biotechnology. Due to a considerable decrease in manufacturing costs in recent years, ceramic membranes have been increasingly used for water purification. A large number of small-scale drinking water treatment processes using ceramic membranes have been deployed in Japan over the past decade (Freyberg, 2014). A bigger drinking water ceramic membrane filtration plant has been recently constructed by PWN Technologies in the Netherlands, with a reported capacity of 120,000 m^3/day (Freyberg,

2014). Further, ceramic membranes have the ability to sustain intermittent operations while maintaining a stable and reliable performance (Fujioka et al., 2014). These types of membranes (ceramic membranes) offer an ideal platform for strategic drinking water supply from compromised water sources for disaster relief and military operations whereby the removal of trace organic chemicals (TrOCs) of significant health risk is an important concern. However, to date, ceramic membrane applications for drinking water treatment have been primarily limited to MF and UF, but efforts are underway to introduce them in nanofiltration (NF). The challenge is the defects that allow larger molecules to pass through.

The rejection of neutral TrOCs by a ceramic and two polymeric (i.e., NF90 and NF270) NF membranes increased in the order of their molecular weight increase (Puhlfürß et al., 2001; Zeidler et al., 2014). It was reported in the study that the NF90 membrane showed the lowest molecular weight cutoff (MWCO) (191 Da), followed by the NF270 membrane (326 Da) and the ceramic NF membrane (345 Da). The ceramic membrane showed the highest MWCO possible because of the defects. However, it is remarkable that the rejection of the smallest compound—NDMA—by the ceramic NF membrane (20%) was higher than that by the NF90 membrane (8%) and the NF270 membrane (4%), and a similar observation was reported for other small molecules. These results points to the fact that while size exclusion is the dominating rejection mechanism for both ceramic and polymeric NF membranes, they respond contrarily to other factors that can also influence organic pollutants rejection. The discrepancy can be the doing of considerable adsorption of hydrophobic TrOCs to the ceramic and polymeric NF membranes (Zeidler et al., 2014).

In 2015 a work was done where the purpose was to develop new membranes made from the natural material kaolinite clay (Meyn et al., 2012). The study used elaborated flat discs that were intended to be used as MF membranes to manufacture porous ceramics that would be applied to the production of consumption water. The membrane was prepared by shaping and pressing of a mineral plastic paste containing natural clay and sawdust as the main components. Separation characteristic of the membrane was investigated by performing filtrations of particles. The reported results confirm that the rejection of particles strongly depends on the operating cycle, and a very high retention rate was reached after the first cycle (Meyn et al., 2012). The intrinsic rejections are found to decrease with additional cycles. The maximum retention rate of particles is found to be 97% after cycle 4 (Meyn et al., 2012). It was concluded in the study that due to the positive MF tests supported by the low-cost membrane, these discs should find applications for economical treatment of consumption water containing toxic and particles in suspension in poor and developing countries (Meyn et al., 2012).

Macroporous ceramics can be developed by ceramic compacts partially sintered that are made by conventional ceramic techniques. In this case the pore structure is controlled by the packing of particles and particle shape. The pore size is commonly controlled by the particle size. Ceramic membranes are convincing principally in terms of high mechanical, chemical, and thermal stability compared to polymeric membranes. These properties enable their use under harsh conditions, and they can tolerate the application of chemicals and steam for membrane and module/plant cleaning. Their use in a wide range of different harsh applications like wastewater treatment and clarification and sterilization of beverages are well known. In principle, ceramic membranes possess unrestricted resistance against organic solvents; hence they are of particular interest for organic solvent NF. Organic solvent NF was established as a result of the development of solvent stable polymeric membranes and offers a high potential for applications in pharmaceutical and chemical industries.

According to Zeidler et al. (2014) the first ceramic NF membrane with an MWCO in water below 1000 g/mol was developed by Puhlfürßetal, followed by a successful application in process-integrated cleaning of wastewater from textile finishing (Puhlfürßetal et al., 2000; Zeidler et al., 2014). A ceramic membrane containing amorphous titanium dioxide with a pore size of 0.9 nm was reported to have a water flux of 20×10^{-5} kg/m^2 h Pa and an MWCO of 450 g/mol measured with polyethylene glycols. The application of this membrane in organic solvents resulted in an MWCO above 10,000 g/mol combined with a toluene flux below 2×10^{-5} kg/m^2 h Pa. The low flux and the bad rejection of the membrane despite the low MWCO indicate that the transport of organic solvents occurs through larger defect pores of the membranes due to the hydrophilic behavior of the titania surface. Similar results were reported in studies done by Guizard et al. (2002), who prepared the following mixed oxides (silicon dioxide/titanium dioxide, aluminum oxide/zirconium dioxide, and silicon dioxide/zirconium dioxide) with pore sizes below 1 nm. The defects result from poor miscibility between the components of the membrane (Guizard et al., 2002).

8.3 Surface Modification of Ceramic Membranes

Membrane surface layer properties have a great impact on the material properties. These properties can be tailored by chemical or physical modification. Many industrial wastewater treatment processes depend on controlling the hydrophobicity of the properties. The most regularly used measure of hydrophobicity is water contact angle. It can be determined by placing a drop of water on the surface of a solid of interest and measuring the angle between the aqueous phase and the membrane surface. However, it is necessary that the solid surface be flat and smooth for reliable results. Properties of ceramic membranes can be determined by direct methods like contact angle measurement and indirect methods such as Van Oss-Chaudhury–Good method. The measurement of a contact angle is one of the methods most commonly used to determine polymer surface properties of materials coated with monolayers or multilayers.

The advantage of hydrophobization is that for a more hydrophobic surface, a lower value of mean roughness is observed. This observation agrees well with observations reported in literature where it was established that the membrane surface roughness decreased after hydrophobization of membranes (Yoshida and Cohen, 2003; Wei et al., 2012). The roughness and level of hydrophobicity have an important impact on the wetting behavior of the membrane. A more hydrophilic membrane wets faster and easier in water than a hydrophobic membrane. Ceramic membranes were efficiently modified with perfluoroalkyl silanes during surface grafting to obtain hydrophobic surfaces with the apparent contact angles higher than 126° (Kujawa et al., 2015). The study reported that the higher hydrophobicity of modified membranes was created at longer grafting.

In an attempt to address the issue of defects in ceramic membranes, different post-treatment methods have been attempted to create a hydrophobic membrane surface. The first method was the silanization of mesoporous zirconium dioxide membranes with an initial pore size of 3 nm, as reported by Dudziak et al. (2010). Membranes produced via this way had the MWCO in toluene measured with polystyrenes reduced down to 600 g/mol. Van Gestel et al. (2003) applied the silanization method to modify mesoporous aluminum oxide/titanium dioxide membranes (Van Gestel et al., 2003).

Commercial ceramic membranes are typically made from metal oxides like alumina, zirconia, and/or titania. These materials generally possess a hydrophilic character due to the presence of surface hydroxyl (OH) groups. Depending on the intended application, a hydrophobization process of membrane surface is required. For instance, the membrane has to be modified to enable the utilization of this type of membrane in pervaporative removal of volatile organic compounds from water. During chemical modification the hydroxyl groups present on the pristine ceramic membranes can react with reactive groups of grafting molecules, for example, ethoxy, methoxy, or chlorine grouping, forming a chemically bonded monolayer.

The authors used 1H,1H,2H,2H-perfluorooctyltriethoxysilane (C6) and 1H,1H,2H,2H-perfluorotetradecyltriethoxysilane (C12) as grafting molecules. The grafting procedure consisted of hydrophobization by dipping membrane samples in the grafting solution (C6 or C12) for a given period of time, among other steps. The evaluation of the grafting efficiency was done by determining the value of the water entry pressure with water (Van Gestel et al., 2003).

8.4 Future Prospects of Ceramic Membranes in Water Treatment

Another water purification technique that has gained interest among researchers for its application in wastewater and water treatment is membrane distillation (MD). MD is a recently emerging, thermally driven technique applied in seawater desalination, wastewater treatment, and separation techniques (Khayet, 2011; Guan et al., 2012; Camacho et al., 2013). MD enables the recovery of high-purity water from contaminated water driven by vapor pressure across a macroporous hydrophobic membrane.

MD has many advantages that can be appealing as a separating technique in wastewater or water treatment processes. These include the efficiency of the technique in producing pure water, a simple membrane structure, low costs associated with preventing membrane clogging, and no chemical requirement for the application of this technique (Aneesh et al., 2016). However, these advantages are accompanied by major drawbacks such as high costs for MD membrane fabrication and application and also sediment clogging, resulting in reduced power penetration (Aneesh et al., 2016). This is an opportunity for researchers to fabricate cost-effective materials that can be used for MD purposes and also modify the application process.

A number of studies have been dedicated to the fabrication of innovative MD membranes with improved performance (Salehi et al., 2016). However, despite the merits of MD technology, the fabrication of ceramic membranes for application in MD has not been extensively explored. Predominantly, the inorganic nature of ceramic membranes suits the potential applicability of these membranes in this field.

An inspection of the state-of-the-art polymeric materials used for the manufacturing of commercial membranes reveals that most of the membranes currently on the market are based on a relatively few polymers that had originally been developed for other engineering applications. This is a highlight to the fact that there is a lot that needs to be done to improve the performance and cost effectiveness of ceramic membrane systems in water treatment. Therefore, it should be possible to foresee membrane manufacturing using or adapting existing technologies or using novel technologies at a competitive cost. It is envisaged that the following general strategies will lead to a higher separation's performance by ceramic membranes:

- Nonporous membranes, composed of a selective transport and a stable matrix phase at an optimal volume ratio along with minimal tortuosity of the transport pathways, thus combining high selectivity and permeability with high stability

- Porous membranes, with a narrow pore size distribution, high porosity, and minimal tortuosity (ideally straight-aligned pores through the barrier)
- Selective interactions controlled by additional functionalities (based on charge, molecular recognition, or catalysis) combined with nonporous or porous membrane barriers
- Membrane surfaces (external, internal, or both) that are inert toward uncontrolled adsorption and adhesion processes to reduce fouling

8.5 Conclusions

Porous barriers like ceramic membranes could be theoretically used for very precise continuous permselective separations based on subtle differences in size, shape, and/or functional groups. It is vital to mention that both membrane permeability and selectivity can be controlled by concentration polarization (due to the enhancement of the concentration of rejected species on the membrane surface as a function of transmembrane flow) or membrane fouling (due to unwanted adsorption or deposition of matter on/in the separation layer of the membrane). Ceramic membranes have found a wide application in MF and UF processes in water treatment. Defect pores in ceramic membranes make them less attractive for NF membranes. To achieve good rejection of small organic contaminants adsorption processes have to dominate size exclusion. In this regard, hydrophobic interaction also influences the rejection of neutral trace organic pollutants by the ceramic membranes. Unfortunately, in a filtration process adsorption is problematic because it enhances membrane fouling. This is the major drawback for ceramic membranes in NF applications. Therefore, a lot of work is required to make these durable types of membranes more attractive and to make them economically viable for NF applications in water treatment.

Acknowledgments

This chapter is part of a project funded by SASOL Group Technology under the SASOL University Collaboration Program.

References

Almandoz MC, Marchese J, Prádanos P, Palacio L, Hernández A. (2004). Preparation and characterization of non-supported microfiltration membranes from aluminosilicates. *J. Membr. Sci.*, **241**: 95–103.

Amin MT, Alazba A, Manzoor U. (2014). A review of removal of pollutants from water/wastewater using different types of nanomaterials: review. *Adv. Mater. Sci. Eng.*, **2014**: 825910(1–24).

Aneesh V, Antony R, Paramasivan G, Selvaraju N. (2016). Distillation technology and need of simultaneous design and control: a review. *Chem. Eng. Process. Process Intensif.*, **104**: 219–242.

Aryanti PTP, Subagjo S, Ariono D, Wenten IG. (2015). Fouling and rejection characteristic of humic substances in polysulfone ultrafiltration membrane. *J. Membr. Sci.*, **1**: 41–45.

Belibi P, Nguemtchouin MMG, Rivallin M, NdiNsami J, Sieliechi J, Cerneaux S, Ngassoum MB, Cretin M. (2015). Microfiltration ceramic membranes from local Cameroonian clay applicable to water treatment. *Ceram. Int.*, **41**: 2752–2759.

Bet-moushoul E, Mansourpanah Y, Farhadi K, Tabatabaei M. (2015). TiO_2 nanocomposite based polymeric membranes: a review on performance improvement for various applications in chemical engineering processes. *Chem. Eng. J.*, **283**: 29–46.

Camacho LM, Dumée L, Zhang J, De Li J, Duke M, Gomez J, Gray S. (2013). Advances in membrane distillation for water desalination and purification applications. *Water (Switzerland)*, **5**(1): 94–196.

Dudziak G, Hoyer T, Nickel A, Puhlfuerss P, Voigt I. (2010). Ceramic nanofiltration membrane for use in organic solvents and method for the production thereof. EP Patent. 160 3663.

Fakhfakh S, Baklouti S. (2010). Elaboration and characterisation of low cost ceramic support membrane. *Adv. Appl. Ceram.*, **109**: 31–38.

Freyberg T. (2014). Large scale ceramic membrane drinking water plant demonstrated in the Netherlands. *WaterWorld*, Penn Well Corporation.

Fujioka T, Khan SJ, McDonald JA, Nghiem LD. (2014). Nanofiltration of trace organic chemicals: a comparison between ceramic and polymeric membranes. *Sep. Purif. Technol.*, **136**: 258–264.

Graves CG, Matanoski GM, Tardiff RG. (2001). Weight of evidence for an association between adverse reproductive and developmental effects and exposure to disinfection by-products: a critical review. *Regul. Toxicol. Pharmacol.*, **34**: 103–124.

Guan G, Wang R, Wicaksana F, Yang X, Fane AG. (2012). Analysis of membrane distillation crystallization system for high salinity brine treatment with zero discharge using Aspen flowsheet simulation. *Ind. Eng. Chem. Res.*, **51**(41): 13405–13413.

Guizard C, Ayral A, Julbe A. (2002). Potentiality of organic solvents filtration with ceramic membranes. A comparison with polymer membranes. *Desalination*, **147**: 275–280.

Harabi A, Zenikheri F, Boudaira B, Bouzerara F, Guechi A, Foughali L. (2014) A new and economic approach to fabricate resistant porous membrane supports using kaolin and $CaCO_3$. *J. Eur. Ceram. Soc.*, **34**: 1329–1340.

Hau FL, Tsang YF, Wang YJ, Chan SY, Chua H, Sin SN. (2007). Performance study of ceramic microfiltration membrane for oily wastewater treatment. *Chem. Eng. J.*, **128**: 169–175.

Hong J, He Y. (2012). Effects of nano sized zinc oxide on the performance of PVDF microfiltration membranes. *Desalination*, **302**: 71–79.

Jeong BH, Hoek EMV, Yan Y, Subramani A, Huang X, Hurwitz G, Ghosh AK, Jawor A. (2007). Interfacial polymerization of thin film nanocomposites: a new concept for reverse osmosis membranes. *J. Membr. Sci.*, **294**(1–2): 1–7.

Jhaveri JH, Murthy ZVP. (2016). A comprehensive review on anti-fouling nanocomposite membranes for pressure driven membrane separation processes. *Desalination*, **379**: 137–154.

Khayet M. (2011) Membranes and theoretical modelling of membrane distillation: a review. *Adv. Colloid Interface Sci.*, **164**(1–2): 56–88.

Kitis M, Karanfil T, Kilduff JE. (2004). The reactivity of dissolved organic matter for disinfection by-product formation. *J. Eng. Environ. Sci.*, **28**(3): 167–179.

Kochkodan V, Hilal N. (2015). A comprehensive review on surface modified polymer membranes for biofouling mitigation. *Desalination*, **356**: 187–207.

Kujawa J, Cerneaux S, Kujawski W. (2015). Highly hydrophobic ceramic membranes applied to the removal of volatile organic compounds in pervaporation. *Chem. Eng. J.*, **260**: 43–54.

Lee SJ, Dilaver M, Park PK, Kim JH. (2013). Comparative analysis of fouling characteristics of ceramic and polymeric microfiltration membranes using filtration models. *J. Membr. Sci.*, **432**: 97–105.

Li Q, Mahendra S, Lyon DY, Brunet L, Liga MV, Li D, Alvarez PJJ. (2008). Antimicrobial nanomaterials for water disinfection and microbial control: potential applications and implications. *Water Res.*, **42**(18): 4591–4602.

Loi-Brügger A, Panglisch S, Buchta P, Hattori K, Yonekawa H, Tomita Y, Gimbel R. (2006). Ceramic membranes for direct river water treatment applying coagulation and microfiltration. *Water Sci. Technol. Water Supply*, **6**: 89–98.

Lorente-Ayza MM, Sánchez E, Sanz V, Mestre S. (2015). Influence of starch content on the properties of low-cost microfiltration ceramic membranes. *Ceram. Int.*, **41**: 13064–13073.

Meder F, Wehling J, Fink A, Piel B, Li K, Frank K, Rosenauer A, Treccani L, Koeppen S, Dotzauer A, Rezwan K. (2013). The role of surface functionalization of colloidal alumina particles on their controlled interactions with viruses. *Biomaterials*, **34**(17): 4203–4213.

Meyn T, Leiknes TO, König A. (2012). MS2 removal from high NOM content surface water by coagulation–ceramic microfiltration for potable water production. *AIChE J.*, **58**: 2270–2281.

Mhlongo SH, Mamba BB, Krause RW. (2009). Monitoring the prevalence of nitrosamines in South African waters and their removal using cyclodextrin polyurethanes. *Phys. Chem. Earth*, **34**(13–16): 819–824.

Michen B, Meder F, Rust A, Fritsch J, Aneziris C, Graule T. (2011). Virus removal in ceramic depth filters based on diatomaceous earth. *Environ. Sci. Technol.*, **46**: 1170–1177.

Ng LY, Mohammad AW, Leo CP, Hilal N. (2013). Polymeric membranes incorporated with metal/metal oxide nanoparticles: a comprehensive review. *Desalination*, **308**: 15–33.

Nothnagel C, Kotte K, Pienaar JJ, Van Zyl PG, Beukes JP. (2008). Disinfection by-products and extractable organic compounds in South African tap water. *J. Transdisciplin. Res. South Afr.*, **4**(1): 175–180.

Osteen C, Gottlieb J, Vasavada U. (2012). *Agricultural Resources and Environmental Indicators*. United States Department of Agriculture, WA.

Palacio L, Bouzerdi Y, Ouammou M, Albizane A, Bennazha J, Hernández A, Calvo JI. (2009). Ceramic membranes from Moroccan natural clay and phosphate for industrial water treatment. *Desalination*, **245**: 501–507.

Puhlfürß P, Voigt A, Weber R, Morbé M. (2000). Microporous TiO_2 membranes with a cut off o500Da. *J. Membr. Sci.*, **174**: 123–133.

Ranade VV, Bhandari VM. (2014). *Industrial Wastewater Treatment, Recycling and Reuse: An Overview*. Elsevier.

Richardson SD. (2005). New disinfection by-product issues: emerging DBP's and alternative routes of exposure. *Global NEST J.*, **7**(1): 43–60.

Salehi SM, Di Profio G, Fontananova E, Nicoletta FP, Curcio E, De Filpo G. (2016). Membrane distillation by novel hydrogel composite membranes. *J. Membr. Sci.*, **504**: 220–229.

Soticha K, Jareeya Y, Sudjit K, Prapat P. (2014). Assessing water quality of rural water supply in Thailand. *J. Clean Energy Technol.*, **2**(3): 226–228.

Sukhishvili SA, Granick S. (2002). Layered erasable polymer multilayers formed by hydrogen-bonded sequential self-assembly. *Macromolecules*, **35**(1): 301–310.

Sun D. (2007). Removing contaminants in wastewater. *Environ. Eng.*, 2–3.

Tewari PK, Singh RK, Batra VS, Balakrishnan M. (2010). Membrane bioreactor (MBR) for wastewater treatment: filtration performance evaluation of low cost polymeric and ceramic membranes. *Sep. Purif. Technol.*, **71**: 200–204.

Ulucan K, Noberi C, Coskun T, Ustundag CB, Debik E, Kaya C. (2013). Disinfection by-products removal by nanoparticles sinetered in zeolite. *J. Clean Energy Technol.*, **1**(2): 120–123.

United Nations. (2012). *The Millennium Development Goals Report 2012*. Available online at: www.un.org/millenniumgoals/pdf/MDG%20Report%202012.pdf (accessed on March 3, 2016).

Van Gestel T, Van de Bruggen B, Buekenhoudt A, Dotremont C, Luyten J, Vandecasteele C, Maes G. (2003). Surface modification of γ-Al_2O_3/TiO_2 multilayer membranes for applications in non-polar organic solvents. *J. Membr. Sci.*, **224**: 3–10.

Wang H, Yuan X, Wu Y, Huang H, Peng X, Zeng G, Zhong H, Liang J, Ren MM. (2013). Graphene-based materials: fabrication, characterization and application for the decontamination of wastewater and wastegas and hydrogen storage/generation. *Adv. Colloid Interface Sci.*, **195**(196): 19–40.

Wei CC, Li K. (2009). Preparation and characterization of a robust and hydrophobic ceramic membrane via an improved surface grafting technique. *Ind. Eng. Chem. Res.*, **48**: 3446–3452.

Weir MR, Rutinduka E, Detellier C, Feng CY, Wang Q, Matsuura T, Le R, van Mao R. (2001). Fabrication, characterization and preliminary testing of all-inorganic ultra-filtration membranes composed entirely of a naturally occurring sepiolite clay mineral. *J. Membr. Sci.*, **182**: 41–50.

Yoshida W, Cohen Y. (2003). Topological AFM characterization of graft polymerized silica membranes. *J. Membr. Sci.*, **215**: 249–264.

Zeidler S, Puhlfürß P, Kätzel U, Voigt I. (2014). Preparation and characterization of new low MWCO ceramic nanofiltration membranes for organic solvents. *J. Membr. Sci.*, **470**: 421–430.

Zhou JE, Chang QB, Wang YQ, Wang JM, Meng GY. (2010). Separation of stable oil-water emulsion by the hydrophilic nano-sized ZrO_2 modified Al_2O_3 microfiltration membrane. *Sep. Purif. Technol.*, **75**: 243–248.

Zhu Y, Wang D, Jiang L, Jin J. (2001). Recent progress in developing advanced membranes for emulsified oil/water separation. *NPG Asia Mater.*, **6**(5): e101.

Chapter 9

Sol-Gel-Based Synthesis of Metal Oxide Nanoparticles for Air and Water Purification

Rohit Bhatia,[a,b] Seema Garg,[c] and Pankaj Attri[d]

[a]*Lady Shri Ram College for Women, University of Delhi, New Delhi 110024, India*
[b]*Division of Sustainable Technology, Rudraksh Prodhyogiki Sangathan, New Delhi 110092, India*
[c]*Department of Chemistry, University of Delhi, New Delhi 110007, India*
[d]*Plasma Bioscience Research Center/Department of Electrical and Biological Physics, Kwangwoon University, Seoul 139-701, Republic of Korea*

chem.pankaj@gmail.com

According to the World Health Organization (WHO), water and air pollution constitutes about 23%–30% of the global burden of disease in the present-day society. It estimates that nearly 7 million people die due to outdoor air pollution every year, whereas indoor air pollution (owing to households cooking over coal, wood, and biomass stoves) can be linked to approximately 4.3 million deaths in 2012. Similarly, WHO estimates that polluted water accounts for almost 1.5 million deaths per year, including 0.36 million deaths of children below 5 years of age in the developing countries. Hence, there is an urgent need for developing

Smart Ceramics: Preparation, Properties, and Applications
Edited by Ajay Kumar Mishra

ISBN 978-981-4774-30-7 (Hardcover), 978-1-351-67164-4 (eBook)
www.panstanford.com

techniques that could tone down the air and water pollution levels so as to improve the quality of life. In recent years, various metal oxide nanoparticles, attributed with unique physical and chemical properties, have emerged as a choice of technology to remove various types of pollutants from air and water. In this chapter, we will discuss the synthesis of various metal oxide nanoparticles using the sol-gel methodology and their application in the removal of various types of air and water pollutants.

9.1 Introduction

In the past few decades, rapid urbanization and industrialization, together with an exponential growth in the world's population, have adversely affected the environment, resulting in rapid degradation of the quality of air and water available for people all around the world. The consumption of polluted air and water leads to many noncommunicable health problems, such as coronary heart diseases, cancers, and chronic respiratory diseases. According to WHO, approximately 12.6 million deaths per annum occur due to the unhealthy environmental conditions (Prüss-Ustün et al., 2016), out of which approximately 8.2 million deaths could be linked to diseases caused by air pollution (including indoor and outdoor air pollution), whereas water pollution is estimated to account for almost 1.5 million deaths every year, including 0.36 million deaths of children below 5 years of age in the developing countries. As a consequence, the United Nation Organization (UNO) has declared access to clean water and air as one of the key development goals of the millennium. Over the years, a plethora of technologies for purifying air and water have been developed. Some of the most common and conventional technologies for cleaning air encompass high-efficiency particulate arrestance (HEPA)-based purifiers, electrostatic air purifiers, and ozone generator–based purifiers. Amongst the three, HEPA-based air purifiers are most efficient as they can remove almost 99.97% of the particulate particles, including dust, sand, allergens, pollens, and others (of size ≤ 0.03 microns); however, their wide-scale application is limited due to their high cost of purchase and maintenance. Electrostatic air purifiers, on the other hand, are cheap, but they have low removal efficiency, ranging only between 30%–35% of the pollutant particles. Ozone-based air purifiers are typically employed in the industrial step-up processes in order to remove airborne particles, chemicals, mold, viruses, bacteria, and odors. They are not recommended for use in homes as ozone generated by them can cause health problems like irritation of the respiratory tract and headaches.

The conventional water purification technology available nowadays involves a series of steps (i.e., coagulation, flocculation, sedimentation, filtration, chlorination, and ozonation) (Keast and Johnston, 2008) for the removal of both chemical and biological pollutants. Though the technology provides inspiring outcomes it suffers from a lot of disadvantages at the same time, thereby deteriorating the overall quality of the purified water. It has been observed that during the coagulation and flocculation steps, water loses many of its beneficial minerals along with pollutants like clay, silt, bacteria, and organic matter, which impedes the overall quality of water. Sedimentation and filtration steps, employed for the removal of floc and sludge generated during coagulation, require regular check on the algal growth and subsequent replacement of filters, thereby increasing the total cost of purification process (WHO, 2005). The chlorination and ozonation steps, used as microcides for the removal of microbes present in the water, (Nieuwenhuijsen, 2000) are known to produce disinfection by-products (DBPs) on contact with natural organic matter (Boorman, 1999; Hrudey and Charrois, 2012), thereby further posing a threat to the quality of purified water. These shortcomings necessitate the quest for a better technology that is cheap to manufacture and easy to maintain to provide clean air and water to people globally at an affordable cost.

In recent years, nanotechnology is unveiled as a better substitute for the conventional technology, providing an effective and low-cost solution for purifying both air and water. In general, nanomaterials offer burgeoning advantages over their bulk counterparts owing to their flexible size and dimensions, high surface-area-to-volume ratio, improved catalytic properties, high electrical conductivity, self-assembling aspects, and antimicrobial properties (Hillie and Hlophe, 2007; Brame et al., 2011; Khin et al., 2012). Among different types of nanomaterials, metal oxide nanoparticles find increased application in the area of water and air purification due to their unique physicochemical properties attributed to their small size, versatile surface chemistry, and availability of more corner or edge surface sites (Fernández-Garcia and Rodriguez, 2009).

Metal oxide nanoparticles can be synthesized using a number of methods, such as coprecipitation, hydrothermal processing, solvothermal methods, and sol-gel chemistry. Among these methods, the sol-gel method is usually preferred over the others as it can be carried out at room conditions with no further requirement of high-end laboratory equipment, which makes it much more feasible and cost effective. In addition, it ensures atomic-level mixing of reagents and hence helps in synthesizing even complex inorganic materials, that is, ternary and quaternary oxides. Furthermore, this method ensures better control over the morphology and size of nanoparticles, thus widening its applicability (Danks et al., 2016).

The sol-gel method can be of two types:

- Aqueous sol-gel chemistry
- Nonaqueous sol-gel chemistry

9.1.1 Aqueous Sol-Gel Chemistry

The aqueous sol-gel method can be defined in simple terms as the process of conversion of a metal precursor solution into an inorganic solid via inorganic polymerization reactions (hydrolysis and condensation) in an aqueous medium. The metal precursors are often inorganic metal salt or metal alkoxide solutions, with the latter being preferred over the former for their ease of reaction with water. In general, the sol-gel methodology consists of five steps, with the first step involving the preparation of a homogeneous metal precursor solution either by dissolving organic precursors in an organic solvent (which are miscible with water) or by dissolving inorganic salts in water. In the second step, the homogeneous precursor solution is converted into a sol by reacting it with a suitable reagent (generally water with or without any acid/base). The sol is then allowed to age in the next step, followed by shaping and finally thermal treatment or sintering in the last step.

In the process, molecular precursors are converted into a highly cross-linked inorganic polymer by a polymerization reaction, that is, hydrolysis and condensation reactions. The hydrolysis reaction leads to the formation of a sol (depicted as the dispersion of colloidal particles in a liquid), and the subsequent condensation reaction results in the formation of a gel, mostly depicted as an interconnected, rigid, and porous inorganic network enclosing a continuous liquid phase. The resulting gel can be dried in two ways: (i) by the removal of the solvent under hypercritical conditions, leading to the formation of aerogels, and (ii) by drying of the gel under ambient conditions, resulting in the formation of xerogel due to shrinkage of pores. One of the major advantages of the sol-gel process is its prospects to shape the material into any desired form (such as films, fibers, or monosized powders), which can be converted into ceramic material, as required, by thermal treatment (Bradley et al., 2001; Turova and Turevskaya, 2002).

9.1.2 Nonaqueous Sol-Gel Chemistry

In the nonaqueous sol-gel method, the conversion of the metal precursor into a metal oxide is carried out in an organic solvent. The list of potential precursors for the nonaqueous sol-gel method is much larger in comparison to those used in the aqueous sol-gel method. It primarily includes inorganic metal salts, metal alkoxides, metal acetates, and metal acetylacetonates. Besides, organometallic compounds are also frequently used as precursors for the synthesis of metal oxide particles. Due to the availability of such a wide range of metal precursors, metal oxide nanoparticles synthesized by nonaqueous processes have grown immensely in the recent years, ranging from simple metal oxides to more complex binary, ternary, multimetal, and doped systems (Inoue, 1991).

9.2 Synthesis of Metal Oxide Nanoparticles Using the Sol-Gel Method for Removing Pollutants from Water and Air

Water and air purification using metal oxide nanoparticles is currently an area of intense scientific research. The literature cites arrays of example depicting synthesis of metal oxide nanoparticle using the sol-gel approach and their application in purifying both air and water. Some of the most significant and result-oriented studies are discussed in this section. In a very recent study, Zaleska et al. (2015) synthesized a titania photocatalyst doped with lanthanides (Er^{3+}, Yb^{3+}, and Er^{3+}/Yb^{3+}) for the removal of phenol from water. They synthesized the photocatalysts via the sol-gel method using titanium isopropoxide as the metal precursor and freshly prepared $Er(NO_3)_3$ and $Yb(NO_3)_3$ (ranging from 0.25 to 10 mol%) as the dopant. The gel so obtained was dried and ground to obtain a powder that was finally subjected to thermal treatment so as to obtain the final doped TiO_2 nanoparticles. The authors observed that the synthesized TiO_2 nanomaterials after excitation by a 978 nm laser beam showed low intensive upconversion emission.

Among the various photocatalysts synthesized, Yb^{3+}-TiO_2 containing 1 mol% of ytterbium showed the highest photocatalytic activity under visible light. It was proposed that visible light sensitization could be attributed to the reduced electron–hole recombination (due to lowered bandgap), availability of higher adsorption sites, and decrease in the crystallite size with a corresponding increase in the surface area, resulting in enhanced photocatalytic activity of the composite (Zaleska et al., 2015).

In another study, absorption by a TiO_2-based photocatalyst in the visible light region due to a reduction of electron-hole recombination was achieved by synthesizing tungsten-oxide-doped TiO_2 nanomaterials using the sol-gel approach. The authors utilized this new photocatalyst to degrade methylene blue (MB) in an aqueous solution. The photocatalyst sol was prepared by first mixing $Ti(O\text{-}Bu)_4$ in a mixture of absolute ethanol, acetic acid, and distilled water. The reaction mixture was then allowed to age for 24 h. After that, an aqueous solution of ammonium tungstate was added dropwise to the sol under vigorous stirring until a WO_x-TiO_2 (x = 1.5%, 3%, 4.5%, and 10%) gel was formed. The gel so obtained was dried, ground, and sintered to obtain the final product. The authors claimed that the photocatalyst with 3% doping showed maximum efficiency in the removal of MB from the aqueous solution.

In another study based on titania nanomaterials, Yang et al. (2012) synthesized TiO_2/carbon nanomembranes employing the sol-gel methodology for its use in a electrocatalytic membrane reactor to purify oily wastewater. They used tetrabutyl titanate as a metal precursor, which was dissolved in ethanol and was stirred for 50 min, after which a mixture of water and acetic acid was added to it, followed by stirring for 24 h so as to obtain the sol. A carbon membrane, pretreated with HNO_3, was then dip-coated into the sol using a dip-coating machine. In the final steps, a TiO_2-coated membrane was sintered for about 2 h in order to obtain the final product. The authors reported that the synthesized membrane was effective in removing oily waste from water under a high oil concentration with low energy consumption. They also observed that the membrane not only functioned as a filter to remove pollutants but also worked as an anode to catalyze and oxidize organic pollutants present in the oily wastewater. They further added that the synthesized membrane has low permeate flux, which resulted in longer residence time, leading to complete degradation of pollutants via oxidation (Yang et al., 2012).

Synthesis of cobalt-doped titania nanoparticles using the sol-gel method has also been reported for the purpose of water and air purification. Hamadanian and group (2010) used titanium isopropoxide and cobalt nitrate as the metal precursors. They observed that synthesized Co/TiO_2 nanoparticles comprised mainly of the anatase phase and doping of cobalt induced the red-shift in TiO_2, which increased with the increasing concentration of the dopant (Hamadanian et al., 2010).

Synthesis of magnetic nanophotocatalysts for the degradation and subsequent removal of pollutants in water was also reported using the sol-gel methodology. Harraz et al. (2014) synthesized titania-silica (TiO_2-SiO_2)-coated cobalt ferrite ($CoFe_2O_4$) nanoparticles for the degradation of MB dye in an aqueous solution. They first synthesized cobalt ferrite nanoparticles, which served as the core on which silica and titania particles were coated to make a core-shell-type nanocomposite. Magnetic properties of the nanocomposite enhanced the reusability of the catalyst, showing recyclability up to six cycles. The study revealed that the photocatalytic degradation of the dye depends on parameters like catalyst loading, solution pH, and the initial dye concentration. The authors achieved the highest dye degradation efficiency of 98.3% using 0.165 g/L catalyst in a 300 mL solution at a dye concentration of 100 parts per million (ppm) in an alkaline pH of 9 during a time span of 40 min (Harraz et al., 2014).

Similarly, a magnetic TiO_2-graphene photocatalyst synthesized using the sol-gel methodology was assessed for the degradation of herbicides present in water. The catalyst was shown to exhibit strong light absorption in the visible region and had high adsorption capacity toward organic pollutants owing to its high surface area. The synthesis of the nanocomposite was accomplished in two steps, with the first step involving the precipitation of a magnetic-Ti core, followed by coating with graphene in the next step. In a typical procedure, first Fe_3O_4 nanoparticles were dispersed in alcohol using ultrasonication, followed by the addition of a mixture of ammonia water, deionized water, and tetraethylorthosilicate. The resulting mixture was allowed to stir for about 1 h, producing silica-encapsulated Fe_3O_4 nanoparticles (SiO_2@TiO_2). The particles so formed were separated magnetically and were added to an alcoholic solution of tetra-n-butyl titanate. After allowing the mixture to age for 5 h, the solid product thus obtained was separated magnetically and calcinated to obtain a magnetite@SiO_2@TiO_2 (MT@SiO_2@TiO_2) nanocomposite. In the next step, the surface of the MT@SiO_2@TiO_2 composite was modified using aminopropyltrimethoxysilane (APS), followed by its addition to the aqueous dispersion of graphene oxide. The reaction mixture was again stirred for another 5 h and then a stochiometric amount of hydrazine hydrate was added in order to reduce the exfoliated graphene oxide, thereby producing the final photocatalyst. Results revealed that the synthesized catalyst could efficiently remove the herbicide 2,4-D from water with a 33% higher efficiency in comparison to its commercial counterparts. Moreover, the authors claimed that the catalyst was highly stable and can be reused for eight cycles without losing much of its efficacy (Tang et al., 2013).

In another study, Li et al. (2013) reported the synthesis of a TiO_2/carbon nanotube (CNT) composite using the sol-gel methodology for the removal of methyl orange dye from water. They, at first, synthesized multiwalled carbon nanotubes (MWCNTs) on a thin graphite plate (by thermal decomposition of hydrogen gas) so as to form an immobilized 3D network of CNTs. TiO_2 was then precipitated onto the CNTs using the sol-gel methodology. For experimentation, butyl titanate in isopropyl alcohol was used as the metal precursor. The 3D mesh of CNT was added to the precursor solution, and the mixture was allowed to stir for 1 h, followed by the addition of a mixture of H_2O and HNO_3. The resulting sol was allowed to age for 24 h and then the solvent was removed to dry the gel. The dried gel was finally annealed to obtain the final catalyst. Transmission electron microscopy (TEM) studies indicated that TiO_2 nanoparticles coated on CNT walls formed a "coral"-linked nanocomposite network immobilized on a graphite substrate. Such morphology offered a high surface area, large pore sizes, and more interspaces, thereby resulting in an enhanced photocatalytic ability of the catalyst. The authors further revealed that the rate of electron-hole recombination was also low in the synthesized photocatalyst and it can completely remove methyl orange from water within 200 min of treatment (Li et al., 2013).

Jaiswal et al. (2016) recently discussed the synthesis of a Co-B-codoped TiO_2 photocatalyst for the degradation of organic water pollutants under visible light using the sol-gel methodology. The authors used titanium butoxide $[Ti(OC_4H_9)_4]$, boric acid (H_3BO_3), and cobalt chloride ($CoCl_2 \cdot 6H_2O$) as precursor salts for Ti, B, and Co, respectively. In a typical experimental procedure, titanium butoxide was first dispersed in ethanol and stirred for about 1 h, followed by the addition of a mixture of distilled water, ethanol, and nitric acid so as to obtain a homogeneous solution. Cobalt-doped TiO_2 and boron-doped TiO_2 were then precipitated by mixing a hydrated cobalt chloride solution and a boric acid solution with the precursor homogeneous solution, respectively. The resulting solution was then aged for 24 h under ambient conditions. The resulting gel was centrifuged and dried to remove the excess of solvent. In the last step, the dried sample was calcinated (at 400°C) so as to obtain the final catalyst. Morphological investigations of the photocatalyst revealed that at a low concentration of boron doping (1 wt%), B occupies the interstitial site and as the concentration of boron increases (2 wt% and 3 wt%), the dopant also occupies the substitutional O position. Catalytic efficiency studies revealed that the presence of B in both lattice locations of TiO_2 resulted in its enhanced photocatalytic activity. The authors attributed such activity enhancement to the effective charge separation obtained by the TiO_2-B_{int} shallow energy level and higher visible light absorption, achieved with TiO_2-$B_{int+sub}$, to the presence of two deep energy levels in the bandgap. Jaiswal et al. (2016) successfully utilized the boron-doped TiO_2 photocatalyst for the degradation of p-nitrophenol (p-NP) and rhodamine B (RhB) in aqueous solutions.

In a similar study, Bettinelli et al. (2007) reported the synthesis of a boron- and vanadium-doped TiO_2 photocatalyst using the modified sol-gel methodology for the purification of wastewater. A V-doped TiO_2 catalyst was prepared using titanium butoxide in ethanol as the precursor solution and vanadium alkoxide as the dopant, whereas a B-doped TiO_2 photocatalyst was prepared using boric acid as the dopant. The resulting gels were dried and annealed so as to obtain the final catalyst. Catalytic efficiency investigations of the synthesized catalyst revealed that 1% and 9% doping of V and B in TiO_2, respectively, led to a significant increase in the photocatalytic activity of the catalyst in comparison to the undoped catalyst (Bettinelli et al., 2007).

In a recent study, Ahmed et al. (2013) reported the synthesis of mesoporous SnO_2/TiO_2 nanoparticles using the sol-gel method, employing polymethylmethacrylate as the template. The catalyst was synthesized by mixing tin chloride and titanium chloride in ethanol, followed by the addition of polymethylmethacrylate (dissolved in acetone) solution to it. The reaction mixture was vigorously stirred for about 3 h, after which a dilute solution of ammonia was added to it till a turbid sol appeared. The sol mixture was then allowed to age for 48 h. The gel so formed was finally dried and calcined so as to obtain the final catalyst. The authors observed that the incorporation of tin led to a remarkable reduction in the particle size of TiO_2 (up to 6 nm) and increased the surface area (up to 161 m^2/g), thereby enhancing the photocatalytic efficiency of the composite. Catalytic activity investigations revealed that a sample with 10% SnO_2 could remove 92% of rhodamine dye within 3 h of treatment (Ahmed et al., 2013).

In another report, Ahmed et al. (2013) synthesized Fe_2O_3/TiO_2 nanoparticles using an almost similar sol-gel methodology employing an octadecylamine surfactant as the template. A morphological assessment of the photocatalyst revealed that doping of Fe_2O_3 led to noteworthy changes in the optical and physical properties of TiO_2. It was observed that TiO_2 doped with 7 wt% Fe_2O_3 exhibited a drastic decrease in its crystallite size (up to 7 nm) and a corresponding increase in the surface area (up to 182 m^2/g), leading to its enhanced photocatalytic activity (Ahmed et al., 2013).

Besides, Pal et al. (1999) used TiO_2/Fe_2O_3 binary mixed oxide nanoparticles, synthesized using the sol-gel methodology, for the removal of cresol from wastewater. They used titanium isopropoxide and ferric acetylacetonate as the organic precursors. The catalyst was synthesized by first suspending titanium isopropoxide in a mixture of isopropanol and HCl under continuous stirring for about 1 h, followed by the addition of a solution of ferric acetylacetonate over a period of 1 h. The resulting mixture was then allowed to age for 24 h under constant stirring. The gel so formed was dialyzed, dried, and sintered at different temperatures. The authors observed that sintering at high temperatures (900°C) produced a rutile phase and a pseudobrookite phase in the final product, whereas sintering at a medium temperature (500°C–700°C) produced a mixture of anatase and rutile phase and thus showed a pronounced effect on the photocatalytical activity of samples. It was also observed that a sample with the anatase and rutile phase exhibited better photocatalytic activity in comparison to the rutile and pseudobrookite phase and the same could be attributed to the small particle size and more iron oxide on the TiO_2 surface in the anatase and rutile phase containing the catalyst (Pal et al., 1999).

The synthesis of a mesoporous-mixed Fe_2O_3/TiO_2 photocatalyst using the sol-gel methodology was also reported by Murugesan et al. (2013). They used titanium isopropoxide and iron nitrate nanohydrate (at variable concentrations) as the metal precursors and pluronic P123 as the soft template for controlling the size and shape of the final nanoparticles. In a typical procedure, pluronic P123 was at first suspended in ethanol, to which the precursor solutions were added and the mixture was stirred for about 2 h. A solution of distilled water and ammonia was then added to the above mixture, followed by its stirring for another 24 h. The solid so formed was separated using centrifugation and dried and calcined at 450°C. Results revealed that incorporation of iron in TiO_2 led to a red shift in the bandgap excitation of TiO_2 such that the synthesized catalyst could also absorb in visible light region, enhancing its photocatalytic abilities. The observed order for the photocatalytic activity of the synthesized catalyst for the photocatalytic oxidation of 4-chlorophenol follows meso-30 wt% Fe_2O_3/TiO_2 > meso-10 wt% Fe_2O_3/TiO_2 > meso-50 wt% Fe_2O_3/TiO_2 > meso-70 Fe_2O_3/TiO_2 > meso-90 wt% Fe_2O_3/TiO_2 > meso-Fe_2O_3 > meso-TiO_2 (Murugesan et al., 2013).

Apart from the single and binary photocatalytic systems, thin films of TiO_2 synthesized using the sol-gel methodology have also been reported for the photocatalytic reduction of chromium present in water. Titanium tetraisopropoxide was used as the organic precursor. Kajitvichyanukula et al. (2005) prepared two types of sols, which were then dip-coated on the glass slide. The gel-coated slides were heated at temperatures ranging from 400–700°C. The authors observed that the anatase phase predominated in films and the concentration of the rutile phase depends upon the number of coating cycles. They also showed that films with a high anatase phase were more effective in reduction of chromium than catalysts containing the rutile phase (Kajitvichyanukula et al., 2005).

Similarly, thin films of TiO_2-ZrO_2 were reported on glass rings for the removal of volatile organic compounds (VOCs) using the sol-gel methodology. $Ti(OPr^i)_4$ and $Zr(OPr^n)_4$ were used as metal precursors. The precursors were suspended in an aqueous solution of nitric acid for 3 days, after which they were subjected to dialysis. After dialysis, the resulting sol with three layers was coated with "Raschig rings" and calcined (350°C) so as to obtain final thin films of catalyst. The authors observed that the incorporation of Zr leads to a distinguished increase in the surface area of the composite with respect to TiO_2 and also stabilized the anatase phase. The synthesized catalyst was used to successfully degrade acetone and methylcyclohexane present in air (Coronado et al., 2006).

Further, rapid synthesis of TiO_2 photocatalyst using a modified sol-gel method was also reported for the removal of formaldehyde. In this approach, synthesis of titania sols involved acid hydrolysis of a $TiCl_4$ precursor instead of titanium alkoxides. The synthesized sol was then modified with doping of Pt, SiO_2, ZrO_2, WO_3, and MoO_3. The modified catalysts were used for the degradation of formaldehyde at physiological temperature. Catalytic efficiency studies revealed that enhanced activity was observed for composites doped with silica or zirconia in comparison to pure TiO_2, while reduction in the activity was observed when Pt and MoO_3 were used as dopant, and no significant change was observed when WO_3 was used as dopant. The best activity was shown by the SiO_2-TiO_2 catalyst, which successfully degraded 94% of the formaldehyde present in water (Zhang et al., 2001). Using this modified approach, Harraz et al. (2013) also reported synthesis of a TiO_2-SiO_2 catalyst for the removal of cyanide and heavy metals ions—Cr(III), Co(II), and Pb(II)—from water. They claimed that cyanide and heavy metal ion removal depend largely upon the weight of the catalyst used: as they increased the weight of the catalyst they achieved a higher percentage of pollutant removal (Harraz et al., 2013).

In another report, polypropylene-supported TiO_2 was used for the degradation of formaldehyde. The authors coated a TiO_2 sol on the polypropylene films, which were then annealed at 40°C to obtain the final catalyst. It was observed that the TiO_2 sol-gel formed an approximately 1.0 μm thick film on polypropylene and predominantly had the anatase phase with a regular surface. The authors utilized these films to successfully remove formaldehyde to its limit of detection. Moreover, it was observed that the films could retain their photocatalytic ability for 360 h and can be reused (Canela et al., 2015).

TiO_2 doped with 2% WO_3, synthesized using the sol-gel methodology. has also been reported for the successful degradation of the pesticide malathion, typically found in wastewater. Tetrabutyl orthotitanate (TBT) and ammonium p-tungstate were used as metal precursors. The catalyst was prepared by first dispersing TBT in 2-butanol with continuous stirring for 2 h, followed by the addition of an aqueous solution of ammonium p-tungstate dropwise over a period of 1 h. The gel so formed was then dried for 48 h and calcined at 500°C for 4 h to obtain the final catalyst. The author claimed that the synthesized catalyst showed a crystalline anatase phase with a mean particle size of about 20 nm. A catalytic efficiency assessment showed that the synthesized catalyst exhibited a better photocatalytic ability owing to its high surface area and reduced charge recombination. It was observed that under visible light, the catalyst can degrade malathion in 2 h, while 63% of TOC abatement was achieved within 5 h (Hernández-Ramíreza, 2013).

The sol-gel methodology is also reported for the synthesis of a SiO_2-TiO_2 aerogel for the degradation of organic pollutants present in contaminated water. Kim et al. (2013) used sodium silicate and titanium oxychloride as low-cost metal precursors. They synthesized a series of catalysts by varying concentrations of the metal precursors. The nanocatalysts so formed were made hydrophobic by functionalizing their surface with trimethylchlorosilane. The authors concluded that the porosity and crystallinity of the aerogels depend largely upon the calcination temperature. They observed that the surface area and textural properties of the nanocatalyst increased up to a temperature 400°C, beyond which (at 500 to 1000°C) it started losing its hydrophobicity. Consequently, hydrophilic aerogels together with TiO_2 or SiO_2 crystals were produced at higher calcination temperatures. Catalytic activity studies suggested that the synthesized catalyst with 16 g of sodium silicate showed most effective photodegradation of MB (Kim et al., 2013).

In a recent example, cobalt-doped TiO_2 nanoparticles synthesized using a modified sol-gel methodology were also reported for the successful degradation of methyl orange present in contaminated water. Sökmen et al. (2016) employed surfactant Triton X100 or Tween 20 to control the final size of particles and also to enhance their photocatalytic ability. In addition, they modified the cobalt-doped TiO_2 catalyst with phthalocyanine (Pc) so as to enhance its electron conduction capability, and hence the photocatalytic ability of the composite. Morphological investigations of the Pc/Co-TiO_2 nanocomposite revealed that Pc did not affect the crystal phase of Co-TiO_2, instead increasing its surface area, especially when prepared with Tween 20 in comparison to Triton X100 as the template. UV-Vis studies of the composite showed that addition of Pc led to an extension of the catalyst's absorption band to the visible region. Catalytic activity studies of the composite indicated that it could efficiently remove nearly 80% of methyl orange within 150 min, which is way higher than the unfunctionalized Co-TiO_2 catalyst (Sökmen et al., 2016).

Similarly, the TiO_2-CeO_2 nanocomposite, synthesized using the sol-gel method, was also evaluated for the removal of methyl orange dye from water. Titanium isopropoxide and cerous nitrate were used as metal precursors whereas cetyl trimethyl ammonium bromide (CTAB) and ammonia were used as capping agent and precipitating agent, respectively. The nanocatalyst was synthesized by first adding a solution of cerous nitrate to an alcoholic solution of CTAB under continuous stirring for about 30 min. Ammonia solution was then added slowly to the reaction mixture till precipitation occurred. The precipitated solid was finally collected using a centrifuge and dried and calcined (at 500°C) for 2 h so as to obtain CeO_2 nanoparticles. The latter were then added (in stoichiometric amounts) to the alcoholic solution of CTAB and titanium isopropoxide under continuous stirring for about 1 h. Again, an ammonia solution was added slowly to the formed suspension in order to precipitate the final nanocomposite. The solid was calcined at 500°C for 2 h. Hankare et al. (2013) reported that the addition of CeO_2 nanoparticles to TiO_2 lowered the rate of electron-hole recombination and thus provided better photocatalytic activity. They also showed that synthesized nanocatalysts could perform effectively in both UV and visible light regions for the efficient removal of dye from water. Amongst the various composites synthesized, the one with TiO_2:CeO_2 in a 7:3 ratio showed the highest photocatalytic degradation of dye in UV (60 min) and real sunlight (90 min) (Hankare et al., 2013).

Synthesis of a nanoabsorbent for the removal of lead, copper, and arsenic using the sol-gel methodology has also been reported recently. Yürüm et al. (2013) synthesized anatase nanoabsorbent employing titanium tetraisopropoxide as the metal precursor. The adsorbant was synthesized by dissolving the precursor in 2-propanol, followed by the addition of a mixture of distilled water and 2-propanol (1:2) to it. The resulting mixture was aged for 2 h, after which the solvent was removed and the solid obtained was annealed at 400°C. The maximum adsorption capacity of the catalyst for Cu(II), Pb(II), and As(III) ions as investigated was found to be 31.25 mg/g, 23.74 mg/g, and 16.95 mg/g, respectively, which makes it an ideal candidate for the removal of heavy metal ion pollutants from water (Yürüm et al., 2013).

In another study, Tobaldi et al. (2013) reported the synthesis of tungsten-and-silver-codoped titanias using an aqueous sol-gel methodology for the removal of textile dyes from water. Titanium isopropoxide, tungstic acid, and silver nitrate were employed as the respective metal precursors. Morphological investigations of the nanocatalysts indicated that Ag- and/or

Ag/W-doped TiO_2 (26 and 30 nm, respectively) were larger in size in comparison to pure TiO_2 (9 nm). Moreover, the calcination temperature had a profound impact on the crystal structure of the nanocatalysts, thereby affecting their photocatalytic activity. The authors claimed that at 450°C rutile, anatase, and brookite phases with a substantial amount of amorphous phase were produced, whereas at 600°C, crystalline phases, that is, anatase and rutile, predominated. Catalytic studies revealed that the catalyst fired at 450°C showed better photocatalytic activity than the catalyst fired at 600°C (Tobaldi et al., 2013).

Silver-doped hollow spheres of TiO_2, as reported by Paria and Boxi (2015), were synthesized using the sol-gel and sacrificial core methodology in order to carry out the photodegradation of antibiotics nitrobenzene (NB) and metronidazole (MTZ), and textile dye MB. The nanocatalyst was prepared by first synthesizing AgBr nanoparticles in aqueous SDBS media, which served as the core for the precipitation of TiO_2. For the same, titanium butoxide was added slowly to the homogeneous suspension of AgBr nanoparticles under constant stirring for about 35 min. The AgBr/TiO_2 core-shell nanoparticles so formed were then isolated and treated with ammonia solution for 48 h so as to allow the dissolution of the AgBr core. The hollow particles were finally separated and calcined at 450°C for about 2 h to improve the crystallinity of the final nanocatalyst. A morphological analysis indicated that during the dissolution process, silver ions were well incorporated in the interstitial sites of TiO_2. It was also found that the synthesized Ag-TiO_2 nanoparticles comprised purely the anatase phase having a mean particle size of about 17.76 nm. The catalyst exhibited a higher surface area in comparison to the solid TiO_2 nanoparticles. A catalytic efficiency examination revealed that the maximum degradation of NB (nearly 95.5%) was obtained under visible light irradiation for 3.5 h. The degradation efficiency of metronidazole, on the other hand, was observed to be 96.55% and 94.77% under the irradiation of visible light for the initial MTZ concentration of 15 and 30 mg L^{-1} and a catalyst dose of 0.5 g L^{-1}. Moreover, the catalyst could be reused for six cycles without any major decrease in its efficiency (Paria and Boxi, 2015).

The photocatalytic degradation of phenol and benzoic acid using zinc oxide nanoparticles, synthesized using the sol-gel method, has also been reported. Benhebal et al. (2013) suspended zinc acetate in ethanol and refluxed the reaction mixture. Oxalic acid (dispersed in ethanol) was then added to this refluxing mixture and was stirred for 1 h. The gel so obtained was dried at 80°C to obtain xerogel, which was finally powdered and calcinated in order to obtain the ZnO catalyst. The authors studied the effect of various parameters such as pH of the solution, catalyst loading, and initial phenol and benzoic acid concentrations on the photodegradation ability of the catalyst. They reported that at pH = 2.5, the ZnO concentration = 1.5 g L^{-1} and the initial phenol concentration = 0.20 g L^{-1}. Sixty percent degradation of phenol was achieved within 120 min of treatment (Benhebal et al., 2013).

In a similar study, zinc oxide nanoparticles have also been evaluated for the removal of cadmium from contaminated water. The catalyst was synthesized by dissolving zinc chloride in distilled water followed by dropwise addition of ammonium hydroxide to it till precipitation occurred. The precipitate was collected, washed, dried, and calcined in a muffle furnace at 450°C for about 2 h so as to obtain the final catalyst. It was found that the synthesized particles were hexagonal in shape and agglomerated with an average particle size of 18–31 nm. Catalytic studies revealed that they could remove 92% of cadmium at an initial concentration of 200 mg/L (Srivastava, 2013).

Fe^{3+}-doped ZnO nanoparticles, synthesized using the sol-gel methodology, were also analyzed for the removal of 2-chlorophenol. Zinc acetate, oxalic acid, and iron nitrate were used as precursor materials. The synthesis was accomplished by first dissolving zinc acetate in alcohol and refluxing it for about 30 min, before a solution of oxalic acid and iron nitrate was slowly added to it till a gel appeared. The gel was then allowed to age for 1 h and subsequently dried in oven for 5 h. In the final step, the dried gel was calcined at 400°C for 2 h. Morphological investigations of the product showed that the catalyst encompassed wurtzite as the dominant phase with significant incorporation of Fe^{3+} ions into the crystal lattice and was hexagonal in shape. Moreover, the concentration of dopant had a significant impact on the particle size since as the concentration of dopant increased from 0.25 wt% to 1 wt%, the particle size of doped ZnO decreased in comparison to undoped ZnO. Furthermore, it was also observed that doping shifted the spectral absorption of ZnO to the visible light region by lowering its bandgap. Hence the overall enhancement in the photocatalytic activity of doped ZnO could be attributed to its small particle size, large surface area, and low bandgap (Ba-Abbad, 2013).

Like titania thin films, ZnO thin films prepared using the sol-gel methodology have also been assessed for the removal of textile dyes from water. Talebian et al. (2013) synthesized these films by first dissolving zinc acetate in monoethanolamine and then coating the solution onto microscopic glass slides (five layers to achieve the desired film thickness) using the dip-coating methodology. The coated slides were dried at 190°C and annealed at 550°C under ambient pressure for 1 h to obtain the final catalyst. A morphological assessment of the synthesized films elucidated that they exhibited a high specific area, purity, better crystal orientation, and modified bandgap energy (E_g) levels suitable for higher UV and visible photoresponsiveness. These characteristics made the thin films superior for removal of diazo dye and monoazo dye over ZnO particles (Talebian et al., 2013).

Similarly using a sol-gel dip-coating methodology, Cu-doped ZnO thin films were reported for the removal of MB dye from water. Zinc acetate and copper acetate were used as the metal precursors. Studies revealed that the grain size and film thickness decreased with the increasing concentration of dopant, with the sample containing 0.5 mol% of copper ions showing the highest photocatalytic activity. Moreover, the absorption data showed that the synthesized film could also absorb in the visible region (400–800 nm) (Amornpitoksuk et al., 2012).

In another study, bimetallic ZnO-TiO_2 nanotubes, synthesized using the sol-gel methodology, were reported by Chakma and Moholkar Vijayanand (2016) for the fast removal of dye from water. The group used an ultrasound-assisted sol-gel technique together with hydrothermal treatment for the synthesis of these tubes. In a typical procedure, titanium isopropoxide was at first suspended in ethanol under sonication, to which a solution of zinc acetate was added. It was again ultrasonicated for about 25 min. The resulting sol-gel was air-dried for about 24 h. The particles so obtained were then suspended in a mixture of NaOH and KOH and refluxed for about 48 h. The resulting powder was washed, dried, and calcined at 300°C for 3 h so as to obtain the final catalyst. Morphological studies of the synthesized nanotubes revealed that nanotubes were open ended and hollow in nature with diameters of approximately 10 nm. Further, they exhibited a high specific surface area and low bandgap energy (2.95 eV) in comparison to pure ZnO and TiO_2. Catalytic efficiency studies showed that the tubes achieved >97% removal of MB within 10 min of treatment (Chakma and Moholkar Vijayanand, 2016).

In a recent report, synthesis of codoped ZnO nanoparticles for the removal of dyes from wastewater using the sol-gel methodology was reported by Pineda et al. (2014). The catalyst was synthesized by suspending zinc nitrate and cobalt nitrate (at different concentrations) an in aqueous solution of polyvinyl alcohol at a maintained ratio of 1:6 for polymer to metal and 1:4 for Co and Zn ions, respectively. The resulting mixture was aged, extracted, and calcined under atmospheric air at 400°C so as to obtain the final catalyst. The authors reported that all the synthesized catalysts exhibited a hexagonal wurtzite structure with an average particle size varying from 25 to 50 nm. The synthesized nanocatalyst also showed variation in its bandgap, which decreased with an increase in the dopant concentrations. The synthesized catalyst effectively removed dye MB from water within 120 min of treatment (Pineda et al., 2014).

Cai et al. (2014) reported amino-functionalized mesostructured Al_2O_3, synthesized using the sol-gel methodology, for the removal of chromium and carbon dioxide from contaminated water. They used aluminum isopropoxide as the metal precursor, Pluronic P123 as the soft support, and tetraethylenepentamine (TEPA) as the functionalization agent. The catalyst was synthesized by first suspending P123 in ethanol at room temperature, followed by the addition of nitric acid (67 wt%) and aluminum isopropoxide to it. The mixture was then allowed to age for about 5 h before the solvent was evaporated to obtain a light-yellow PA composite. Further functionalization was carried out by adding 2 g of composite to the TEPA suspension (in ethanol) under continuous stirring for 24 h. The solid so formed was isolated and dried for 12 h at 100°C. Results indicated that the synthesized composite was highly efficient in purifying water as it could easily remove 92% of chromium within 50 min of treatment with an adsorption capacity of 59.5 mg g^{-1} and could also absorb

0.7 mmol g^{-1} of carbon dioxide. The authors attributed the high photocatalytic efficiency of the composite to its high surface area and presence of surface active amino groups (Cai et al., 2014).

Magnetite (Fe_3O_4) nanoparticles, synthesized using the sol-gel methodology, were utilized to remove dye congo red from wastewater. The synthesis of nanoparticles was accomplished by first dissolving ferric nitrate in ethylene glycol and heating it for about 2 h to obtain a brown-colored gel. The obtained gel was then aged for 1 h and dried to obtain xerogel. The dried xerogel were annealed to obtain the final catalyst. It was observed that the average particle size of synthesized catalyst was 61.24 nm and it could remove 90% of the dye within 4 h of treatment (Shaker et al., 2013).

Degradation of p-hydroxybenzoic acid (p-HBZ) using the catalytic wet hydrogen peroxide oxidation approach, employing Fe/TiO_2 and $Ru\text{-}Fe/TiO_2$ as catalysts, has been recently reported by Medina et al. (2015). They used the sol-gel methodology to synthesize the bimetallic catalyst. The catalyst (Fe/TiO_2) was synthesized by first dissolving titanium isopropoxide in ethanol, followed by the addition of ferric nitrate solution to it. The resulting mixture was stirred for half an hour, after which a mixture of ethyl acetoacetate and nitric acid was added to it. The gel so formed was dried under supercritical conditions so as to obtain an aerogel. The aerogel was then calcined at 500°C for 3 h so as to obtain the final catalyst. In case of the catalyst $Ru\text{-}Fe/TiO_2$, ruthenium nitrosyl nitrate was used together with ferric nitrate to impregnate titanium isopropoxide solution. A series of nanocatalysts were synthesized by varying the concentrations of Fe and Ru. It was observed that among the various catalysts synthesized, $0.5Ru\text{-}3Fe/TiO_2$ showed the best catalytical ability as it removed 71% of p-HBZ within 10 min of treatment (Medina et al., 2015).

To remove the boron present in water, Semiat and Zelmanov (2014) developed iron (Fe^{+3}) oxide/hydroxide-based nanoparticles (NanoFe) and impregnated them into a granular-activated carbon (GAC) adsorbent using the sol-gel methodology. The catalyst was synthesized by first preparing the iron sol through hydrolysis of ferric chloride. The resulting gel was air-dried or freeze-dried by vacuum and calcined to obtain iron nanoparticles. These nanoparticles were then used to impregnate activated carbon. Catalytic efficiency studies revealed that the average particle size of synthesized particles was about 50 nm, showing 95–98% boron recovery. The authors claimed that boron adsorption onto the NanoFe-impregnated GAC adsorbent followed pseudo-second-order reaction kinetics and the Langmuir isotherm model. Moreover, the recovered iron and Fe-impregnated activated particles can be easily reused without significant loss in their absorption activity (Semiat and Zelmanov, 2014).

Iron-doped TiO_2 for degradation of phenol was also reported. Akhlaghian and Sohrabi (2016) used titanium isopropoxide and iron nitrate as metal precursors. They reported that the as synthesized catalysts were not effective for the degradation of phenol; however, their activity was enhanced in the presence of oxidizing agents, that is, H_2O_2 (Akhlaghian and Sohrabi, 2016).

Nickel oxide (NiO) nanosheets, synthesized using the sol-gel methodology, were also reported for the removal of carbon monoxide from air. Naskar et al. (2016) used nickel nitrate as a metal precursor. They first dissolved nickel nitrate in ethanol, followed by the addition of bis(2-ethyl hexyl) sulfosuccinate sodium salt (AOT) solution to it. The resulting mixture was stirred for 1 h at room temperature before a solution of tartaric acid was added to it. The mixture was then stirred overnight to obtain a greenish product. The solid was isolated, dried, and calcined at 450°C for 2 h. Morphological investigations revealed that synthesized nanosheets were 80–100 nm thick, in which nanograins (10–20 nm) were assembled. Among the series of catalysts synthesized at the varying concentrations of AOT, the sample containing 1 mmol of AOT showed the best results. It exhibited a surface area of 88.23 m^2g^{-1} and showed better CO oxidation at a lower temperature in comparison to its other counterparts—T50 (50% conversion) and T100 (100% conversion) were at 245°C and 288°C, respectively (Naskar et al., 2016).

NiO-loaded $InTaO_4$ nanocatalyst, synthesized using the sol-gel methodology, was reported for the photoreduction of CO_2 in a recent study. Mul et al. (2010) at first prepared $InTaO_4$ sol by mixing $In(NO_3)_3 \cdot H_2O$ and $Ta(OC_4H_9)_5$ in equal (moles) amounts, followed by the addition of a nickel nitrate solution to it. The resulting mixed $NiO/InTaO_4$ sol was stirred for about 48 h. The gel so obtained was dried and calcined at 1100°C for 12 h.

Studies revealed that the synthesized catalyst successfully converted CO_2 into methanol under visible light and real sunlight with a methanol production rate of 11.1 μmol/g in a light intensity of 327mW/cm^2 at 25°C (Mul et al., 2010).

Nitrogen-doped bismuth vanadate was reported for the photodegradation of organic contaminants as a novel photocatalyst. The synthesis was accomplished by first dissolving bismuth nitrate in HNO_3, followed by the addition of citric acid to it so as to achieve a molar ratio of 2:1 (for citric acid and $Bi(NO_3)_3$, respectively). In another flask, a mixture of NH_4VO_3 (dissolved in distilled water) and citric acid in a molar ratio of 2:1 was separately formed. The two solutions were subsequently mixed in a ratio of 1:1 under continuous stirring for about 1 h, followed by the addition of hexamethylenetetramine. The resulting solution was stirred in the absence of sunlight till a dark-blue $BiVO_4$ sol-gel was obtained. The resulting gel was dried and calcined in air at different temperatures, ranging from 350–500°C, for about 5 h. It was observed that $BiVO_4$ achieved a monoclinic structure easily due to N-doping and it also resulted in a slight expansion of the crystal lattice. The authors attributed such behavior to the intertwining nature of N atoms into the $BiVO_4$ interstitial structure, which possibly replaced an O atom and formed a different O–Bi–N–V–O bond. It was also reported that nitrogen doping led to a red shift in $BiVO_4$ and thus slightly affected its photocatalytic activity (Wang et al., 2013).

9.3 Conclusion

The exposure to polluted air and water is responsible for many health-related problems, which lead to the deaths of a million people every year globally. In recent years a number of technologies were developed for purification of air and water, such as reverse osmosis, ultrafilters, and HEPA-based air purifiers. But due to their high cost of manufacturing and maintenance there is a need to develop new materials that are cheap, easy to synthesize, and environmentally friendly. In last two decades nanotechnology has emerged as a viable alternative to conventional technology for the synthesis of materials. Among the various nanoparticles reported, metal oxide nanoparticles find increasing application in the area of water and air purification due their physical and chemical properties, which are attributed to their small size and a high density of the corner or edge surface sites. The metal oxide nanoparticles can be synthesized using a number of methods, such as coprecipitation, hydrothermal processing, solvothermal methods, and sol-gel chemistry. The sol-gel method is preferred over others due to its low cost and the fact that it is easy to handle at room temperature. The metal oxide nanoparticles synthesized using a sol-gel methodology were reported to remove pollutants such as phenol, dyes, 4-chlorophenol, formaldehyde, CO, CO_2, and NOx successfully from water and air. Hence, using these particles low-cost and environment friendly air and water purifiers can be manufactured to provide clean water and air to the growing population.

References

Abdel-Messih MF, Ahmed MA, El-Sayed AS. (2013). Photocatalytic decolorization of rhodamine B dye using novel mesoporous SnO_2–TiO_2 nano mixed oxides prepared by sol–gel method. *J. Photochem. Photobiol. A*, **260**: 1–8.

Ahmed MA, El-Katori EE, Gharni HZ. (2013). Photocatalytic degradation of methylene blue dye using Fe_2O_3/TiO_2 nanoparticles prepared by sol–gel method. *J. Alloys Compd.*, **553**: 19–29.

Altýn Ý, Sökmen M, Býyýklýoðlu Z. (2016). Sol-gel synthesis of cobalt doped TiO_2 and its dye sensitization for efficient pollutant removal. *Mater. Sci. Semicond. Process.*, **45**: 36–44.

Ba-Abbad MM, Kadhum HAA, Mohamad AB, Takriff SM, Sopian K. (2013). Visible light photocatalytic activity of Fe^{3+}-doped ZnO nanoparticle prepared via sol–gel technique. *Chemosphere*, **91**: 1604–1611.

Benhebal H, Chaib M, Salmon T, Geens J, Leonard A, Lambert DS, Crine M, Heinrichs B. (2013). Photocatalytic degradation of phenol and benzoic acid using zinc oxide powders prepared by the sol–gel process. *Alexandria Eng. J.*, **52**: 517–523.

Bettinelli M, Dallacasa V, Falcomer D, Fornasiero P, Gomba V, Montini T, Romano L, Speghini A. (2007). Photocatalytic activity of TiO_2 doped with boron and vanadium. *J. Hazard. Mater.*, **146**: 529–534.

Boorman GA. (1999). Drinking water disinfection byproducts: review and approach to toxicity evaluation. *Environ. Health Persp.*, **107**(Suppl 1): 207–217.

Bose P, Ghosh S, Basak S, Naskar KM. (2016). A facile synthesis of mesoporous NiO nanosheets and their application in CO oxidation. *J. Asian Ceram. Soc.*, **4**: 1–5.

Boxi SS, Paria S. (2015). Visible light induced enhanced photocatalytic degradation of organic pollutants in aqueous media using Ag doped hollow TiO_2 nanospheres. *RSC Adv.*, **5**: 37657.

Bradley DC, Mehrotra RC, Rothwell IP, Singh A. (2001). *Alkoxo and Aryloxo Derivatives of Metals*. Academic Press, London.

Brame J, Li Q, Pedro JJ. (2011). Alvarez nanotechnology-enabled water treatment and reuse: emerging opportunities and challenges for developing countries. *Trends Food Sci. Technol.*, **22**: 618–624.

Cai W, Tan L, Yu J, Jaroniec M, Liu X, Cheng B, Verpoort F. (2014). Synthesis of amino-functionalized mesoporous alumina with enhanced affinity towards Cr(VI) and CO_2. *Chem. Engg. J.*, **239**: 207–215.

Chakma S, Moholkar Vijayanand S. (2016). Synthesis of Bi–metallic oxides nanotubes for fast removal of dye using adsorption and sonocatalysis process. *J. Ind. Eng. Chem.*, **37**: 84–89.

Curcio SM, Michel PO, Waldman RW, Sánchez B, Canela MC. (2015). TiO_2 sol-gel for formaldehyde photodegradation using polymeric support: photocatalysis efficiency versus material stability. *Environ. Sci. Pollut. Res.*, **22**(8): 800–809.

Danks AE, Hall SR, Schnepp Z. (2016). The evolution of "sol-gel" chemistry as a technique for materials synthesis. *Mater. Horiz.*, **3**: 91–102.

Dong Y, Tang D, Li C. (2013). Photocatalytic oxidation of methyl orange in water phase by immobilized TiO_2–carbon nanotube nanocomposite photocatalyst. *Appl. Surf. Sci.*, **296**: 1–7.

Fernández-García M, Rodriguez JA. (2009). Metal oxide nanoparticles. In *Encyclopedia of Inorganic Chemistry*. Wiley, USA.

Hamadanian M, Reisi-Vanani A, Majedi A. (2010). Sol-gel preparation and characterization of Co/TiO_2 nanoparticles: application to the degradation of methyl orange. *J. Iran. Chem. Soc.*, **7**: S52–S58.

Hammedi T, Triki M, Ksibi Z, Ghorbe A, Medina F. (2015). Catalytic wet hydrogen peroxide oxidation of p-hydroxybenzoic acid over Fe/TiO_2 and $^{0.5}Ru$–$^{3}Fe/TiO_2$. *J. Sol-Gel Sci. Technol.*, **76**(3): 679–685.

Harraz AF, Abdel-Salam OE, Mostafa AA, Mohamed RM, Hanafy M. (2013). Rapid synthesis of titania–silica nanoparticles photocatalyst by a modified sol–gel method for cyanide degradation and heavy metals removal. *J. Alloys Compd.*, **551**: 1–7.

Harraz FA, Mohamed RM, Rashad MM, Wang Y, Sigmund W. (2014). Magnetic nanocomposite based on titania–silica/cobalt ferrite for photocatalytic degradation of methylene blue dye. *Ceram. Int.*, **40**: 375–384.

Hernandez-Alonso MD, Tejedor-Tejedor I, Coronado JM, Soria J, Anderson MA. (2006). Sol–gel preparation of TiO_2–ZrO_2 thin films supported on glass rings: influence of phase composition on photocatalytic activity. *Thin Solid Films*, **502**: 125–131.

Hernández-Ramíreza A. (2013). Synthesis by sol–gel of WO_3/TiO_2 for solar photocatalytic degradation of malathion pesticide. *Catal. Today*, **209**: 35–40.

Hillie T, Hlophe M. (2007). Nanotechnology and the challenge of clean water. *Nat. Nanotechnol.*, **2**: 663–664.

Hrudey SE, Charrois JWA. (2012). Disinfection by-products (DBPs) as a public health issue. In *Disinfection By-Products and Human Health*. IWA, London, pp. 1–10.

Inoue M, Kominami H, Otsu H, Inui T. (1991). Synthesis of microcrystalline titania in organic media. *Nippon Kagaku Kaishi*, 1364–1366.

Jaiswal R, Patela N, Dashora A, Fernandes R, Yadav M, Edla R, Varma SR, Kothari DC, Ahuja BL, Miotello A. (2016). Efficient Co-B-codoped TiO_2 photocatalyst for degradation of organicwater pollutant under visible light. *Appl. Catal. B*, **183**: 242–253.

Jongnavakit P, Amornpitoksuk P, Suwanboon S, Ndiege N. (2012). Preparation and photocatalytic activity of Cu-doped ZnO thin films prepared by the sol–gel method. *Appl. Surf. Sci.*, **258**: 8192– 8198.

Kajitvichyanukula P, Ananpattarachaia J, Pongpom S. (2005). Sol–gel preparation and properties study of TiO_2 thin film for photocatalytic reduction of chromium(VI) in photocatalysis process. *Sci. Technol. Adv. Mater.*, **6**: 352–355.

Keast G, Johnston R. (2008). *UNICEF Handbook on Water Quality*. UNICEF, New York.

Khin MM, Nair AS, Babu VJ, Murugan R, Ramakrishna S. (2012). A review on nanomaterials for environmental remediation. *Energy Environ. Sci.*, **5**: 8075–8109.

Kim YN, Shao GN, Jeon SJ, Imran SM, Sarawade BP, Kim HT. (2013). Sol–gel synthesis of sodium silicate and titanium oxychloride based TiO_2–SiO_2 aerogels and their photocatalytic property under UV irradiation. *Chem. Engg. J.*, **231**: 502–511.

Li XZ, Li FB, Yang CL, Ge WK. (2001). Photocatalytic activity of WOx-TiO_2 under visible light irradiation. *J. Photochem. Photobiol. A*, **141**: 209–217.

Lima MK, Fernandes DM, Silva MF, Baesso ML, Neto AM, Morais GR, Nakamura CV, Caleare AO, Hechenleitner AAW, Pineda EAG. (2014). Co-doped ZnO nanoparticles synthesized by an adapted sol–gel method: effects on the structural, optical, photocatalytic and antibacterial properties. *J. Sol-Gel Sci. Technol.*, **72**(2): 301–309.

Nieuwenhuijsen MJ. (2000). Chlorination disinfection byproducts in water and their association with adverse reproductive outcomes: a review. *Occup. Environ. Med.*, **57**: 73–85.

Özlem Z, Atakl K, Yürüm Y. (2103). Synthesis and characterization of anatase nanoadsorbent and application in removal of lead, copper and arsenic from water. *Chem. Engg. J.*, **225**: 625–635.

Pal B, Sharon M, Nogami G. (1999). Preparation and characterization of TiO_2/Fe_2O_3 binary mixed oxides and its photocatalytic properties. *Mater. Chem. Phys.*, **59**: 254–261.

Palanisamy B, Babu CM, Sundaravel B, Anandan S, Murugesan V. (2013). Sol–gel synthesis of mesoporous mixed Fe_2O_3/TiO_2 photocatalyst: application for degradation of 4-chlorophenol. *J. Hazard. Mater.*, **252–253**: 233–242.

Prüss-Ustün A, Wolf J, Corvalán C, Bos R, Neira M. (2016). Preventing disease through healthy environments: a global assessment of the burden of disease from environmental risks. WHO, Geneva.

Ramos-Delgado NA, Hinojosa-Reyes L, Guzman-Mar IL, Gracia-Pinilla MA. (2013). Solar photocatalytic activity of TiO_2 modified with WO_3 on the degradation of an organophosphorus pesticide. *J. Hazard. Mater.*, **263**: 36–44.

Reszczyńskaa J, Grzyb T, Sobczak WJ, Lisowski W, Gazda M, Ohtani B, Zaleska A. (2015). Visible light activity of rare earth metal doped (Er^{3+}, Yb^{3+} or Er^{3+}/Yb^{3+}) titania photocatalysts. *Appl. Catal. B*, **165**: 40–49.

Shaker S, Zafarian S, Chakra S, Rao VK. (2013). Preparation and characterization of magnetite nanoparticles by sol-gel method for water treatment. *Int. J. Innovative Res. Sci. Eng. Technol.*, **2**(7): 2969–2973.

Sohrabi S, Akhlaghian F. (2016). Surface investigation and catalytic activity of iron-modified TiO_2. *J. Nanostruct. Chem.*, **6**(1): 93–102.

Srivastava V, Gusain D, Sharma CY. (2013). Synthesis, characterization and application of zinc oxide nanoparticles. *Ceram. Int.*, **39**, 9803–9808.

Talebian N, Nilforoushan RM, Najimeh M. (2013). Ultraviolet to visible-light range photocatalytic activity of ZnO films prepared using sol–gel method: the influence of solvent. *Thin Solid Films*, **527**: 50–58.

Tang Y, Zhang G, Liu C, Luo S, Xu X, Chen L, Wang B. (2013). Magnetic TiO_2-graphene composite as a high-performance and recyclable platform for efficient photocatalytic removal of herbicides from water. *J. Hazard. Mater.*, **252–253**: 115–122.

Tobaldi DM, Pullar RC, Gualtieri AF, Seabra MP, Labrincha JA. (2013). Sol–gel synthesis, characterisation and photocatalytic activity of pure, W-, Ag- and W/Ag co-doped TiO_2 nanopowders. *Chem. Eng. J.*, **214**: 364–375.

Turova NY, Turevskaya EP. (2002). *The Chemistry of Metal Alkoxides*. Kluwer Academic, Boston.

Wandre TM, Gaikwad PN, Tapase AS, Garadkar KM, Vanalakar SA, Lokhande PD, Sasikala R, Hankare PP. (2015). Sol–gel synthesized TiO_2–CeO_2 nanocomposite: an efficient photocatalyst for degradation of methyl orange under sunlight. *J. Mater. Sci. Mater. Electron.*, **27**: 825–833.

Wang M, Liu Q, Che Y, Zhang L, Zhang D. (2013). Characterization and photocatalytic properties of N-doped $BiVO_4$ synthesized via a sol–gel method. *J. Alloys Compd.*, **548**: 70–76.

Wang ZY, Chou HC, Wu CSJ, Tsai DP, Mul Gu. (2010). CO_2 photoreduction using $NiO/InTaO_4$ in optical-fiber reactor for renewable energy. *Appl. Catal. A*, **380**: 172–177.

WHO/UNICEF. (2005). *Water for Life: Making It Happen*. WHO and UNICEF, Geneva.

Yang Y, Wang H, Li J, He B, Wang T, Liao S. (2012). Novel functionalized nano-TiO_2 loading electrocatalytic membrane for oily wastewater treatment. *Environ. Sci. Technol.*, **46**: 6815–6821.

Zelmanov G, Semiat R. (2014). Boron removal from water and its recovery using iron (Fe^{+3}) oxide/hydroxide-based nanoparticles (nanoFe) and nanoFe-impregnated granular activated carbon as adsorbent. *Desalination*, **333**: 107–117.

Zhang Y, Xiong G, Yao N, Yang W, Fu X. (2001). Preparation of titania-based catalysts for formaldehyde photocatalytic oxidation from $TiCl_4$ by the sol–gel method. *Catal. Today*, **68**: 89–95.

Chapter 10

Ceramic Nanofibers and Their Applications

Sanjay R. Dhakate

CSIR-National Physical Laboratory, Dr. K. S. Krishnan Marg, Pusa, New Delhi 110012, India
Academy of Scientific and Innovative Research (AcSIR), NPL Campus, New Delhi 110012, India

dhakate@nplindia.org

Ceramic materials are normally hard and inert and therefore well known for their excellent mechanical and thermal properties, as well as excellent resistance against corrosion and chemical erosion. The nanostructured ceramic can replace all bulk ceramic materials with improved properties. A large number of approaches have been developed for processing ceramics into nanostructured materials. Among the different approaches, electrospinning is a simple and versatile technique for preparing the various types of nanofibers with controlled morphology and structure on an industrial scale and has the merits of simplicity, high efficiency, low cost, and high reproducibility. In this chapter different types

Smart Ceramics: Preparation, Properties, and Applications
Edited by Ajay Kumar Mishra

ISBN 978-981-4774-30-7 (Hardcover), 978-1-315-16359-8 (eBook)
www.panstanford.com

of ceramic nanofibers developed by the electrospinning technique are reported with their prospective applications in the various fields.

10.1 Introduction

Nanoscience and nanotechnology have attracted a lot of attention of the scientific community across the globe. Controlling the chemical composition and geometry, such as the morphology and particle size, is of great importance in the engineering of nanostructured materials for successful realization. Nanomaterials exhibit significantly improved properties because of their small size and minimal defects, which are almost close to the corresponding atoms and molecules. One-dimensional nanostructured materials such as nanofibers (NFs), nanowires (NWs), nanotubes (NTs), and nanorods (NRs) continue to be at the forefront of nanoscience and nanotechnology (Yaman, et al., 2011; Palacios, 2012) due to their unique physical and chemical characteristics. NTs belong to the class of nanomaterials with diameters <100 nm and aspect ratio >1000, while in the case of NFs the diameter varies up to a few microns.

Ceramic materials are normally hard and inert and therefore well known for their excellent mechanical and thermal properties, as well as excellent resistance against corrosion and chemical erosion. These attributes make ceramic materials widely used in many applications, including the fabrication of electronic and photonic devices, sensors, electrodes, catalyst supports, and drug delivery components. Nanostructured ceramic oxides are distinguished for their stability and green chemistry, unlike some polymers that are toxic and find diverse technical applications. Nanostructured ceramic materials could nearly replace all bulk ceramics in applications such as catalysis, fuel cells, solar cells, membranes, hydrogen storage batteries, structural applications requiring high mechanical strength, biology for tissue engineering, biomolecular machines, and biosensors due to their high value addition. Besides, nanostructured ceramic oxides are attractive materials that find potential uses ranging from simple everyday applications like paints and pigments to sophisticated ones such as bioimaging, nanoelectronics, optical, sensors, resonators, and opto- and magnetoelectronic devices due to the physical properties these materials possess on account of their electronic structure. The inability to economically synthesize nanoscale ceramic structures on a large scale and simultaneously achieve precise control of their size has restricted their real-time application (Ju et al., 2007). The potential of ceramic nanostructures for various applications has necessitated their production in a uniform way.

Over the past several decades, a large number of approaches have been developed for processing ceramics into nanostructured materials, with distinguished examples including those based on sol-gel chemistry (Fujita, 2007), pyrolysis (Gupta and Kumar, 2008), and solvothermal and hydrothermal treatment (Takamura et al., 1999). These methods do not provide directly continuous NFs buy yield in to only nanopowder. To get morphologies such as rods, wires, thin films, or fibers at the nanoscale, novel fabrication techniques need to be realized. These have to allow for the necessary chemical reactions that enable metal oxide formation while allowing for particular shapes or structures to form with precise size control. Among different methods, electrospinning is an efficient process that can fabricate NFs on an industrial scale and has the merits of simplicity, high efficiency, low cost, and high reproducibility. Electrospinning was first developed as a patent to produce continuous fibers in 1934 (Formalas, 1934). However, the process was largely forgotten until the 1990s. Electrospinning is currently the only technique that allows the fabrication of continuous fibers with diameters down to a few nanometers. Fibers with complex architectures, such as ribbons, porous fibers, core-shell fibers, or hollow fibers, can be produced by special electrospinning methods (McCann et al., 2005). It can also produce structures ranging from single fibers to ordered arrangements of fibers. This technique is not only limited to research institutes and university laboratories but also applied in industry. The scope of applications, in fields as diverse as optoelectronics, sensor technology, catalysis, filtration, and medicine, is very broad (Li and Xia, 2004b). However, electrospinning was earlier largely limited to the fabrication of NFs of organic polymer materials, including synthetic and natural polymers, polymer alloys, and polymers loaded with nanoparticles, or active agents, as it is relatively convenient to prepare a polymer solution or melt with appropriate rheological properties required for electrospinning (Li and Xia, 2004b). Metals and ceramics are usually not possible to spin by the electro-spinning technique directly (Dersch et al., 2007).

It is only possible to obtain ceramic NFs by directly electro-spinning from a sol solution that contains an inorganic precursor (metal alkoxides or metal salts) and a solvent. But the inappropriate rheological properties and the rapid hydrolysis rates of metal alkoxides or metal salts make it very hard to control the electrospinning process (Li, McCann, and Xia, 2006). To overcome this problem, one has to introduce a polymer into the solution as a matrix to adjust the rheological properties, as well as an additive, such as a catalyst, to control the hydrolysis rate of the precursor. Hence, a typical spinnable precursor solution contains an alkoxide or metal salt precursor, a polymer, an additive, and a relatively volatile solvent such as ethanol, water, isopropanol, chloroform, and dimethyl formaldehyde (DMF). Polyvinyl pyrolidone (PVP) is one of the most popular polymers employed as a matrix due to its high solubility in ethanol or water and its good compatibility with many metal alkoxides and salts.

In addition, other polymers, such as polyvinyl alcohol (PVA) (Yang et al., 2005), polyvinyl acetate (PVAc) (Viswanathamurthi et al., 2004), polyacrylonitrile (PAN) (Zhang et al., 2009), polymethylmethacrylate (PMMA) (Ban, 2009), and polyacrylic acid (PAA) (Chen, 2009), are widely used. Some additives, such as catalysts and salts, are usually added into the solution to stabilize the precursor and facilitate the electrospinning process. Although only a small amount is required, these additives play an important role in stabilizing the solution as well as the jet. A catalyst, such as acetic acid (Li and Xia, 2003), hydrochloric acid (Choi et al., 2003), or propionic acid (Viswanathamurthi et al., 2004), can be used to adjust both the hydrolysis and gelation rates, preventing the solution from blocking the spinneret, thus ensuring a continuous spinning. A salt, such as sodium chloride or tetramethylammonium chloride, can increase the charge density on the liquid jet and thus eliminate the formation of beads, which is a common problem in electrospinning (Li and Xia, 2004b).

10.2 Electrospinning

Electrospinning is a technique to synthesized polymeric NFs by applying a high electric field to a polymeric solution. It is a simple, versatile, and cost-effective technique to produced NFs with diameter ranges from a few microns to nanometers. It is found to be unique in producing continuous NFs several hundred micrometers in length, flexible, and with ease of fiber production. It is one of the most efficient techniques other than self-assembly, phase separation, drawing, template synthesis, and sol-gel synthesis to obtain continuous NFs. Figure 10.1 shows the schematic representation of electrospinning assembly. It consists of a syringe pump, a high-voltage DC source, a collector, etc. The spinning solution is kept in a vertical syringe with a stainless steel needle having a small orifice. The needle is electrically kept in contact with a positive high voltage. The collector is placed below the syringe and it is grounded.

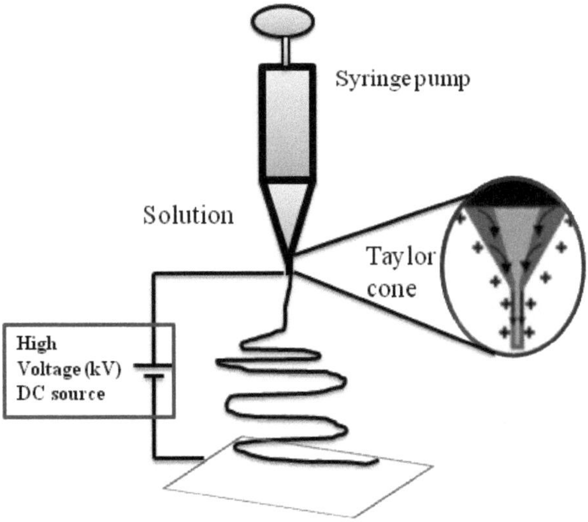

Figure 10.1 Schematic of an electrospinning apparatus.

The stationary collectors on this random NF are collected. In the electrospinning process, an electric force acts on the elements of charged fluid. The polymeric chains are polarized and consequently attracted by the electric field, creating the Taylor cone. At this point the pendant droplet of the polymer solution at the needle tip is deformed into a conical shape, typically referred to as the Taylor cone. This causes the ejection of a charged jet from a capillary tube, which is elongated and accelerated by an external electric field. The polymer solution forms a continuous NF when the electric forces overcome the surface tension of the fluid. If the voltage surpasses a critical value, the electrostatic force overcomes the surface tension and a fine charged jet is ejected in the form of NFs. The morphology of the fiber depends on the number of parameters, which may be divided into two categories, solution parameters (molecular weight, concentration, surface tension, viscosity, and conductivity of the polymer) and processing parameters (Dhakate, 2010, 2011). In contrast, the parameters that can be manipulated easily in experimental processing include the applied voltage, the flow rate, and the distance between syringe tip and collector. The resultant fiber diameter determines the mechanical and electrical properties of electrospun fiber mats. The mechanical and electrical properties of the membrane or mat of electrospun fibers are also diameter sensitive (Yao et al., 2014). The decrease in the fiber diameter of the fibers decreases the porosity but increases fiber density and mechanical strength. The electrospun fibers are classified according to morphology, as given in the following sections.

10.2.1 Monolithic Fibers

These fibers are synthesized by electrospinning a solution composed of only a single polymer without any additives. Monolithic fibers are simpler to manufacture as they can be electrospun by dissolving the polymer solution in an organic solvent, and some of the polymers can be dissolved in water. The chemical properties of drawn fibers are the same as those of the polymer. However, the fiber has a very high surface area and mechanical properties as compared to bulk polymer. Figure 10.2a shows an example of monolithic PAN NFs (Kaur et al., 2016).

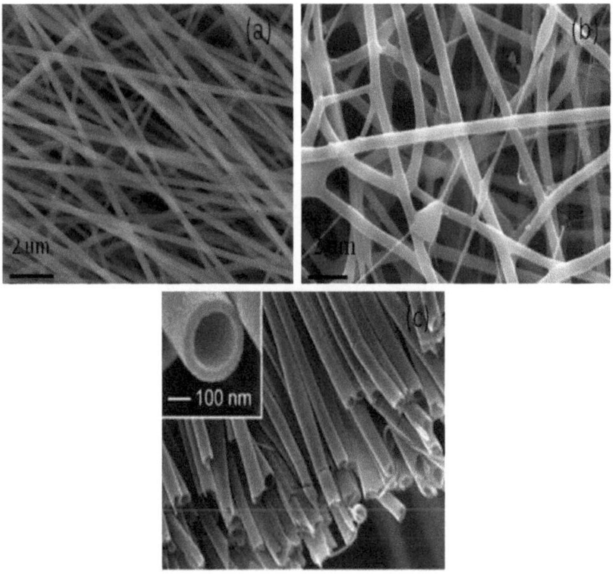

Figure 10.2 (a) Monolithic PAN nanofibers, (b) PVA-sodium alginate composites, and (c) core-shell TiO_2/PVP hollow nanofibers (Li et al., 2003; Kaur et al., 2016). (c) Reproduced from Sharma A, Gupta A, Rath G, Goyal A, Mathur RB. (2013). Electrospun composite nanofiber-based transmucosal patch for anti-diabetic drug delivery. *J. Mater. Chem. B*, **1**: 3410–3418, with permission from The Royal Society of Chemistry.

10.2.2 Composite or Blend Fibers

The blend or composite fibers are produced from a solution containing a blend of more than one polymers or a polymer with additives, nanoparticles, cells, drugs, biomolecules, ceramic particles, etc. The additives are homogeneously dispersed across the fiber's length, forming a composite structure. An essential criterion for obtaining homogeneous composite fibers is that both components should be miscible in a single solvent system, without any phase separation during the electrospinning process. Figure 10.2b shows an example of composite NFs of PVA and sodium alginate (Sharma et al., 2013).

10.2.3 Core-Shell Fibers

The electrospun fiber has a core-shell morphology in which core and shell are separately distinguishable when observed under an electron microscope. It is possible to isolate the core from the shell. This property is desirable for designing advanced nanostructures such as NWs, nanofluidic channels, and nano-optic fibers. The core-shell structure can act as an entrapment matrix, bioactive agents, drugs, live cells, cellular organelles, etc. Figure 10.2c shows core-shell hollow fibers produced by electrospinning of a TiO_2/PVP solution (Li and Xia, 2004a). The technique is not only capable of forming different types of NFs but also able to develop tubes (Xie et al., 2005), flowers (Kumar et al., 2012), and rods (Ostermann et al., 2006) by controlling the solution chemistry and feed rate strategy.

10.3 Electrospinning of Ceramic Nanofibers

The electrospinning technique has been recognized as a fabrication method for polymer and carbon NFs (Huang et al., 2003). Since 2002, electrospinning has been further explored as a high-efficiency method for the generation of 1D ceramic NFs (Li and Xia, 2003). Ceramic NFs of different sizes, compositions, and morphologies are fabricated by the combination of two conventional techniques, electrospinning and sol-gel (Li and Xia, 2004a). Generally, ceramic NFs are made by the electrospinning of ceramic precursors in the presence of polymers, followed by calcination at higher temperatures. To generate well-controlled and high-quality ceramic NFs by electrospinning, one typically has to use the following procedure: Figure 10.3 shows the schematic ceramic NF processing at different stages.

(1) Preparation of an electrospinning solution containing a polymer and a sol-gel precursor to the ceramic material.
(2) Electrospinning of the solution under appropriate conditions to produce precursor NFs containing inorganic precursor and polymer assistant materials.
(3) Heat treatment of the precursor NFs at a high temperature to remove polymers and obtain the ceramic phase.

So far more than a hundred ceramic NFs have been prepared using the electrospinning technique, and their probable applications are listed in Table 10.1.

Table 10.1

Ceramic	Precursor	Polymer/solvent	Application	References
Al_2O_3-B_2O_3	Aluminum acetate, boric acid	PVA,[a] water	Electronic devices	(Dai et al., 2002)
$NiFe_2O_4$	Iron_III_ethylhexano isopropoxide, nickel ethylhexano isopropoxide	Acetic acid, I isopropanol, PVP[b]	Storage devices	(Li, Herricks, and Xia, 2003)
Al_2O_3	Aluminum di-sec-butoxide	Ethanol, H_2O + HCl	Filter membranes	(Larsen et al., 003)
Co_3O_4	Cobalt acetate	H_2O, PVA	Biomarkers	(Guan et al., 2003)
Nb_2O_5	Niobium ethoxide	PVAc,[c] ethanol, acetic acid	Superconductors	(Viswanathamurthi et al., 2003)
Fe_3O_4	Iron _II_ chloride + Iron _III_chloride	Graft copolymer, PEO,[e] or PVA	Catalyst, carriers	(Wang A. et al., 2004)
$MgTiO_3$	Magnesium ethoxide, titanium isoproproxide 2-ethoxyethanol	PVAc in DMF[d]	Catalyst, dielectric	(Dharmaraj et al., 2004)
$NiTiO_3$	Nickel acetate, ethanol, HNO_3, titanium isopropoxide	PVAc in DMF	Tribological coatings, pigments	(Dharmaraj et al., 2004)
ZrO_2	Zirconium oxychloride	PVA, H_2O	Electroceramics	(Shao, Guan, Liu, Gong et al., 2004)

(Continued)

Table 10.1 (*Continued*)

Ceramic	Precursor	Polymer/solvent	Application	References
GeO_2	Germanium isopropoxide, isopropanol, H_2O, CH_3CH_2COOH	PVAc in acetone	Semiconductors	(Viswanathamurthi et al., 2004)
Mn_2O_3-Mn_3O_4	Manganese acetate	PVA, H_2O	Catalysts	(Shao, Guan, Liu, Li et al., 2004)
TiO_2	Titanium butoxide	P-123-EO-PO-EO copolymer	Photocatalysis solar cell	(Madhugiri et al., 2004)
NiO/ZnO	Nickel acetate, zinc acetate	PVA, H_2O	Smart materials, semiconductors	(Shao, Yang et al., 2004)
SnO_2	Dimethyldineodecanoate tin	PEO, $CHCl_3$	Sensors	(Wang Y. et al., 2004)
CeO_2	Cerium nitrate	PVA, H_2O	Biomedical and catalyst	(Yang et al., 2005)
TiO_2	Titanium isopropoxide	PVP	Solar cells	(McCann et al., 2005)
SiO_2	Tetraethylorthosilicate	Ethanol	Drug carrier, reinforcements	(Zhang et al., 2005)
$BaTiO_3$	Barium acetate, titanium isopropoxide, ethanol, acetic acid	PVP, ethanol, acetic acid	Piezo and microwave material	(Yuh et al., 2005)

Ceramic	Precursor	Polymer/solvent	Application	References
Erbia-modified TiO_2	Titanium isopropoxide	PVP, ethanol	Fuel cells	(Tomer et al., 2005)
Y_2O_3-stabilized ZrO_2	Zircony chloride, yttrium nitrate	PVP, ethanol	Automobile thermal barrier coatings	(Azad et al., 2005)
$LiCoO_2$	Lithium acetylacetonate, cobalt acetate	PVP, ethanol, acetic acid	Batteries, alternative energy sources	(Fu et al., 2005)
$LiMn_2O_4$	Lithium chloride, manganese acetate	PVA, H_2O	Storage devices	(Yu et al., 2005)
TiO_2	Titanium isopropoxide	PVP, ethanol		(Lee et al., 2005)
CuO	Copper acetate	PVAc, H_2O	Catalysis, semiconductors	(Wu et al., 2006)
ZnO	Zinc acetate	PVP, ethanol	Photocatalysts	(Bender et al., 2006)
$BaTiO_3$	Barium titanium ethylhexano-propoxide	PVP, acetyl acetone	Ferroelectric and piezoelectric sensors	(McCann et al., 2006)
Cr_2O_3	Chromium chloride	PVP	Pigments	(Hao et al., 2006)

(*Continued*)

Table 10.1 (*Continued*)

Ceramic	Precursor	Polymer/solvent	Application	References
WO_3	Tungsten powder, 30%	PVP, ethanol, H_2O_2	Field emission instruments	(Lu et al., 2006)
WO_3	Tungsten isopropoxide	PVAc, DMF		(Wang et al., 2006)
ZnO_2	Zinc acetate	Zinc acetate	Drug formulation-transport	(Wu et al., 2006)
$NaCo_2O_4$	Sodium acetate trihydrate, cobalt_III_acetate tetrahydrate	PAN[f] in DMF	Thermoelectric sensors	(Maensiri and Nuansing, 2006)
Y_2O_3-stabilized ZrO_2	Zirconyl chloride, yttrium nitrate	PVP, ethanol	Fuel cells	(Azad AM, 2006)
Ta_2O_5	Tantalum isopropoxide	PVAc, DMF, acetic acid, ethanol	Rectifiers, biomedical implants, dielectric	(Dharmaraj, 006)
Transparent Al_2O_3	Aluminum 2,4-pentanedionate	PVP, ethanol, acetone	Pyrotechnics, rocket propellant	(Azad A-Z, 2006)
$La_2Zr_2O_7$	Zirconium oxychloride, lanthanum nitrate	PVP	Fuel cells, catalysts	(Li, Dai et al., 2006)
TiO_2-pyrrole cable	Tetrabutyl titanate	PVP, pyrrole, $FeCl_3$	Sensor	(Lu et al., 2006)

Ceramic	Precursor	Polymer/solvent	Application	References
V_2O_5 nanorods TiO_2 nanofibers	Titanium isopropoxide, vanadium oxytriisopropoxide	PVP	Fuel cells, SOFC	(Ostermann et al., (2006)
TiO_2	Titanium propoxide	PVAc	Gas sensors	(Kim, Rothschild et al., 2006)
CeO_2-ZrO_2	Ceric nitrate, zirconyl oxychloride	PVP	Selective membranes	(Zhang, 2007)
$BaTiO_3$	Barium acetate, $[(CH_3)_2CHO]_4Ti$	PVP	Ferroelectric transducers	(Yuh et al., 2007)
Hydroxy apatite on TiO_2	Titanium isopropoxide	PVAc, simulated body fluid	Surgical implants	(Zhang, 2007)

[a]PVA: Polyvinyl alcohol
[b]PVP: Polyvinyl pyrrolidone
[c]PVAc: Polyvinyl acetate
[d]DMF: Dimethyl formaldehyde
[e]PEO: Polyethylene oxide
[f]PAN: Polyacrylonitrile

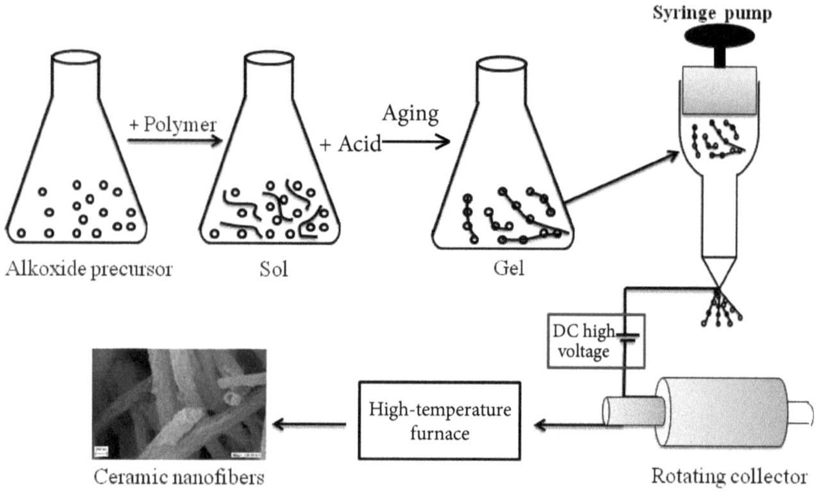

Figure 10.3 Schematic of processing of ceramic NFs at different stages.

10.3.1 Ceramic Nanofibers

Titania is one of the important materials used in a number of applications related to environmental cleaning and protection, photocatalysis, gas sensing, and fabrication of solar cells and batteries (Hoffmann et al., 1995). Titanium oxide (TiO_2) is an important semiconductor material for use in a wide range of applications, including photocatalysis, environmental pollution control, and solar energy conversion (Wang et al., 2007). However, there are some drawbacks of conventional powder catalysts, that is, low efficiency of light use and difficulty of separation after reaction (Zhang et al., 2009). These disadvantages of TiO_2 powders result in low efficiency of photocatalytic activity. On the other hand, TiO_2 NFs demonstrated a high surface area and efficient degradation of some organic compounds in the photocatalytic process (Lee et al., 2005). The titania NFs can be conveniently prepared by electrospinning an alcohol solution that contains a polymer of high molecular weight ad a titanium alkoxide precursor. The calcination in a subsequent step leads to the formation of polycrystalline NFs.

The synthesis of TiO_2 NF (Li and Xia, 2003) by electrospinning was reported, in which titanium tetraisopropoxide (a sol-gel precursor to titania) is directly added to an alcohol solution containing high-molecular-weight polyvinyl pyrrolidone (PVP) to prepare TiO_2/PVP composite NFs. The PVP is selected as a base polymer because of its good solubility in alcohols and water and its compatibility with some titania precursors.

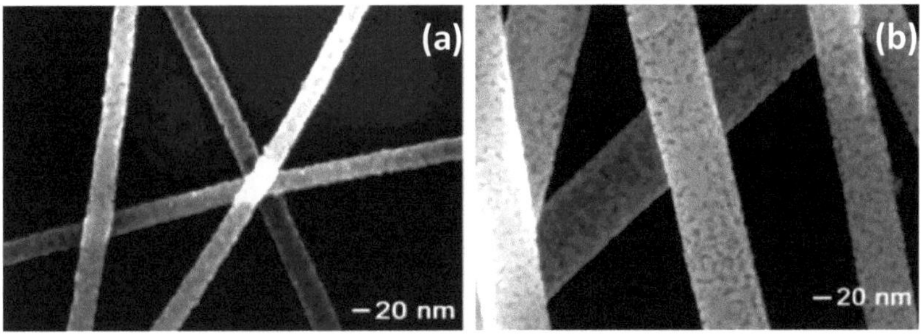

Figure 10.4 SEM micrographs of TiO_2/PVP composite nanofibers (a) with 0.025 g/mL and (b) 0.15 g/mL TiO_2 precursor $Ti(OiPr)_4$ after it is calcined in air at 500°C for 3 h. Reprinted with permission from Li D, Xia Y. (2003). Fabrication of titania nanofibers by electrospinning. *Nano Lett.*, **3**(4): 555–560. Copyright (2003) American Chemical Society.

Figure 10.4 shows the scanning electron microscopy (SEM) image of TiO_2/PVP composite NFs (Fig. 10.4a) with 0.025 g/mL and (Fig. 10.4b) 0.15 g/mL TiO_2 precursor $Ti(OiPr)_4$ after it is calcined in air at 500°C for 3 h. Both images indicate that each NF is formed through the sintering of TiO_2 nanoparticles that are ~10 nm in diameter and voids exist between adjacent nanoparticles. This suggests that the porous structures on the surfaces of titania NFs can be manipulated by controlling the hydrolysis rate of $Ti(OiPr)_4$ in the fibers.

The TiO_2 or TiO_2 composite NFs are produced by combining sol-gel and electrospinning techniques (Tekmen et al., 2008). The process of sol-gel during fiber formation makes the electrospinning solution unstable, which needs more time due to hydrolysis of the precursors. To overcome these problems, Gao et al. (2002) prepared the TiO_2 NFs without the sol-gel process, directly from the solution of PMMA and tetrabutyl titanate (TBT) via electrospinning technique and subsequent calcination. These TiO_2 fibers possess a very rough surface due to a great deal of mesopores distributed on the surface of TiO_2 fibers.

The photocatalytic degradation of TiO_2 fibers on methyl orange (MeO) in an aqueous solution shows that the degradation rate reached almost 90% after 5 h. TiO_2 fibers demonstrated excellent photocatalytic properties for organic dye. This is because the porous fibers with a high specific surface area led to the generation of a large number of hydroxyl free radicals on the surface of TiO_2 fibers in the solution, which cause oxidation of the dyestuff. The H_2O gets easily absorbed on the TiO_2 fiber surface and reacts with the valence-band hole to produce a great number of hydroxyl free radicals. The photogenerated holes had a direct reaction with the dyestuff, and fibers with a high surface area worked well to concentrate MeO around the TiO_2 fibers, which also facilitated the photo-catalytical degradation (Gao et al., 2002).

Copper oxide (CuO) is a p-type semiconductor that exhibits a narrow bandgap of 1.2 eV and possesses a number of other interesting properties (Musa et al., 1998). There has been increasing interest in developing 1D nanostructures of CuO for diverse applications such as high-critical-temperature superconductors (MacDonald, 2001), gas sensors (Chowdhuri et al., 2004), and lithium ion electrode materials. But it is a challenge to synthesize and get the oriented assembly of CuO NFs with lengths of over 20 μm by the traditional nonmaterial synthesis process. It is only possible by the electrospinning technique to produce the CuO NFs from a viscous solution of copper acetate/PVA (Wu et al., 2006). After calcination at 500°C ultimate CuO fibers possess a diameter of 60 nm and a length of several hundreds of micrometers (Fig. 10.5).

The CuO fibers shown in Fig. 10.5b are randomly oriented on the substrate. Therefore, it is difficult to handle and measure their electrical properties. To measure the electrical properties of NFs, a special type of fiber collector is designed for collecting aligned fibers. The electrospun CuO NF is deposited bridging two parallel electrodes. Electrical measurement is conducted between the two electrodes. The conductivity of the individual CuO NF is measured to be 3×10^{-3} S/cm. The field-effect transistor (FET) behavior in a single CuO NF is also observed, showing that a CuO NF is an intrinsic p-type semiconductor. It is demonstrated that semiconductive ceramic NFs can be directly assembled into FETs using the simple and versatile electrospinning method. The NF transistors should be useful in building low-cost logic and switching circuits, as well as highly sensitive chemical and biological sensors with reduced device dimensions.

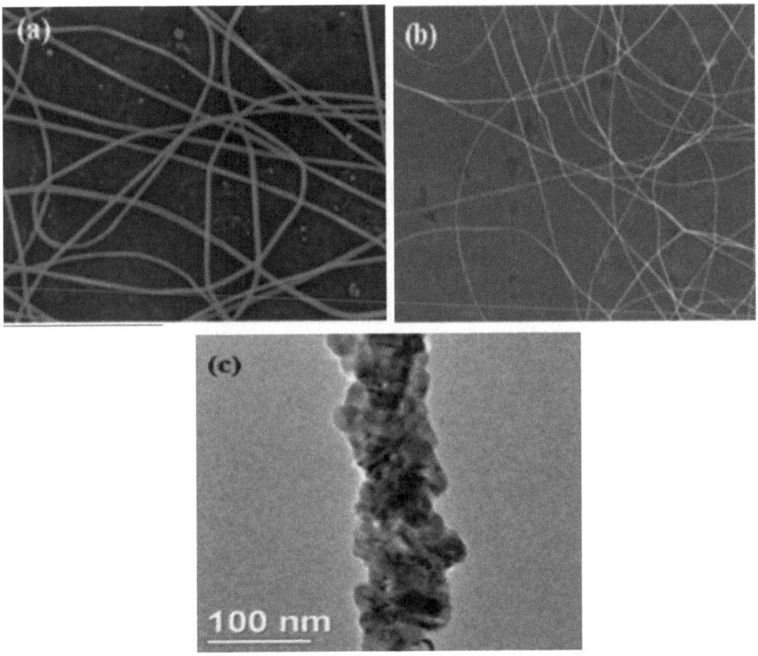

Figure 10.5 FESEM image of the electrospun (a) copper acetate/PVA composite fibers and (b) CuO fibers and (c) a TEM image of CuO. Reprinted from Wu H, Lin D, Pan W. (2006). Fabrication, assembly, and electrical characterization of CuO nanofibers. *Appl. Phys. Lett.*, **89**: 133125, with the permission of AIP Publishing.

Silicon dioxide (SiO_2) NFs are an interest of research because it is a well-studied material for a number of applications (Shao et al., 2002; Choi et al., 2003). Electrospun SiO_2 NFs can significantly outperform glass particles because they are amorphous in nature and possess the desired morphological and structural properties such as a small fiber diameter and a large aspect ratio and strength/modulus. These can be used as a reinforcement in dental composites due to their outstanding properties. The glass fibers can support in resisting the crack opening in dental matrix by a bridging mechanism (Xo et al., 1999) due to their NF structure. Liu et al. (2008) reported the preparation of SiO_2 NFs from tetraethylorthosilicate (TEOS) as the alkoxide precursor; PVP as the polymer; and EtOH, DMF, and HAc (acetic acid) as three of the solvents. The electropsun NFs undergo pyrolysis at different temperatures (from 600°C to 1400°C), and it is reported that the NFs retained their overall fiber morphology after being subjected to vigorous ultrasonic vibrations.

Liu et al. 2008 reported that as-spun TEOS/PVP fibers are uniform and without beads. But the average diameter is 2.0 μm and after pyrolysis at 600°C, the diameter decreases to 1.4 μm. The fiber diameter is very high. To reduce the diameter the PVP concentration is varied, and it is found that the optimum concentration of TEOS and PVP is 1:1 to get fibers in the nanorange. In another case, DMF is used as the solvent with a mixture of acetic acid (HAc). The beadless fibers have a diameter of 700 nm and after pyrolysis at 600°C, the fiber diameter is in the range of 500 nm. These NFs are subjected to vigorous ultrasonic vibrations to check the durability and stability of the fiber morphology. It is reported that fibers pyrolyzed up to 1000°C retain the fiber morphology. This reveals that in SiO_2 NFs, SiO_2 nanoparticles are not an agglomeration but are assembled in o-SiO_2 NFs.

Zinc oxide (ZnO), being nontoxic and photoactive and possessing a wide bandgap, that is, 3.4 eV at room temperature, is the most important oxide material with a large exciton binding energy of 60 meV (Lee et al., 2003). For high photocatalytic activity the size of ZnO plays a significant role. In the case of large particle size photo induced electron holes recombine rapidly, which lowers the photocatalytic activity. On the other hand, nanoparticles have a problem of aggregation and even repollute the water, while in the case of ZnO thin films the yield is very low to meet the industrial requirement for the water treatment applications. Among all the forms, due to a large aspect ratio and a high specific surface area, 1D nanostructures have been gaining more attention among researchers. These are very stable and do not aggregate or repollute the water; as a result they are proven to be better photocatalytic agents.

Gupta et al. (2015) reported the synthesis of ZnO NFs by the electrospinning technique from a mixture of PVA and zinc acetate dihydrate, $Zn(ac)_2 \cdot 2H_2O$. Figure 10.6 shows the morphology of PVA-$Zn(ac)_2$ electrospun NFs observed by SEM. There is a wide variation in the diameter of NFs and some of the $Zn(ac)_2$ particles also visible on the surface of NFs. After calcination at 600°C, a self-assembled ZnO nanofibrous structure is derived from the electrospun PVA/$Zn(ac)_2$ NFs (Fig. 10.6). These ZnO NFs have a very rough surface and diameters in the range of 160–180 nm. During calcination ZnO is formed from the $Zn(ac)_2$ dehydrated salt, in which differently shaped ZnO particles bond with each other to form ZnO NFs. These are formed through self-assembly directly through the minimization of Gibbs free energy. The minimization of free energy is a result of weaker intermolecular forces between self-assembled ZnO particles. So the forces acting between ZnO nanoparticles are weak intermolecular forces like van der Waals or weak polar forces of attraction. There is some balance of attractive and repulsive forces, which allows them to organize in the form of fibers. The ZnO NFs are used for photodegradation of acid fuchsin present in wastewater under solar radiation. When solar radiation of the desired energy (wavelength) falls on ZnO NFs, electrons get excited to the conduction band, leaving holes (positively charged carriers) in the valence band.

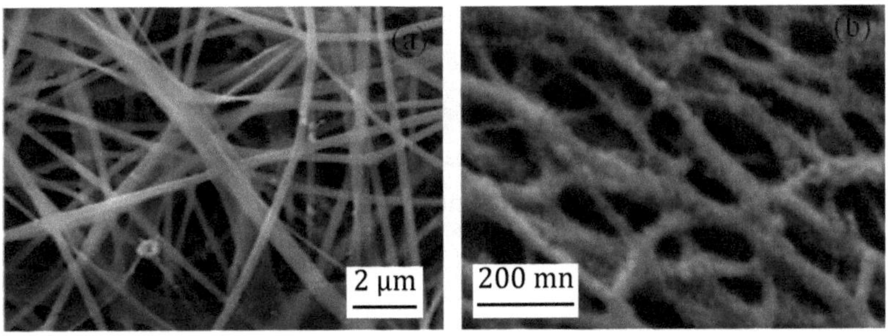

Figure 10.6 SEM image of electrospun PVA-$Zn(ac)_2$ and self-assembled ZnO nanofibers (Gupta Nandanwar, and Dhakate, 2015).

These electrons and holes generated during photocatalysis can degrade almost all types of organic, inorganic, and microbial contaminants due to their high redox potentials. Holes react with hydroxyl ions (OH^-) to form hydroxyl free radicals (OH). In the same way photogenerated electrons help to form superoxide anion free radicals (O_2). But these may form H_2O_2 when reacting with water and evolve out of the reaction medium. Free radicals OH and O_2 produced in the solution then react with organic pollutants (dyes) and degrade the conjugated bonds and as a result decolorize the dyes. The H_2O and CO_2 are produced as by-products, which are harmless. In this way water purification occurs with ZnO NFs. The schematic of the mechanism of degradation of fuchsin dye is illustrated in Fig. 10.7. It demonstrates 99% degradation within 2.5 h. Thus a self-assembled ZnO nanofibrous structure can be a promising cost-effective and sustainable solar photocatalyst for different applications.

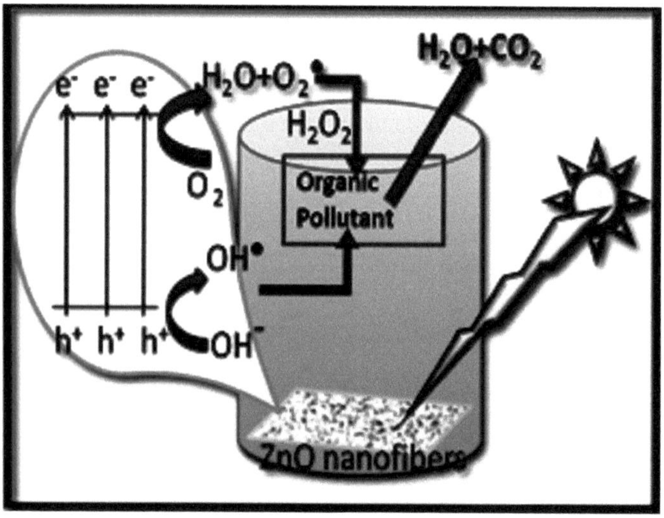

Figure 10.7 Schematic of ZnO nanofiber activation and dye degradation mechanism (Gupta Nandanwar, and Dhakate, 2015).

Tin dioxide (SnO_2) is an environmentally friendly n-type, broad bandgap (3.6 eV) oxide semiconductor. Nowadays, morphology and structure modification of nano-SnO_2 have attracted much attention because they enable the utilization of the unique optical, electronic, and chemical properties of SnO_2 in different applications (Ji et al., 2010; Cheng et al., 2011).

Lots of effort has been made to develop its nanostructures with diverse morphologies and aspect ratios with an aim to enhance the specific surface area and further improve its properties. Nonconventional morphologies of SnO_2, such as tetrapods (Fouad et al., 2008) and flowers (Kumar et al., 2012), have been synthesized from colloidal solutions with improved properties compared to their particulate or 1D analogs. The SnO_2 NWs and NTs are also developed using the electrospinning technique with improved properties (Gubbala et al., 2008). Wali et al. (2014) reported that various 1D nano-morphologies, such as multiporous nanofibers (MPNFs), porous nanofibers (PNFs), and NWs of SnO_2, are synthesized using the electrospinning technique by controlling the tin precursor concentration.

As a typical procedure, 3 g of PVP is dissolved in an equal volume ratio (1:1) of ethanol and DMF. To this solution, various concentrations of the tin precursor are dissolved at room temperature until the solution becomes transparent. The concentrations are 5.5, 7.0, 8.5, 10, and 11.5 mM and are labeled as C0, C1, C2, C3, and C4, respectively. These solutions are electrospun by controlling processing parameters to get different morphologies of NFs (Wali et al., 2014). The solid NFs are collected from the collector and annealed at 600°C for 3 h at a heating rate of 0.5°C per minute. It is observed that as-spun polymeric fibers are 200–600 nm in diameter, which upon annealing result in NWs, MPNFs, and PNFs. Figure 10.8 shows field-emission scanning electron microscopy (FESEM) images of annealed NFs. For the sample C0, porous solid NWs are formed with diameters ranging from 100 to 170 nm. A small increase in the precursor concentration (7 mM, C1) results in the formation of MPNFs with slightly larger diameters (120–190 nm), which upon further increments transform into PNFs. The number of channels in MPNFs vary from 2 to 4, with their inner diameters ranging from 30 to 70 nm. Further increase in the precursor concentration (samples C2 and C3) results in the formation of PNFs with outer and inner diameters in the range of 160–220 and 50–100 nm, respectively. This reveals that an increase in the precursor concentration leads to significant decreases in the tube diameter and at the highest concentration, the tube structure vanishes. Excessive concentration (C4) results in varying diameters of NWs (between 210 and 340 nm). It is reported that the Brunauer–Emmett–Teller (BET) specific surface area increases from C0 to C2. The surface area of NWs is 7–78 m^2 g^{-1} and that of MPNFs is six to eight times higher compared to PNFs. This is attributed to multichannels inside a single wall and the smaller grains and pores in MPNFs. The minimum surface area is of NWs (C0). This type of high-surface-area 1D SnO_2 nanomaterial is used as the electrode material in lithium ion batteries (LiBs). Although having a high capacity, Sn and SnO_2 suffer from rapid decay in capacity (i.e., a poor cycle life) due to the volume change and pulverization of the electrode during alloying-de-alloying. To overcome these problems, another approach is adapted to improve the cyclability and maintain the high capacity of SnO_2 as the anode material in LiB. Carbon is used in the SnO_2 matrix, which acts as a buffer to reduce the strain during lithiation-delithiation.

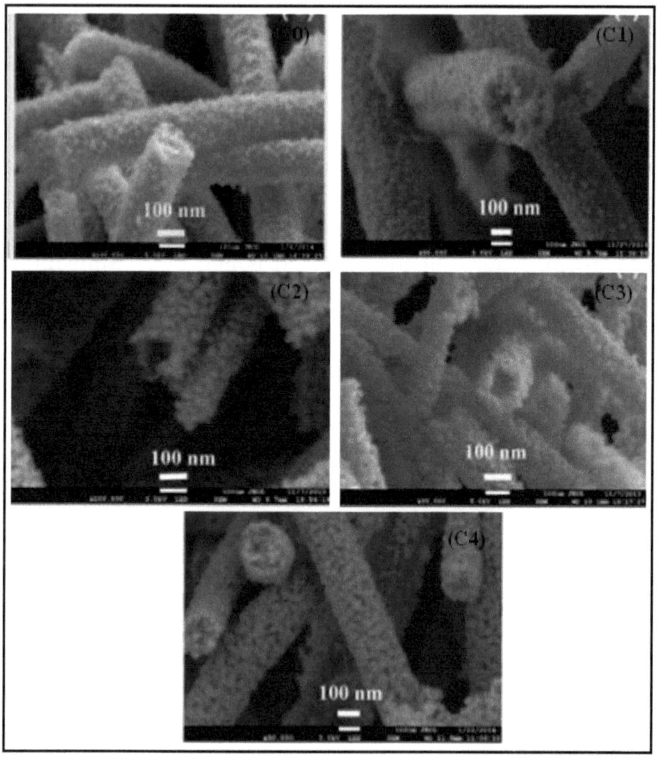

Figure 10.8 FESEM images of annealed nanofibers (C0 to C4). Reproduced from Wali Q, Fakharuddin A, Ahmed I, Ab Rahim MH, Ismail J, Jose R. (2014). Multiporous nanofibers of SnO_2 by electrospinning for high-efficiency dye-sensitized solar cells. *J. Mater. Chem. A*, **2**: 17427, with permission from The Royal Society of Chemistry.

In this direction recently Gupta et al. (2017) developed SnO_2-carbon NFs from a solution of PAN and tin chloride by electrospinning followed by calcination at 600°C in an oxidizing atmosphere to control the carbon content in the SnO_2 matrix. The higher content of carbon is not favorable. Gupta et al. (2017) demonstrated controlled carbon content (less than 10%) in a SnO_2 matrix and studied its performance for 100 cycles as the anode material.

Figure 10.9 shows the FESEM images of SnO_2-carbon NFs formed after heat treatment at 600°C. The image shows the formation of hierarchicaly porous NFs in a self-supporting network having diameters in the range 150–180 nm. At low and high magnification a clearly demonstrated mesoporous structure of SnO_2-carbon NFs is observed. The elemental mapping of SnO_2-carbon NFs reveals the presence of C, Sn, and O.

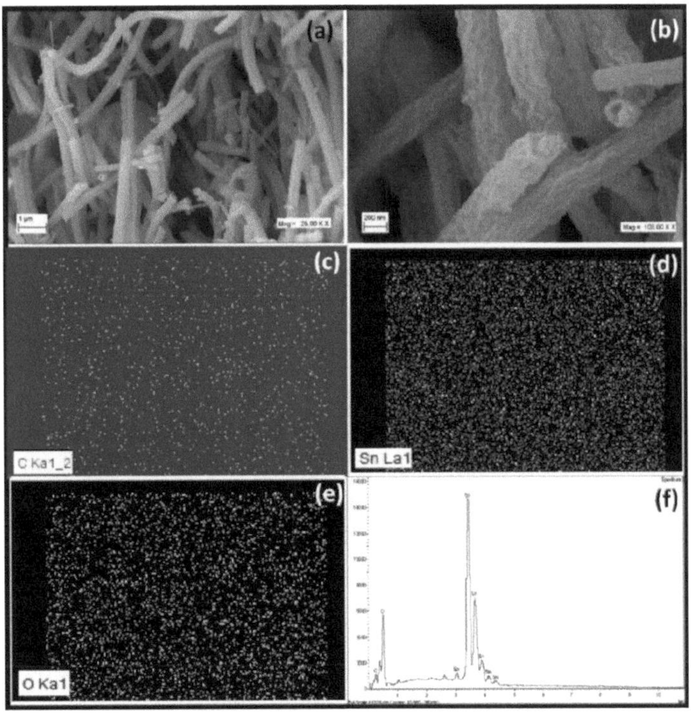

Figure 10.9 FESEM images of SnO_2-carbon nanofibers and corresponding elemental mapping formed after heat treatment at 600°C (Gupta et al., 2017).

10.4 Applications of Ceramic Nanofibers

A gas sensor is a device that detects the presence of different gases in an area, especially those gases that might be harmful to humans or animals. The development of gas sensor technology has received significant attention in recent years for monitoring environmental pollution. It is well known that chemical gas sensor performance features such as sensitivity, selectivity, time response, stability, durability, reproducibility, and reversibility are largely influenced by the properties of the sensing materials used (Kong et al., 2000).

The sensors based on semiconducting oxides, such as WO_3, TiO_2, SnO_2, ZnO, and other wide bandgap metal oxides, are known for their ability to detect trace concentrations of various gases in air via charge transfer interactions between the sensor and chemisorbed species that changes the resistance of oxides (Kong et al., 2000) and favor low-cost detectors for reducing gases. Nanomaterials have a unique advantage for sensor applications due to their size-dependent behavior. The smaller size of nanomaterials enhances not only sensitivity but also selectivity against the target analyte. The reduction of the crystallite size, that is, from micron to nano, can result in the greater proportion of crystallite surface atoms, which are more prone to react with the surrounding ambient or to exhibit unusual structural features. The NFs are preferred for sensor applications because they provide a large surface-to-volume ratio as well as continuity of crystallite and alignment in a particular direction as compared to other 1D nanostructures. Compared to other nanostructure materials, 1D NFs are easily configurable and have a high compatibility with conventional microelectronic devices.

Tungsten oxide (WO_3) is a well-known catalyst. It has a number of structural polymorphs, which exhibit manifold properties, such as photochromic and electrochromic effects (Salje, 1974). These effects are bound up in the features that make a material sensitive to gases by modifying the surface conduction bands due to a change in the number of free electrons when a gas molecule interacts with a surface layer (Gouma, 2003).

Wang et al. (2006) reported electrospun tungsten oxide ceramic NFs as an ammonia gas sensor application. The tungsten oxide ceramic NFs fabricated from a tungsten isopropoxide sol-gel precursor are incorporated into PVAc/DMF and sol-gel-derived solution electrospun. The as-spun composite NFs are calcined at 500°C to obtain pure tungsten oxide NFs with controllable diameters of around 100 nm.

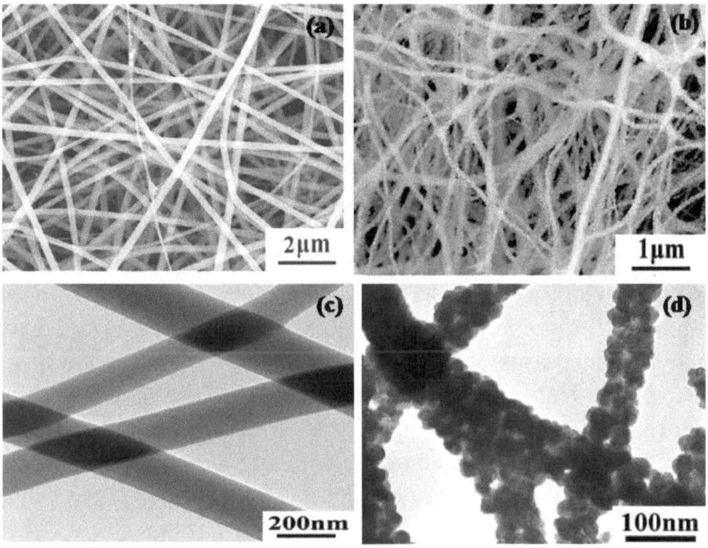

Figure 10.10 (a) SEM image of tungsten oxide/PVAc composite fibers of a sample with 25% W(*i*Pr)6; (b) after calcination in air at 500°C for 3 h; and (c, d) TEM images before and after calcination. Reprinted with permission from Wang G, Ji Y, Huang X, Yang X, Gouma P-I, Dudley M. (2006). Fabrication and characterization of polycrystalline WO_3 nanofibers and their applications for ammonia sensing. *J. Phys. Chem. B*, **110**: 23777–23782. Copyright (2006) American Chemical Society.

Figure 10.10a shows a SEM image of the as-prepared WO_3/PVAc composite NFs 25% W(*i*Pr)$_6$ before calcination. Each individual fiber is uniform in cross section, and the average diameter is around 200 nm. Calcination in air at 500°C (Fig. 10.10b) results in metal oxide NFs of continuous structures but with a reduced diameter, which is less than 100 nm. The size reduction is due to the loss of PVAc from the NFs and the crystallization of tungsten oxide. Figure 10.10c,d displays the TEM images before and after calcination of WO_3/PVAc composite NFs. TEM images show that the darkness along the fiber is uniform because the tungsten oxide sol-gel was distributed uniformly within the polymer template, so it is possible to obtain continuous ceramic NFs after removal of the polymer template. After calcination, Fig. 10.10d shows a grained fiber structure with relatively uniform diameters. The grains are crystalline tungsten oxide with grain sizes <20 nm.

Tungsten trioxide is weakly sensitive toward NH_3 compared with a superior sensing performance toward NO*x* (Moseley and Crocker, 1996; Ponzoni et al., 2006). However, the sensing performance of the material can be strongly affected by the material microstructure. On the other hand, Wang et al. (2006) reported that tungsten oxide NFs had a very quick response toward NH_3 in very low concentrations at an operating temperature of 350°C. Figure 10.11 illustrates the dynamic gas response of the tungsten oxide ceramic NFs to NH_3 pulses with concentrations of 50, 100, 200, and 500 parts per million (ppm). The resistance of the material decreases obviously with respect to the baseline in each NH_3 pulse. This result is in agreement with the model developed (Moseley and Crocker, 1996) that predicted the reduction of resistance of an n-type semiconductor in the presence of reductive gases (such as NH_3). It is demonstrated that the response time of a tungsten-oxide-NF-based sensor is less than 20 s. This is attributed to factors such as porosity of the surface and the large surface-to-volume ratio tungsten oxide electrospun NF after calcination, which provides good accessibility of the gases to the sensing layers. This makes the adsorption of ammonia much easier. Another reason is the high purity of the material processed by sol-gel chemistry and the fact that after calcination there exists only the pure orthorhombic phase as the sensing material.

Nguyen et al. (2011) reported tungsten oxide NFs that have a rough surface morphology and an average diameter of around 40 nm are found formed by the axial agglomeration of prolate spheroid-like tungsten oxide nanoparticles with monoclinic crystalline phases. Gas-sensing measurements of the polycrystalline tungsten oxide NF mats demonstrated n-type sensing response and sensitive NH_3 detection up to 10 ppm with a well-defined relationship between the concentration and detection response at an operating temperature of 300°C. The NFs are formed by thermal treatment of electrospun composite NFs consisting of PVP and tungstic acid at 500°C in air. The results are interpreted by applying the space-charge layer model used in the semiconducting metal oxide sensor systems.

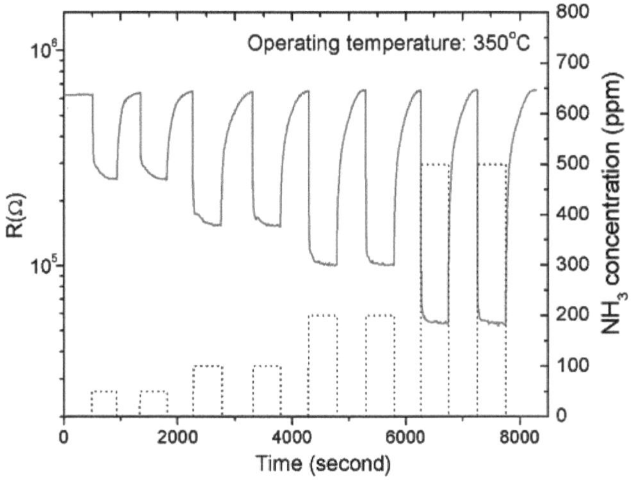

Figure 10.11 Dynamic gas response of the tungsten oxide ceramic nanofibers to NH_3 pulses with concentrations of 50, 100, 200, and 500 ppm. Reprinted with permission from Wang G, Ji Y, Huang X, Yang X, Gouma P-I, Dudley M. (2006). Fabrication and characterization of polycrystalline WO_3 nanofibers and their applications for ammonia sensing. *J. Phys. Chem. B*, **110**: 23777–23782. Copyright (2006) American Chemical Society.

An ultrasensitive chemiresistor based on electrospun TiO_2 NFs (Kim, Rothschild et al., 2006) is fabricated by directly depositing TiO_2/PVAc composite NF mats onto interdigitated Pt electrode arrays with subsequent hot pressing at 120°C and calcining at 450°C, as shown in Fig. 10.12. Figure 10.12 shows SEM images of electrospun TiO_2 NFs. The as-spun TiO_2/PVAc composite fibers possess diameters in the range of 200 to 500 nm (Fig. 10.12a).

In the case of not-hot-pressed pretreatment, after calcination at 450°C in air for 30 min to remove the PVAc, the TiO_2 fibers demonstrate a bundle structure composed of sheaths of 200–500 nm diameter cores filled with ~10 nm thick fibrils, as shown in Fig. 10.12c. As spun onto the substrate, the TiO_2 fiber mats have poor adhesion to the substrates and to the Pt electrodes. Therefore, these are hot-pressed to drive the PVAc above its glass transition temperature, resulting in better adhesion. In contrast to untreated fibers, the hot-pressed fibers did not peel off of the substrate during handling and testing. An interconnected morphology of the TiO_2/PVAc composite fibers is obtain, as depicted in Fig. 10.12b, due to the partial melting of PVAc. Subsequent calcination at 450°C for 30 min results in the extremely high surface area structures shown in Fig. 10.12d. The mechanical pressure applied during the hot pressing at 120°C served to break the outer sheaths, thereby exposing the ~10 nm thick fibrils. This unique morphology results in high gas accessibility to the exceptionally high surface-to-volume ratio structure. This type of morphology is highly advantageous for gas sensors. Hot pressing thus serves two important purposes: that of enhancing adhesion between the TiO_2 fiber mats and the substrates and of creating an accessible large active area for the gas-sensing interactions.

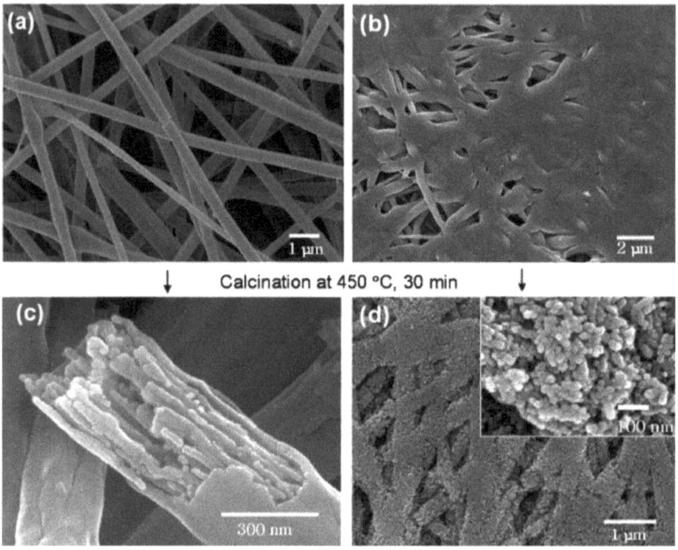

Figure 10.12 (a) SEM micrographs of the as-spun TiO_2/PVAc composite fibers fabricated by electrospinning from a DMF solution. (b) SEM image of TiO_2/PVAc composite fibers after hot pressing at 120°C for 10 min. (c) SEM image of unpressed TiO_2 NFs after calcination at 450°C. (d) SEM image of hot-pressed TiO_2 NFs after calcinations at 450°C (Kim, 2009). Reprinted with permission from Kim ID, Rothschild A, Lee BH, Kim DY, Jo SM, Tuller HL. (2006). Ultrasensitive chemiresistors based on electrospun TiO_2 nanofibers. *Nano Lett.*, **6**: 2009. Copyright (2006) American Chemical Society.

It is reported that the hot-pressed TiO_2 fibers have a much higher specific surface area (138 m^2/g) than unpressed fibers (31 m^2/g) and lead to exceptionally high gas sensitivity. The excellent results of the resistance response of one of the sensors upon cyclic exposure to NO_2 at different concentrations and operating temperatures are registered in Fig. 10.13.

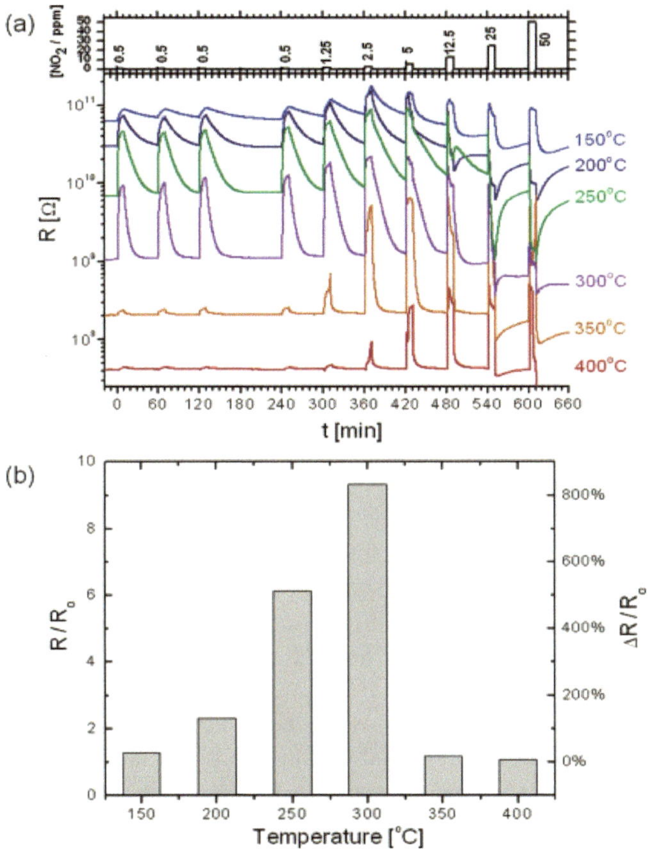

Figure 10.13 (a) The resistance response during cyclic exposure to 10 min pulses with increasing concentrations of NO_2 mixed in dry air at various operating temperatures. (b) ensitivity-versus-temperature histogram upon exposure to 500 ppb NO_2 in dry air. (R_0 designates the baseline resistance in dry air). Reprinted with permission from Kim ID, Rothschild A, Lee BH, Kim DY, Jo SM, Tuller HL. (2006). Ultrasensitive chemiresistors based on electrospun TiO_2 nanofibers. *Nano Lett.*, **6**: 2009. Copyright (2006) American Chemical Society.

It is reported that resistance increases monotonically with NO_2 concentrations of 500 parts per billion (ppb) to 12.5 ppm and recovered reasonably fast, returning to the baseline resistance in dry air following each pulse of NO_2-enriched air. The sensitivity versus temperature histogram reaches through a peak at 300°C. The histogram's shape results from the competition between slow kinetics at low temperatures and enhanced desorption at high temperatures. The remarkable response patterns are registered at high NO_2 concentrations (>12.5 ppm), consistent with n-to-p inversion of the surface-trap limited conduction facilitated by the high surface-to-volume ratio of TiO_2 NF.

Qi et al. (2009) reported improved NH_3-, C_2H_5OH-, and CH_3COCH_3-sensing properties of electrospun SnO_2 NFs by adding block copolymer P123 [poly(ethylene oxide)-block-poly(propylene oxide)-block-poly(ethane oxide)]. The SEM micrographs of SnO_2 NFs derived via a simple electrospinning of a tin chloride solution with (a) SnO_2 NFs with block copolymer and (b) SnO_2 NFs without the block copolymer are depicted in Fig. 10.14. In both cases, diameters of NFs range from 80 to 150 nm and lengths are several tens of micrometers. The Fig. 10.14A inset shows a TEM image revealing that SnO_2 NFs become coarse and loose after adding block copolymer. The porous features of NFs are investigated by the nitrogen adsorption method. Nitrogen sorption of type IV isotherm according to the International Union of Pure and Applied Chemistry (IUPAC) classification is reported. The BET surface areas of NFs with and without block copolymer are 113.261 and 18.025 $m^2 g^{-1}$. These results indicate that NFs with block copolymer have a much higher surface-to-volume ratio than NFs without block copolymer. The pore size distribution suggests that NFs with block copolymer possess a relative narrow pore size distribution and a maximum pore volume of pore size 5.3 nm (Qi et al., 2009).

The responses of SnO_2 NFs (a) and (b) to 500 ppm NH_3, C_2H_5OH, and CH_3COCH_3 as a function of operating temperature are shown in Fig. 10.15. For each curve, the response increases and reaches its maximum and later on decreases rapidly with a further increase in the temperature. The maximum responses of samples to NH_3, C_2H_5OH, and CH_3COCH_3 appear at 280°C, 300°C, and 340°C, which are defined as the optimum operating temperatures for the detection of NH_3, C_2H_5OH, and CH_3COCH_3, respectively.

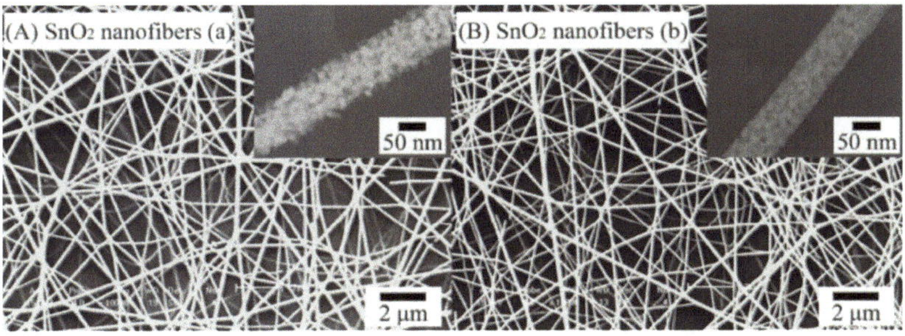

Figure 10.14 SEM image of (A) SnO_2 nanofibers with (inset a) TEM image of block copolymer. (B) SEM image of SnO_2 nanofibers without (inset b) block copolymer. Reprinted from Qi Q, Zhang T, Liu L, Zheng X, Lu G. (2009). Improved NH_3, C_2H_5OH, and CH_3COCH_3 sensing properties of SnO_2 nanofibers by adding block copolymer P123. *Sens. Actuators B*, **141**: 174–178, Copyright (2009), with permission from Elsevier.

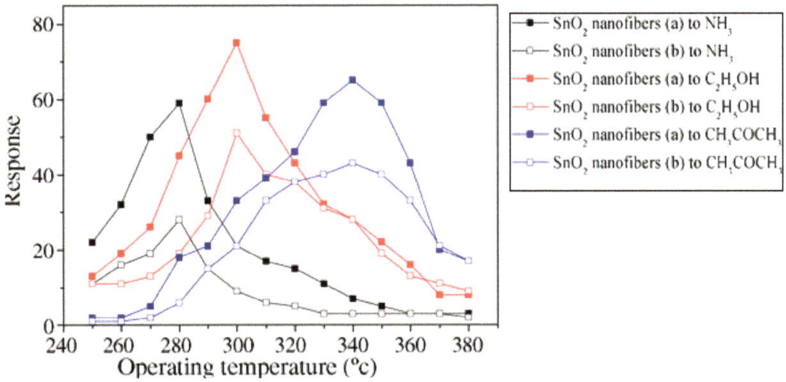

Figure 10.15 Responses of SnO_2 nanofibers (a) and (b) to 500 ppm NH_3, C_2H_5OH, and CH_3COCH_3 at different operating temperatures. Reprinted from Qi Q, Zhang T, Liu L, Zheng X, Lu G. (2009). Improved NH_3, C_2H_5OH, and CH_3COCH_3 sensing properties of SnO_2 nanofibers by adding block copolymer P123. *Sens. Actuators B*, **141**: 174–178, Copyright (2009), with permission from Elsevier.

The sensing measurements are repeated after 10 days for 100 days to know the stability of SnO_2 NFs with a block copolymer. As shown in Fig. 10.16, SnO_2 NFs with a block copolymer exhibit nearly constant electrical signals to the target gases in

100 days, validating the excellent stability of SnO_2 NFs with a block copolymer.

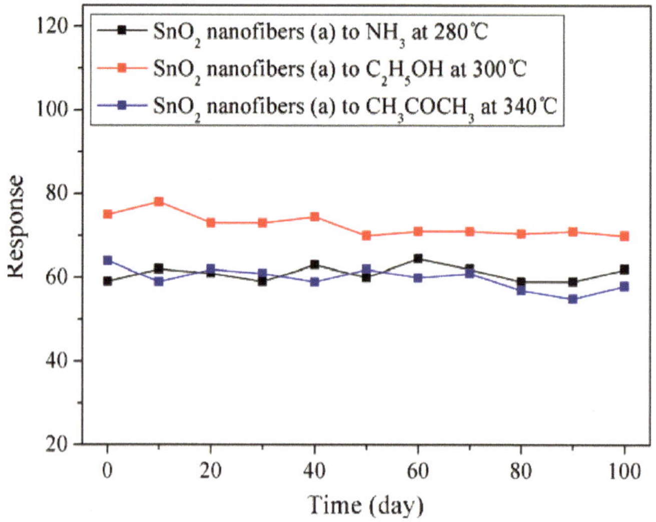

Figure 10.16 Stability of SnO_2 nanofibers (a) to 500 ppm NH_3, C_2H_5OH, and CH_3COCH_3 at 280°C, 300°C, and 340°C, respectively. Reprinted from Qi Q, Zhang T, Liu L, Zheng X, Lu G. (2009). Improved NH_3, C_2H_5OH, and CH_3COCH_3 sensing properties of SnO_2 nanofibers by adding block copolymer P123. *Sens. Actuators B*, **141**: 174–178, Copyright (2009), with permission from Elsevier.

The sensing mechanism in SnO_2 NFs is because the oxygen vacancy in SnO_2 NFs acts as an electron donor to provide electrons to the conduction band of SnO_2 and makes the SnO_2 NFs be an n-type metal oxide semiconductor (Barsan et al., 2007). The SnO_2 NF sensor is surrounded by air, oxygen molecules will adsorb on the fiber surface to generate chemisorbed oxygen species (O_{2-}, O^{2-}, or O^-), and O^- is believed to be dominant (Windischmann and Mark, 1979). Consequently, depletion layers are developed in the surface area of SnO_2 NFs, leading to a decrease of carrier concentration and electron mobility (Li and Xia, 2004a), and making the NFs less conductive. While the sensor is exposed to reducing gases (such as NH_3, C_2H_5OH, and CH_3COCH_3), these reducing molecules may react with O^- and release the trapped electron back to the conduction band, increasing the carrier concentration and electron mobility and resulting in the reducing of fiber resistance. Compared with NFs (b), SnO_2 NFs (a) exhibit an enhanced response value, a higher saturated-detection concentration, and a lower minimum-detection limit, which can be explained by their morphologic characteristics and porous features. The SnO_2 NFs without block copolymer possess a coarse and loose morphology with a high BET surface area, resulting in more target molecules being adsorbed on the fiber surface (Shimizu et al., 2005), which eventually improves the gas-sensing properties of SnO_2 NFs.

10.4.1 Nanofiber Membranes for Filtration

Ceramic membranes are of interest for various separation processes because they can be used under severe conditions owing to their chemical and thermal stability (DeVos and Verweij, 1998). They are able to function impassively within organic and biological systems as well as at high temperatures. These can be readily cleaned by using steam treatment and exhibit long operational lives. Because of low energy consumption and absence of potentially harmful chemical agents in the separation processes using ceramic membranes, they can be used for both gas and liquid systems (Verweij, 2003). The conventional porous ceramic membranes have an unsymmetrical layered structure consisting of a macroporous (pore size > 400 nm) support, a porous intermediate layer, and a nanoporous top layer (Nunes et al., 1995). The conventional ceramic separation membranes encounter major difficulties because of the formation of pinholes and cracks during the drying and calcination process. Then there is a dramatic loss of flux when pore sizes are reduced to increase selectivity (Sun et al., 2006).

To achieve ceramic membranes with high efficiency radical changes have to be introduced in the membrane texture. In this direction Ke et al. (2007) developed ceramic nanoporous filters with a hierarchically structured separation layer on a porous substrate using larger titanate and smaller boehmite NFs. The randomly oriented titanate NFs can completely cover the rough surface of the porous substrate of micrometer α-alumina particles, leaving no pinholes or cracks. On top of this titanate fiber layer, a layer of alumina fibers is formed using boehmite NFs. Boehmite NFs, which can be converted to γ-alumina fibers by heating at temperatures above 773 K (Zhu, Gao, Song et al., 2004), and titanate NFs (Zhu, Gao, Lan et al., 2004) are used to make the web structures. The components of hierarchical layers of randomly oriented fibers (LROFs) are depicted in Fig. 10.17. Figure 10.17A is a schematic profile of the ceramic membranes with LROFs, Fig. 10.17B is a SEM image of the alumina substrate particles of small (2–3 μm) and large (>10 μm) α-alumina particles, Fig. 10.17C shows a TEM image of titanate fibers 20–30 μm long and 40–100 nm thick, and Fig. 10.17D shows a TEM image of small boehmite fibers 60–100 nm long and 2–5 nm thick.

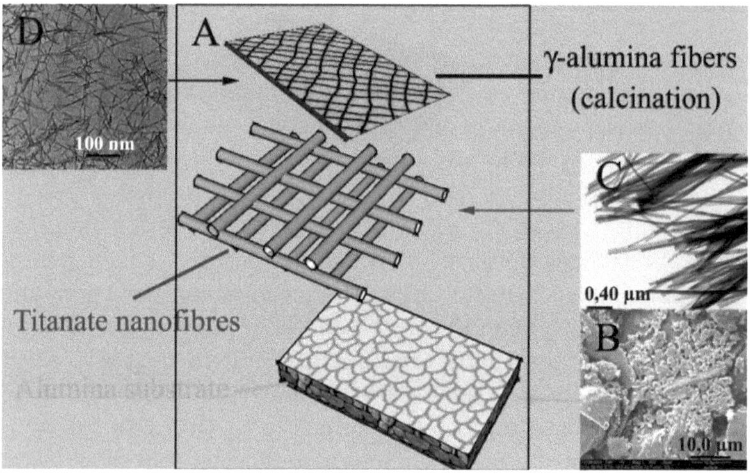

Figure 10.17 The components of a hierarchical LROF structure. (A) Schematic profiles of the ceramic membranes with LROF structures of titanate and α-alumina NFs. (B) An SEM image of the α-alumina substrate surface. (C) A TEM image of titanate NFs. (D) A TEM image of the boehmite NFs, which convert into γ-alumina NFs when heated at 773 K. Reproduced from Ke XB, Zhu HY, Gao XP, Liu JW, Zheng ZF. (2007). High-performance ceramic membranes with a separation layer of metal oxide nanofibers. *Adv. Mater.*, **19**: 785–790, with permission from Wiley.

The resulting membranes can effectively filter out species larger than 60 nm at flow rates orders of magnitude greater than with conventional membranes, and they do not have the structural deficiencies of conventional ceramic membranes. The use of ceramic NFs or nanorods (NRs), instead of particulates with irregular shapes, to fabricate ceramic membranes is a new direction in developing high-performance ceramic membranes. A mesh structure of threads should be the most efficient structure for filtration, achieving high selectivity and maintaining much higher flux than membranes of other forms. It is possible to tailor the thickness of boehmite fibers from 3 to 10 nm and the length from 40 to 100 nm by controlling the electrospinning processing parameters (Zhu, Gao, Song et al., 2004) or tailor the thickness of titanate fibers from 10 nm to 100 nm and the length from 100 nm to 20–30 nm (Zhu, Gao, Lan et al., 2004). These NFs can be readily dispersed in aqueous or alcohol solutions.

This allows constructing LROFs on a porous ceramic substrate as the separation layer by applying thin layers of dispersed NFs. The NFs are able to filter out species with a size of tens of nanometers. The membranes having the ability to filter out species of 50–60 nm or larger have attracted significant interest because of the variety of important potential applications for them (Sun et al., 2006). For example, most viruses and waterborne pathogens are larger than 50–60 nm. The viruses are a major concern to the international community, specifically the severe acute respiratory syndrome (SARS) corona virus, which is 80–200 nm in size, and the avian flu virus, which is 80–120 nm in diameter (Noda et al., 2006). Such membranes are able to effectively filter out these dangerous viruses from air, water, and even blood (Yang et al., 2006). A double-layered nanoporous membrane with cylindrical pores is successfully reported for virus filtration (Yang et al., 2006). It can exhibit high sensitivity and flux for the separation of the pathogen of the common cold in humans.

Recently, Argonide Inc., in Pittsburgh, Pennsylvania, manufactured alumina NFs Nanoceram® that are about 2 nm in diameter for filtration applications (http://www.argonide.com). The Nanoceram® alumina fibers are electropositive and attract dust. These bind and zap viruses and bacteria. The nanosize enables higher flux, at least 2 orders of magnitude compared to other membranes, and does not get clogged by submicron particles, which is a major problem in commercial membranes. Figure 10.18 shows the TEM image of the alumina fibers used in the Nanoceram® filters. The Nanoceram® filter module is also depicted in Fig. 10.18.

Figure 10.18 (Left) TEM image of alumina nanofibers and filter modulus (Nanoceram®) and (right) 4.2 nanofiber separators for lithium ion batteries. Reprinted from Ramaseshan R, Sundarrajan S, Jose R. (2007). Nanostructured ceramics by electrospinning. *J. Appl. Phys.*, **102**(11), doi: http://dx.doi.org/10.1063/1.2815499, with the permission of AIP Publishing.

The separator is a critical component in liquid electrolyte batteries and is located between the positive electrode and the negative electrode to prevent physical contact of the electrodes while enabling free ionic transport and isolating electronic flow. Lithium ion rechargeable batteries are used by all handheld electronic devices. In such batteries the anode and cathode are separated using separators. These separators can be broadly divided into three groups: microporous polymer membranes, nonwoven fabric mats, and inorganic composite membranes. The microporous polymer membranes are characterized by their thinness and thermal shutdown properties. The nonwoven mats have a high porosity and a low cost, while the composite membranes have excellent wettability and exceptional thermal stability. Among numerous battery separators, the thermal shutdown and ceramic separators such as alumina and titania are of special importance in enhancing the safety of LiBs. Zhang (2007) demonstrated the applicability of ceramic nanostructures in LiB separators.

The ceramic separator consists of nanosize ceramic materials coated on two sides with a flexible and highly porous nonwoven matrix, as shown in Fig. 10.19, which enhances safety by retaining extremely stable dimensions even at very high temperatures to prevent direct contact of the electrode. Alumina and titania are the earliest ceramics used to fabricate ceramic NFs by the electrospinning process. The currently used complex membrane fabrication procedure consists of a coating of nanoceramics on nonwovens. It is possible to use just ceramic NF membranes produced by the one-step electrospinning process. The membrane integrity can be maintained by blending a small amount of polymer.

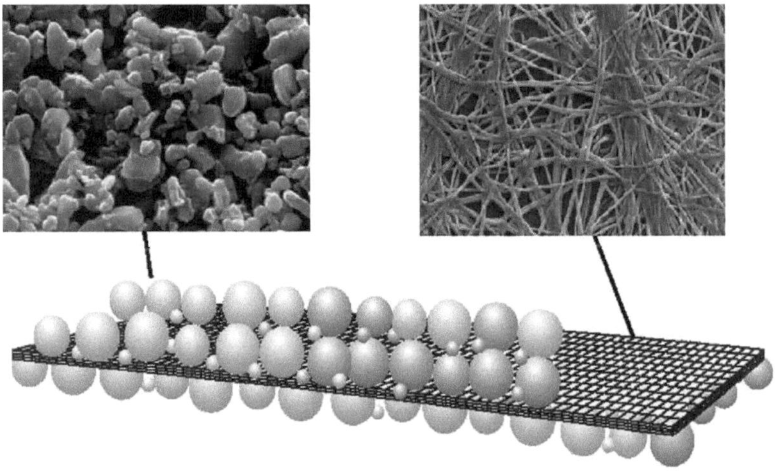

Figure 10.19 Ceramic separator configuration in Li ion batteries. Reprinted from Zhang SS. (2007). A review on the separators of liquid electrolyte Li-ion batteries. *J. Power Sources*, **164**: 351–364, Copyright (2007), with permission from Elsevier.

Due to their thermal and chemical stability, hybrid ceramic polymer-electrolyte membranes derived from organic materials make a commendable option to a conventional polymer electrolyte membrane in view of high temperature application in energy generation devices like the direct methanol fuel cell (DMFC).

The Separion (a trade name) separator membrane is prepared by combining the uniqueness of polymeric nonwoven and ceramic materials (Augustin et al., 2002). The Separion separator is a flexible perforated nonwoven mat coated with a porous ceramic layer on each side, as illustrated in Fig. 10.19. The pore size of the Separion separator is selectively set by an appropriate choice of the ceramic materials. Generally, the ceramic materials used in these separators are alumna, silica, zirconia, or their mixture, and their particle size is required to be of nanosize. To keep the excellent hydrophilic properties of the ceramic materials, an inorganic binder is used for the production of such separators. The inorganic binder sol is prepared by hydrolyzing a mixture of tetraethoxysilane, methyltriethoxysilane, and (3-glycidyloxypropyl) trimethoxysilane in the presence of a hydrochloric acid aqueous solution (Hennige et al., 2006). The resulting sol is used to suspend aluminum oxide powders, and then the homogenized suspension is coated on a porous nonwoven polyethylene terephthalate (PET), followed by drying at 200°C to obtain the separator.

The CREAFILTER® membrane is a commercial membrane (Augustin et al., 2002) that combines the advantages of both polymeric membranes and ceramic membranes. Another application of this type of membrane is in the low-temperature fuel cell. Fuel cell technology is getting more and more important because as fuel only a renewable fuel is used, such as methanol or hydrogen. The CREAFILTER® separator and membrane works at temperatures higher than 80°C, even at reduced relative humidities. Such type of membrane is suitable for both fuel cells and separators for LiB.

10.4.2 Ceramic Nanofibers for Photovoltaic Cells

To utilize the huge resource offered by the sun in the form of solar energy without any investment, throughout the world researchers have actively investigated the solar cell based on photovoltaic devices. Since the first report, by O'Regan and Gratzel (1991), dye-sensitized solar cells (DSSCs) have been intensively explored because of their low cost, belonging to the group of thin-film solar cells, and promising commercial efficiency. A DSSC is based on a semiconductor formed between a photosensitized anode and an electrolyte, a photoelectrochemical system.

This device consists of an optically transparent film of semiconducting (TiO_2) nanoparticles, a monolayer of dye as a sensitizer to harvest light, and an electrolyte containing the tri-iodide/iodide redox couples (O'Regan and Gratzel, 1991). The grain boundaries between nanocrystals are a major issue in this configuration because they act as electron traps and as a consequence rapid recombination of electrons occurs (Jiu et al.,

2006). The 1D ceramic NFs consist of fused or connected tiny nanocrystals. Therefore fewer grain boundaries are anticipated and their high surface area and porosity are also helpful. Recently, several researchers have tried to directly electrospin ceramic NFs (such as TiO_2, SnO_2, and ZnO) onto conductive substrates, as photoelectrodes for the DSSCs, and achieved interesting performance (Song et al., 2004). Jo et al. (2005) used porous electrospun TiO_2 NFs in the quasi-solid-state DSSCs and achieved a photocurrent generation of over 90% of the liquid-electrolyte DSSCs (Jo et al., 2005). The NFs are composed of 20 nm thick fibrils with an islands-in-a-sea morphology, as shown in Fig. 10.20A. To enhance the adhesion and conductivity between the TiO_2 fibrous membranes and the substrates, electrospun TiO_2 membranes are treated with the $TiCl_4$ aqueous solution to implant rutile crystals on the fibers' surface (Fig. 10.20B). As a result, the photocurrent increases by more than 30% (Song et al., 2004). Figure 10.20A shows a NRs-on-nanoparticles configuration, developed by Fujihara et al. (2007). In this TiO_2 NFs are mechanically grinded to yield TiO_2 NRs and they spread onto a layer of TiO_2 nanoparticles. Figure 10.20B shows a cross section of the resulting electrode. This method can potentially solve the poor adhesion problem in DSSCs and, therefore, achieve an energy conversion efficiency of 5.8% (Fujihara et al., 2007).

Wali et al. (2014) reported the fabrication of DSSC using different nanomorphologies of SnO_2 as the photoanode. The PNFs and MPNFs of SnO_2 are depicted in Fig. 10.21. Figure 10.21 shows SEM images of the cross-sectional view of the photoanodes fabricated using PNFs and MPNFs. The thickness of the films is 17 ± 0.5 μm. There is no agglomeration of PNFs and MPNFs in the fabricated photoanode film, ensuring high porosity available for efficient electrolyte penetration (Zhang and Cao, 2011). The dye-loading on high-surface-area PNF- and MPNF-based photoanodes is calculated by a desorption test. The calculated dye-loading of MPNFs is 3.38×10^{-7} mol cm^{-2}, which is four times larger than that of PNFs (0.83×10^{-7} mol cm^{-2}). The significant enhancement in dye-loading is attributed to the higher surface area of MPNFs. These results are compared with the control electrode of TiO_2, which is fabricated using a commercial P25 paste. The dye-loading of the P25-based electrode (7.02×10^{-7} mol cm^{-2}) is twice that of the MPNF-based electrodes. The current-voltage analysis of the

DSSCs is given in Fig. 10.22. As expected from the high surface area in MPNFs, a high short-circuit photocurrent density (J_{sc}) of ¼ 18 mA cm^{-2} and power conversion efficiency (PCE) of ~4.3% are registered for MPNF DSSCs, while for PNF DSSCs, J_{sc} remains 10 mA cm^{-2} and the PCE is 2%.

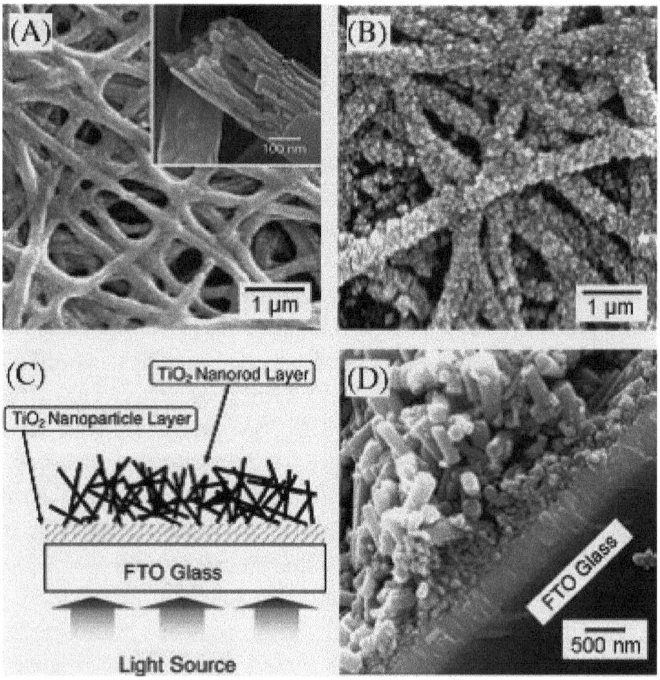

Figure 10.20 SEM images of anatase TiO_2 electrodes derived from electrospun nanofibers: (A) after calcining the as-spun TiO_2/PVAc composite nanofibers (as shown in the inset, each TiO_2 fiber consists of a bundle of fibrils roughly 20 nm in diameter) and (B) after treatment with an aqueous $TiCl_4$ solution to deposit an additional layer of rutile to improve the performance. (C) Schematic of a nanorod-on-nanoparticle electrode prepared by mechanically grinding electrospun TiO_2 nanofibers and then electrospraying the resultant TiO_2 NRs onto a fluorine-doped tin oxide (FTO) substrate. (D) SEM image of a cross section of such an electrode. (A, B) Adapted from Song MY, Kim DK, Ihn KJ, Jo SM, Kim DY. (2004). Electrospun TiO_2 electrodes for dye-sensitized solar cells, *Nanotechnology*, **15**: 1861–1865. © IOP Publishing. Reproduced with permission. All rights reserved. (C, D) Adapted from Fujihara K, Kumar A, Jose R, Ramakrishna S, Uchida S. (2007). Spray deposition of electrospun TiO_2 nanorods for dye-sensitized solar cell. *Nanotechnology*, **18**: 365709. © IOP Publishing. Reproduced with permission. All rights reserved.

DSSCs are nearly half (J_{SC} = ¼ 10 mA cm^{-2} and PCE ≈ 2%). The open-circuit voltage (V_{oc}) in MPNF DSSCs demonstrates an increment of 0.57 mV as a result of the enhanced J_{sc}. The fill factors (FFs) of both DSSCs remain practically the same (0.47 in MPNFs and 0.45 in PNFs). As the thickness of both electrodes is the same, the presence of multichannels in MPNFs and their smaller grain size compared to PNFs result in up to eight times increase in the surface area and up to five times larger dye-loading, attributed to the higher J_{sc}. On the other hand, it is reported that low V_{oc} and FF are the main issues in SnO_2-based DSSCs due to its lower conduction band edge, which results in high electron recombination with the electrolyte species. Therefore, to suppress the charge recombination, a thin layer of TiO_2 is introduced onto the SnO_2 PNF and MPNF photoanodes by $TiCl_4$ treatment (the treated samples were named T-PNFs and TMPNFs, respectively) (Ito et al., 2005). The $TiCl_4$-treated photoelectrode demonstrates a significant increase in photovoltaic parameters (Fig. 10.22). The V_{oc}, J_{sc}, and PCE in MPNF-based

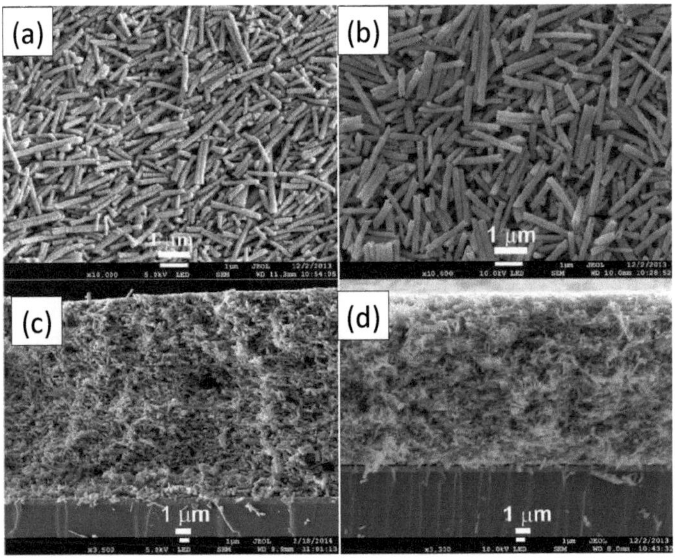

Figure 10.21 (a, b) The top views of coated PNF and MPNF photoanodes and (c, d) the cross-sectional view. Reproduced from Wali Q, Fakharuddin A, Ahmed I, Ab Rahim MH, Ismail J, Jose R. (2014). Multiporous nanofibers of SnO_2 by electrospinning for high-efficiency dye-sensitized solar cells. *J. Mater. Chem. A*, **2**: 17427, with permission from The Royal Society of Chemistry.

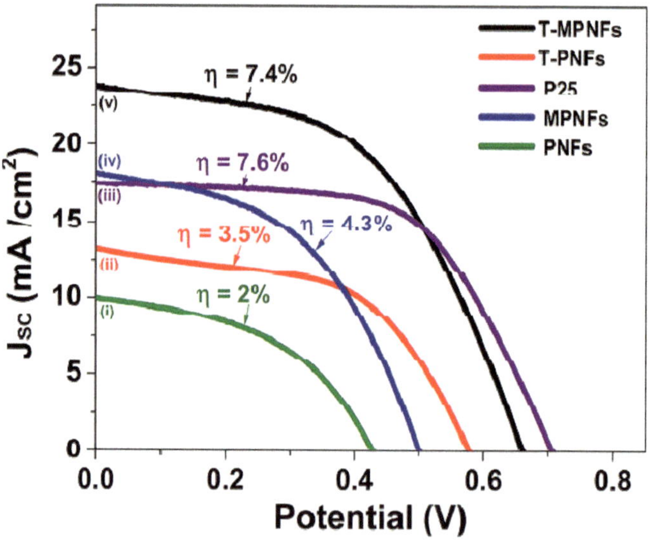

Figure 10.22 Current density versus voltage (*J–V*) curves of (i) pure PNFs; (ii) pure MPNFs; and (iii) P25, (iv) T-PNF, and (v) MPNF DSSCs at AM 1.5 G conditions. Reproduced from Wali Q, Fakharuddin A, Ahmed I, Ab Rahim MH, Ismail J, Jose R. (2014). Multiporous nanofibers of SnO_2 by electrospinning for high-efficiency dye-sensitized solar cells. *J. Mater. Chem. A*, **2**: 17427, with permission from The Royal Society of Chemistry.

DSSCs increase from 0.51 to 0.66 V, 18 to 24 mA cm^{-2}, and 4.3% to 7.4%, respectively. The observed 30% increase in V_{oc} and 33% increase in J_{sc} results in ~80% increment in the final PCE upon $TiCl_4$ treatment. A similar increase in photovoltaic parameters is achieved for PNF DSSCs. The MPNFs excel compared to conventional NFs and NWs when employed as photoanodes in DSSCs, and a significantly higher short-circuit current is achieved in MPNF DSSCs compared to the conventional devices in which a commercial TiO_2 paste is used. The MPNFs enabled >80% charge collection (45% higher than spherical P25 TiO_2 particles) and also demonstrated an approximately three times longer electron lifetime and electron diffusion length. Owing to the scalability of the electrospinning technique, desirable device characteristics, and huge commercial potential of SnO_2 nanostructures, the present materials are expected to be utilized in commercial devices.

10.5 Conclusion

This chapter brings about the fact that there are huge possibilities for the synthesis of ceramic NFs from a variety of ceramic precursors in combination with polymer and solvent by the electrospinning technique. Electrospinning is a simple and versatile technique for preparing the various types of NFs with a controlled morphology and structure. Ceramic NFs have reached an unprecedented level of success. There is no doubt that electrospinning has become one of the most favored techniques for fabricating 1D ceramic NFs in a cost-effective and controllable fashion. By taking advantage of their attractive features, such as high porosity, large surface area, and membrane-like morphology, people have also started to apply them to a broad range of applications, especially in the areas of catalysis, environment science, energy technology, etc.

References

Argonide Advanced Filtration Technologies, Nanoceram Filtration Solutions, http://www.argonide.com.

Augustin S, Hemmige V, Horpel G, Hying C. (2002). Ceramic but flexible: new ceramic membrane foils for fuel cells and batteries. *Desalintion*, **146**: 23–28.

Azad AM, Matthews T, Swary J. (2005). Processing and characterization of electrospun Y_2O_3-stabilized ZrO_2 (YSZ) and Gd_2O_3-doped CeO_2 (GDC) nanofibers. *Mater. Sci. Eng. B*, **123**: 252–258.

Azad AM. (2006). Fabrication of transparent alumina nanofibers by electrospinning. *Mater. Sci. Eng. A*, **468**: 435–436.

Azad A-Z. (2006). Fabrication of yttria-stabilized zirconia nanofibers by electrospinning. *Mater. Lett.*, **60**: 67–72.

Ban C, Chernova NA, Whittingham, MS. (2009). Electrospun nano-vanadium pentoxide cathode. *Electrochem. Commun.*, **11**: 522.

Barsan N, Koziej D, Weimar U. (2007). Metal oxide-based gas sensor research: how to? *Sens. Actuators B*, **121**: 18–35.

Bender ET, Katta P, Lotus A, Park SJ, Chase GG, Ramsier RD. (2006). Electrospinning for advanced energy and environmental applications. *Chem. Phys. Lett.*, **423**: 302–305.

Chen WS, Huang DA, Chen HC, Shie TY, Hsieh CH, Liao JD, Kuo C. (2009). Fabrication of polycrystalline ZnO nanotubes from the electrospinning of Zn^{2+}/poly(acrylic acid). *Cryst. Growth Des.*, **9**: 4070–4077.

Cheng GE, Chen JY, Ke HZ, Shang J, Chu R. (2011). Synthesis, characterization and photo catalysis of SnO_2 nanorods with large aspect ratios. *J. Mater. Lett.*, **65**(21/22): 3327–3329.

Choi SS, Lee SG, Im SS, Kim SH, Joo YL. (2003). Silica nanofibers from electrospinning/sol-gel process. *J. Mater. Sci. Lett.*, **22**: 891–893.

Chowdhuri A, Gupta V, Sreenivas K, Kumar R, Mozumdar S, Patanjali PK. (2004). Response speed of SnO_2-based H_2S gas sensors with CuO nanoparticles. *Appl. Phys. Lett.*, **84**: 1180.

Dai HQ, Gong J, Kim H, Lee D. (2002). A novel method for preparing ultra-fine alumina-borate oxide fibres via an electrospinning technique. *Nanotechnology*, **13**: 674–677.

Dersch R, Graeser M, Greiner A, Wendorff JH. (2007). Electrospinning of nanofibres: towards new techniques, functions, and applications. *Aust. J. Chem.*, **60**: 719–728.

DeVos RM, Verweij H. (1998). High-selectivity, high-flux silica membranes for gas separation. *Science*, **279**: 1710–1711.

Dhakate SR, Gupta A, Choudhari A, Tawale J, Mathur RB. (2011). Morphology and thermal properties of PAN copolymer based electrospun nanofibers. *Synth. Met.*, **161**: 411–419.

Dhakate SR, Singla B, Uppal M, Mathur RB. (2010). Effect of processing parameters on morphology and thermal properties of electrospun polycarbonate nanofibers. *Adv. Mater. Lett.*, **1**(3): 200.

Dharmaraj N, Park HC, Kim CH, Viswanathamurthi P, Kim HY. (2006). Nanometer sized tantalum pentoxide fibers prepared by electrospinning. *Mater. Res. Bull.*, **41**: 612.

Dharmaraj N, Park HC, Kim CK, Kim HY, Lee DR. (2004). Nickel titanate nanofibers by electrospinning. *Mater. Chem. Phys.*, **87**: 5–9.

Dharmaraj N, Park HC, Lee BM, Viswanathamurthi P, Kim HY, Lee D. (2004). Solid state gas sensors: industrial application. *Inorg. Chem. Commun.*, **7**: 431–433.

Formalas A. (1934). Process and apparatus for preparing artificial threads. US Patent, US 1975504.

Fouad OA, Glaspell G, El-Shall MS. (2008). Growth and characterization of ZnO, SnO_2 and ZnO/SnO_2 nanostructures from the vapor phase. *Top. Catal.*, **47**: 84–96.

Fu ZW, Ma J, Qin Q-Z. (2005). Nanostructured LiCoOZ and LiMnzQ, fibers fabricated by a high frequency electrospinning. *Solid State Ionics*, **176**: 1635–1640.

Fujihara K, Kumar A, Jose R, Ramakrishna S, Uchida S. (2007). Spray deposition of electrospun TiO_2 nanorods for dye-sensitized solar cell. *Nanotechnology*, **18**: 365709.

Fujita M, Inukai K, Sakida S, Nanba T, Ommyoji J, Yamaguchi A, Miura Y. (2007). Sintering of Al_2O_3-Cr_2O_3 powder prepared by sol-gel process. *J. Soc. Mater. Sci. Jpn.*, **56**: 526–530.

Gao D, Wang Q, Qiao H, Cai Y, Huang F, Wei Q. (2002). Preparation and characterization of porous TiO_2 fibers and their photocatalytic activity. *J. Eng. Fibers Fabr.*, **7**(2): 94–98.

Gouma PI. (2003). Nanostrutured polymorphic oxides for advanced chemosensors. *Rev. Adv. Mater. Sci.*, **5**: 147–154.

Guan HY, Shao CL, Wen SB, Chen B, Gong J, Yang XH. (2003). An introduction to electrospinning and nanofibers. *Mater. Chem. Phys.*, **82**: 1002.

Gubbala S, Chakrapani V, Kumar V, Sunkara MK. (2008). Band-edge engineered hybrid structures for dye-sensitized solar cells based on SnO_2 nanowires. *Adv. Funct. Mater.*, **18**: 2411–2418.

Gupta A, Dhakate SR, Gurunathan P, Ramesha K. (2017). Electrospun hierarchically porous tin oxide(IV)-carbon composite nanofibers as anode material for lithium ion batteries. *Appl. Nanosci.*, (to be updated).

Gupta A, Nandanwar DV, Dhakate SR. (2015). Electrospun self-assembled ZnO nanofibers structures for photocatalytic activity in natural solar radiations to degrade acid fuchsin dye. *Adv. Mater. Lett.*, **6**(8): 706–710.

Gupta R, Kumar A. (2008). Bioactive materials for biomedical applications using sol-gel technology. *Biomed. Mater.*, **3**: 034005.

Hao R, Yuan J, Peng Q. (2006). Fabrication and sensing behavior of Cr_2O_3 nanofibers via in situ gelation and electrospinning. *Chem. Lett.*, **35**: 1248–1249.

Hennige V, Hying C, Horpel G, Novak P, Vetter J. (2006). Separator provided with asymmetrical pore structures for an electrochemical cell. US Patent Appl. 20,060,078,791.

Hoffmann, MR, Martin, ST, Choi W, Bahnemann DW. (1995). Environmental applications of semiconductor photocatalysis. *Chem. Rev.*, **95**: 69–96.

Huang ZM, Zhang YZ, Kotaki M, Ramakrishna S. (2003). A review on polymer nanofibers by electrospinning and their applications in nanocomposites. *Comp. Sci. Technol.*, **63**: 2223–2253.

Ito S, Liska P, Comte P, Charvet R, Pechy P, Bach U, Schmidt-Mende L, Zakeeruddin SM, Kay A, Nazeeruddin MK, Gratzel M. (2005). Control of dark current in photoelectrochemical ($TiO_2/I^-–I_3^-$) and dye-sensitized solar cells. *Chem. Commun.*, **34**: 4351–4353.

Ji L, Lin Z, Guo B, Medford AJ, Zhang ZW. (2010). Assembly of carbon–SnO_2 core–sheath composite nanofibers for superior lithium storage. *Chemistry*, **16**(38): 11543–11548.

Jiu J, Isoda S, Wang F, Adachi M. (2006). Dye-sensitized solar cells based on a single-crystalline TiO_2 nanorod film. *J. Phys. Chem.*, **110**: 2087–2092.

Jo SM, Song MY, Ahn YR, Park CR, Kim DY. (2005). Nanofibril formation of electrospun TiO_2 fibers and its application to dye-sensitized solar cells. *J. Macromol. Sci. A Pure Appl. Chem.*, **42**: 1529–1540.

Ju SY, Facchetti A, Xuan Y, Liu J, Ishikawa F, Ye P, Zhou CW, Marks TJ, Janes DB. (2007). Fabrication of fully transparent nanowire transistors for transparent and flexible electronics. *Nat. Nanotechnol.*, **2**: 378–384.

Kaur N, Kumar V, Dhakate SR. (2016). Synthesis and characterization of multiwalled CNT–PAN based composite carbon nanofibers via electrospinning. *Springer Plus*, **5**(1): 483.

Ke XB, Zhu HY, Gao XP, Liu JW, Zheng ZF. (2007). High-performance ceramic membranes with a separation layer of metal oxide nanofibers. *Adv. Mater.*, **19**: 785–790.

Kim ID, Rothschild A, Lee BH, Kim DY, Jo SM, Tuller HL. (2006). Ultrasensitive chemiresistors based on electrospun TiO_2 nanofibers. *Nano Lett.*, **6**: 2009.

Kong J, Franklin N, Zhou C, Chapline M, Peng S, Cho K, Dai H. (2000). Nanotube molecular wires as chemical sensors. *Science*, **287**: 622–625.

Kumar EN, Jose R, Archana PS, Vijila C, Yusoff MM, Ramakrishna S. (2012). High performance dye-sensitized solar cells with record open circuit voltage using tin oxide nanoflowers developed by electrospinning. *Energy Environ. Sci.*, **5**: 5401–5407.

Larsen G, Velarde-Ortiz R, Minchow K, Barrero A, Loscertales IG. (2003). A method for making inorganic and hybrid (organic/inorganic) fibers and vesicles with diameters in the submicrometer and micrometer range via sol-gel chemistry and electrically forced liquid jets. *J. Am. Chem. Soc.*, **125**: 1154–1155.

Lee SH, Tekmen C, Sigmund WM. (2005). Three-point bending of electrospun TiO_2 nanofibers. *Mater. Sci. Eng. A*, **398**: 77–81.

Lee, JH, Ko, KH, Park, BO. (2003). Electrical and optical properties of ZnO transparent conducting films by the sol–gel method. *J. Cryst. Growth*, **247**: 119–125.

Li D, Herricks T, Xia YN. (2003). Magnetic nanofibers of nickel ferrite prepared by electrospinning. *Appl. Phys. Lett.*, **83**: 4586.

Li D, McCann JT, Xia YN. (2006). Electrospinning: a simple and versatile technique for producing ceramic nanofibers and nanotubes. *J. Am. Ceram. Soc.*, **89**: 1861.

Li D, Wang YL, Xia YN. (2003). Electrospinning of polymeric and ceramic nanofibers as uniaxially aligned arrays. *Nano Lett.*, **3**: 1167–1171.

Li D, Xia Y. (2003). Fabrication of titania nanofibers by electrospinning. *Nano Lett.*, **3**(4): 555–560.

Li D, Xia YN. (2004a). Direct fabrication of composite and ceramic hollow nanofibers by electrospinning. *Nano Lett.*, **4**: 933–938.

Li D, Xia YN. (2004b). Electrospinning of nanofibers: reinventing the wheel? *Adv. Mater.*, **16**: 1151–1170.

Li JY, Dai H, Li Q, Zhong XH, Ma XF, Meng J, Cao XQ. (2006). Lanthanum zirconate nanofibers with high sintering-resistance. *Mater. Sci. Eng. B*, **133**: 209.

Li QH, Liang YX, Wan Q, Wang HT. (2004). Oxygen sensing characteristics of individual ZnO nanowire transistors. *Appl. Phys. Lett.*, **85**: 6389–6391.

Liu Y, Sagi S, Chandrasekar R, Zhang L, Hedin NE, Fong H. (2008). Preparation and characterization of electrospun SiO_2 nanofibers. *J. Nanosci. Nanotechnol.*, **8**(3): 1528–1536.

Lu X, Liu X, Zhang W, Wanga C, Wei Y. (2006). Synthesis and characterization of tungstate ($WO_3 \cdot H_2O$) nanoleaves and nanoribbons. *J. Colloid Interface Sci.*, **298**: 996.

Lu X, Zhao Q, Liu X, Wang D, Zhang W, Wang C, Wei Y. (2006). Preparation and characterization of polypyrrole/TiO_2 coaxial nanocables. *Macromol. Rapid Commun.*, **27**: 430.

MacDonald AH. (2001). Copper oxides get charged up. *Nature*, **414**: 409–410.

Madhugiri S, Sun B, Smirniotis PG, Ferraris JP, Balkus KJ. (2004). Fabrication of refining mesoporous silica nanofibers via electrospinning. *Microporous Mesoporous Mater.*, **69**: 77.

Maensiri S, Nuansing W. (2006). Thermoelectric oxide $NaCo_2O_4$ nanofibers fabricated by electrospinning. *Mater. Chem. Phys.*, **99**: 104.

McCann JT, Chen JIL, Li D, Ye Z-G, Xia Y. (2006). Nanotechnology for the energy challenge. *Chem. Phys. Lett.*, **424**: 162.

McCann JT, Li D, Xia YN. (2005). Electrospinning of nanofibers with core-sheath, hollow, or porous structures. *J. Mater. Chem.*, **15**: 735–738.

Moseley PT, Crocker AJ. (1996). *Sensor Materials*. Institute of Physics, Philadelphia.

Musa AO, Akomolafe T, Carter MJ. (998). Production of cuprous oxide, a solar cell material, by thermal oxidation and a study of its physical and electrical properties. *Sol. Energy Mater. Sol. Cells*, **51**: 305–316.

Nguyen TA, Park S, Kim JB, Kim TK, Seong GH, Choo J, Kim YS. (2011). Polycrystalline tungsten oxide nanofibers for gas-sensing applications. *Sens. Actuators B*, **160**(1): 549–554.

Noda T, Sagara H, Yen A, Takada A, Kida H, Cheng RH, Kawaoka Y. (2006). Architecture of ribonucleo protein complexes in influenza A virus particles. *Nature*, **439**: 490–492.

Nunes SP, Sforça ML, Peinemann K. (1995). Dense hydrophilic composite membranes for ultrafiltration. *J. Membr. Sci.*, **106**: 49–56.

O'Regan B, Gratzel M. (1991). A low-cost, high-efficiency solar cell based on dye-sensitized colloidal TiO_2 films. *Nature*, **353**: 737–740.

Ostermann R, Li D, Yin Y, McCann JT, Xia Y. (2006). V_2O_5 nanorods on TiO_2 nanofibers: a new class of hierarchical nanostructures enabled by electrospinning and calcination. *Nano Lett.*, **6**: 1297–1302.

Palacios T. (2012). Applied physics nanowire electronics comes of age. *Nature*, **481**: 152–153.

Ponzoni A, Comini E, Ferroni M, Sberveglieri G. (2005). Nanostructured WO_3 deposited by modified thermal evaporation for gas-sensing applications. *Thin Solid Films*, **490**: 81.

Ponzoni A, Comini E, Sberveglieri G, Zhou J, Deng SZ, Xu NS, Ding Y, Wang ZL. (2006). Ultrasensitive and highly selective gas sensors using three-dimensional tungustan oxide nanowires networks. *Appl. Phys. Lett.*, **88**: 203101.

Qi Q, Zhang T, Liu L, Zheng X, Lu G. (2009). Improved NH_3, C_2H_5OH, and CH_3COCH_3 sensing properties of SnO_2 nanofibers by adding block copolymer P123. *Sens. Actuators B*, **141**: 174–178.

Ramaseshan R, Sundarrajan S, Jose R. (2007). Nanostructured ceramics by electrospinning. *J. Appl. Phys.*, **102**(11), doi: http://dx.doi.org/10.1063/1.2815499.

Salje EJ. (1974). A new type of electro-optic effect in semiconducting WO_3. *Appl. Crystallogr.*, **7**: 615.

Shao CL, Guan HY, Liu YC, Gong J, Yu N, Yang XH. (2004). Atomically-precise methods for synthesis of solid catalysts. *J. Cryst. Growth*, **267**: 380.

Shao CL, Guan HY, Liu YC, Li XL, Yang XH. (2004). Preparation of Mn_2O_3 & Mn_3O_4 nanofibres via an electrospinning technique. *J. Solid State Chem.*, **177**: 2628–2631.

Shao CL, Kim H, Gong J, Lee D. (2002). A novel method for making silica nanofibres by using electrospun fibres of polyvinyl alcohol/silica composite as precursor. *Nanotechonology*, **13**: 635–637.

Shao CL, Yang XH, Guan HY, Liu YC, Gong H. (2004). Electrospun nanofibers of NiO/ZnO composite. *Inorg. Chem. Commun.*, **7**: 625–627.

Sharma A, Gupta A, Rath G, Goyal A, Mathur RB, Dhakate SR. (2013). Electrospun composite nanofiber-based transmucosal patch for anti-diabetic drug delivery. *J. Mater. Chem. B*, **1**: 3410–3418.

Shimizu Y, Jono A, Hyodo T, Egashira M. (2005). Preparation of large mesoporous SnO_2 powder for gas sensor application. *Sens. Actuators B*, **108**: 56–61.

Song MY, Kim DK, Ihn KJ, Jo SM, Kim D. (2005). New application of electrospun TiO_2 electrode to solid-state dye-sensitized solar cells. *Synth. Met.*, **153**: 77–80.

Song MY, Kim DK, Ihn KJ, Jo SM, Kim DY. (2004). Electrospun TiO_2 electrodes for dye-sensitized solar cells, *Nanotechnology*, **15**: 1861–1865.

Song MY, Kim DK, Jo SM, Kim DY. (2005). Enhancement of the photocurrent generation in dye-sensitized solar cell based on electrospun TiO_2 electrode by surface treatment. *Synth. Met.*, **155**: 635–638.

Sun DH, Chang C, Li S, Lin L. (2006). Near-field electrospinning. *Nano Lett.*, **6**: 839–842.

Takamura N, Taguchi K, Gunji T, Abe Y. (1999). Preparation of silicon oxycarbide ceramic films by pyrolysis of polymethyl- and polyvinylsilsesquioxanes. *J. Sol-Gel Sci. Technol.*, **16**: 227–234.

Tekmen C, Suslu A, Cocen U. (2008). Titania nanofibers prepared by electrospinning. *Mater. Lett.*, **62**: 4470–4472.

Tomer V, Mensah RT, Tokash JC, Stojilovic N, Kataphinan N, Evans EA, Chase GG, Ramsier RD, Smith DJ, Reneker DH. (2005). Selective emitters for thermophotovoltaics: erbia-modified electrospun titania nanofibers. *Sol. Energy Mater. Sol. Cells*, **85**: 477–478.

Verweij H. (2003). Ceramic membranes: morphology and transport. *J. Mater. Sci.*, **38**: 4677–4695.

Viswanathamurthi P, Bhattarai N, Kim HY, Khil MS, Lee DR, Suh EK. (2004). GeO_2 fibers: preparation, morphology and photoluminescence property. *J. Chem. Phys.*, **121**: 441.

Viswanathamurthi P, Bhattarai N, Kim HY, Khil MS, Lee DR, Suh EK. (2004). GeO_2GeO_2 fibers: preparation, morphology and photoluminescence property. *J. Chem. Phys.*, **121**: 441–445.

Viswanathamurthi P, Bhattarai N, Kim HY, Lee DR, Kim SR, Morris M. (2003). Preparation and morphology of niobium oxide fibres by electrospinning. *Chem. Phys. Lett.*, **374**: 79–84.

Wali Q, Fakharuddin A, Ahmed I, Ab Rahim MH, Ismail J, Jose R. (2014). Multiporous nanofibers of SnO_2 by electrospinning for high efficiency dye-sensitized solar cells. *J. Mater. Chem. A*, **2**: 17427.

Wang A, Singh H, Hatton TA, Rutledge GC. (2004). Synthesis techniques for polymer nanocomposites. *Polymer*, **45**: 5505–5514.

Wang CH, Tong YB, Sun ZY, Xin Y, Yan EY, Huang ZH. (2007). New and future developments in catalysis: catalysis by nanoparticles. *Mater. Lett.*, **61**: 5125–5128.

Wang G, Ji Y, Huang X, Yang X, Gouma P-I, Dudley M. (2006). Fabrication and characterization of polycrystalline WO_3 nanofibers and their applications for ammonia sensing. *J. Phys. Chem. B*, **110**: 23777–23782.

Wang Y, Aponte M, Leon N, Ramos I, Furlan R, Evoy S, Santiago-Aviles JJ. (2004). Synthesis and characterization of tin oxide microfibres electrospun from a simple precursor solution. *Semicond. Sci. Technol.*, **19**: 1057.

Windischmann H, Mark P. (1997). A model for the operation of a thin films tin oxide conductance modulation carbon monoxide sensor. *J. Electrochem. Soc.*, **126**: 627–630.

Wu H, Lin D, Pan W. (2006). Fabrication, assembly, and electrical characterization of CuO nanofibers. *Appl. Phys. Lett.*, **89**: 133125.

Wu H, Pan W. (2006). Preparation of zinc oxide nanofibers by electrospinning. *J. Am. Ceram. Soc.*, **89**: 699.

Xie XL, Mai YW, Zhou XP. (2005). Dispersion and alignment of carbon nanotubes in polymer matrix: a review. *Mater. Sci. Eng. R*, **49**: 89–112.

Xo HK, Martin IA, Antonucci JM, Eichmiller FC. (1999). Ceramic whisker reinforcement of dental resin composites. *J. Dent. Res.*, **78**: 706–712.

Yaman M, Khudiyev T, Ozgur E, Kanik M, Aktas O, Ozgur EO, Deniz H, Korkut E, Bayindir M. (2011). Arrays of indefinitely long uniform nanowires and nanotubes. *Nat. Mater.*, **10**: 494–501.

Yang SY, Ryu I, Kim HY, Kim JK, Jang SK, Russell TP. (2006). Nanoporous membranes with ultrahigh selectivity and flux for the filtration of viruses. *Adv. Mater.*, **18**: 70912.

Yang XG, Shao CL, Liu YC, Mu RX, Guan HY. (2005). Nanofibres of CeO_2 via an electrospinning technique. *Thin Solid Films*, **478**: 228–231.

Yao J, Bastiaansen CWM, Peijs T. (2014). High strength and high modulus electrospun *nanofibers. Fibers*, **2**: 158.

Yu N, Shao C, Liu Y, Guan H, Yang X. (2005). Nanofibers of $LiMn_2O_4$ by electrospinning. *J. Colloid Interface Sci.*, **285**: 163–166.

Yuh J, Nino JC, Sigmund W. (2005). Nanoparticulate materials: synthesis, characterization, and processing. *Mater. Lett.*, **59**: 3645–3647.

Yuh J, Perez L, Sigmund WM, Nino JC. (2007). Electrospinning of complex oxide nanofibers. *Physica E*, **37**: 254–259.

Zhang G, Kataphinan W, Teye-Mensah R, Katta P, Khatri L, Evans EV, Chase CG, Ramsier RD, Reneker DH. (2005). Electrospun nanofibers for potential space-based applications. *Mater. Sci. Eng. B*, **116**: 353–358.

Zhang Q, Cao G. (2011). Nanostructured photoelectrodes for dye-sensitized solar cells. *Nano Today*, **6**: 91–109.

Zhang SS. (2007). A review on the separators of liquid electrolyte Li-ion batteries. *J. Power Sources*, **164**: 351–364.

Zhang XW, Xu SY, Han GR. (2009). Fabrication and photocatalytic activity of TiO_2 nanofiber membrane. *Mater. Lett.*, **63**: 1761–1763.

Zhang Y, Li J, Li Q, Zhu L, Liu X, Zhong X, Meng J, Cao X. (2007a). Functional nanofibers and their applications. *J. Colloid Interface Sci.*, **307**: 567.

Zhang Y, Li J, Li Q, Zhu L, Liu X, Zhong X, Meng J, Cao X. (2007b). Preparation of In_2O_3 ceramic nanofibers by electrospinning and their optical properties. *Scr. Mater.*, **56**: 409–412.

Zhang Z, Li X, Wang C, Wei L, Liu Y, Shao C. (2009). ZnO hollow nanofibers: fabrication from facile single capillary electrospinning and applications in gas sensors. *J. Phys. Chem. C*, **113**: 19397.

Zhu HY, Gao XP, Lan Y, Song DY, Xi YX, Zhao JC. (2004). Hydrogen titanate nanofibers covered with anatase nanocrystals: a delicate structure achieved by the wet chemistry reaction of the titanate nanofibers. *J. Am. Chem. Soc.*, **126**: 8380.

Zhu HY, Gao XP, Song DY, Bai YQ, Ringer SP, Gao Z, Xi YX, Martens W, Riches JD, Frost RL. (2004). Growth of boehmite nanofibers by assembling nanoparticles with surfactant micelles. *J. Phys. Chem. B*, **108**: 4245.

Chapter 11

Corrosion-Resistant Ceramic Nanomaterial Systems Derived through Sol-Gel Technology

S. C. Mojaki, A. K. Mishra, and S. B. Mishra

University of South Africa, Nanotechnology and Water Sustainability Research Unit, College of Science Engineering and Technology, Science Campus, Florida, Johannesburg, South Africa

bhards@unisa.ac.za

Structural destructions of materials as a result of a corrosive environment are unpleasant phenomena to deal with. The destructions may not only be economically costly but also be threatening to the safety and operations of equipment and systems, which, if not managed, could lead to fatal or near-fatal incidents, at worst. Ceramic-based nanomaterials have benefited the corrosion industry due to their modifiable chemical and subsequently physical properties in order to meet the demands of performing materials for corrosion protection or inhibition applications. Protective systems containing ceramic nanomaterials

Smart Ceramics: Preparation, Properties, and Applications
Edited by Ajay Kumar Mishra

ISBN 978-981-4774-30-7 (Hardcover), 978-1-315-16359-8 (eBook)
www.panstanford.com

ought to display active and/or passive means of protection on metal materials. The performance of ceramic nanomaterials is measured on the ability of the protective system to reduce the corrosion rate significantly, to minimal or zero level, thereby extending the life span of metal materials to a longer period under corrosive conditions. Control over aspects such as chemical compositions, particle size, specific particle shape, particle size distribution, homogeneity, and purity of natural products and synthetic materials is important for the properties of the final ceramic material to be produced. This chapter provides a review of the concept of corrosion, protective measures, and types of ceramic nanomaterials applied for corrosion protection. Additionally, the chapter presents the synthetic procedure and characterization techniques of these ceramic nanomaterials.

11.1 Introduction

The design of critical structures and their components requires more than just attention to the mechanical properties alone, especially where safety and economic performance of the materials are of vital consideration. Structural destruction has made it clear that it is imperative to know more about the selection of material properties and the types and effects of environment those materials interact with, especially when dealing with corrosion. Economically, corrosion is expensive. It poses a threat to the safety and operations of equipment and systems, such as industries, municipalities, or the military (Pietersen et al., 2004; Elahee, 2010; Revie and Uhlig, 2011; Popoola et al., 2013; Shaw et al., 2013). Other corrosion costs besides structural repair include costs of education and further information on the subject, applied research, technical development, hazardous waste material disposal, and sharing of knowledge and technology (Koch et al., 2002; Pietersen et al., 2004; Bardal, 2007; Sherif et al., 2007; Thompson et al., 2007; Demo et al., 2010). From the areas mentioned, a number of advantageous uses of corrosion may also be drawn in growing and sharing the knowledge and technology of corrosion through research and development strategies.

The use of nanomaterials has benefited the corrosion industries in enhancing materials' performance (Brinker et al., 1999; Nalwa, 2001; Saji and Thomas, 2007; Maia et al., 2012; Roberge, 2013) in addressing corrosion control. Nanomaterials have morphological features on a nanoscale that is smaller than 100 nm in at least one dimension and are known for exhibiting uniquely enhanced chemical, physical, and biological properties (Thostenson et al., 2005). They are lighter in weight and have higher mechanical strength, increased control of the light spectrum, and greater chemical reactivity than their larger counterparts (Zäch et al., 2006). The chemical properties of nanomaterials depend on the structuring and arrangements of atoms and molecules. The decrease in particle size of materials leads to a greater number of atoms being on the surface of the particle as well as an increase in the surface area. This signifies higher chemical reactivity occurring on the particle's surface. Numerous properties that are of importance in a material's performance, such as chemical (corrosion, temperature insulation, biocompatibility, wettability, etc.), mechanical (wear, erosion, etc.), optical (absorption, color, reflection, transmission, etc.), and transport (electrical and thermal conductivity), can be improved at this level.

11.2 Corrosion

Some main concepts, such as environment, degradation, reaction, and material, are worth mentioning in defining corrosion. Thus corrosion is defined as the gradual destruction of material when chemically reacting with its environment under various conditions (Roberge, 2000). Not all corrosions are gradual, and the resultant effects are often detrimental to the usage of the metal material considered. Materials susceptible to corrosion include among others wood, polymers (rubber, plastics, etc.), metals, composites, and cement. However, wood may split or decay, polymers crack or swell in the presence of solvents, granite erodes, while cement may leach away. Metals, on the other hand, are relatively active and get corroded through the same basic mechanisms but different from those other materials. Some metals do exhibit high corrosion resistance, which could be accredited to their chemical components or the nature of electrochemical reaction, among others. Common corrosive environments for metals and alloys are aqueous solutions, gases, soil, and nonaqueous liquids. On the basis of the mechanism it follows, corrosion of metals could be classified as physical (Lee et al., 2012), chemical, or electrochemical. In an aqueous environment, corrosion is governed by electrochemical reactions (illustrated in Fig. 11.1), which consist of the following components: an anode, a cathode, an electrolyte (an aqueous solution that is in contact with both the cathode and the anode, providing an ionic conduction path), and an electrical connection between the cathode and the anode. The anodic component undergoes dissolution, releasing electrons that participate in the cathodic reaction as they enter the electrolyte as positively charged metal ions. Electrons released from the anodic site get consumed as they react with reduced electrolytes, in the cathodic component. The flow of electrons within the metal and ions flowing within the electrolyte causes the corrosion current to flow between anode and cathode (Bradford, 1993).

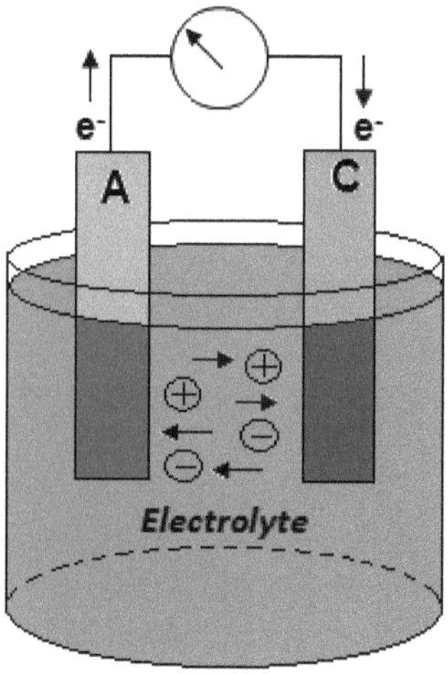

Figure 11.1 Components of an electrochemical corrosion cell (Bradford, 1993).

Corrosion is a natural process that converts the elemental metal to its oxide or salt state. If it is not correctly managed, important structural properties of the materials, such as strength, appearance, and permeability (of moistures and gasses), deteriorate in the process. Besides nature, kinetics, and thermodynamics as aspects affecting the corrosion environment, corrosion or chemical resistance depends on factors such as effect of temperature, oxygen and oxidizers, corrosive concentration, velocity (fluid flow), galvanic coupling, and metallurgical factors (Popoola et al., 2014). Employing suitable approaches in retarding or suppressing anodic, cathodic, or both sites can aid in minimizing the corrosion process. Fundamental laws of electrochemical reaction kinetics are important to assist in improving corrosion protective methods and in developing corrosion-resistant metals or alloys (West, 1986).

11.3 Corrosion Protection

Although some metals have been found to be intrinsically resistant to corrosion than others, it is still imperative to use additional means of surface protection in order to enhance the life span of the metal material. The role of protective systems is to reduce the corrosion rate to a level that will allow metal materials to attain an extended lifetime by preserving essential physical properties of the metal materials such as strength, magnetic property, and electrical and thermal conductivity. To control and manage corrosion, several methods that are of much interest for investigation and used to avoid or prevent corrosion of a metal surface, such as coating (Wang et al., 2000; Carbajal et al., 2001; Voevodin et al., 2003; Zheludkevich et al., 2007; Gopi et al., 2014; Bahrami et al., 2015), inhibition (Pieterse et al., 2006; Umoren et al., 2006; Moutarlier et al., 2008; Wang and Bierwagen, 2009; Harvey et al., 2011; Huang et al., 2012; Kirkpatrick et al., 2014), and cathodic and anodic protection (France, Jr., and Greene, Jr., 1968; Zhong et al., 2006; Riggs, 2012), have been investigated and applied.

11.3.1 Protective Methods

11.3.1.1 Coating protection

The evidence and importance of coating for the purpose of protection and preservation or decoration were always points of consideration by ancient people (Egyptians, Chinese, western

Europeans) when they used varnishes on their artifacts. Currently, protection of metallic substrates from their possible aggressive environments is an active and important area of research in material science. Several barrier coatings and materials capable of preventing corrosion, such as organic coatings (paints, varnish, resin), inorganic coatings (glass linings, enamels), and metallic coatings (through surface treatments), have been employed. Active and passive corrosion protection methods are approaches that can be used in reducing the corrosion process. The former protection mode is carried out by an inhibitive medium or species that functions by decreasing the corrosive activity of the corrosive environment, while the latter protection is accompanied by the formation of a barrier film that will prohibit the corrosive medium or species from contact with the surface of the metal.

11.3.1.2 Inhibition protection

The application of inhibitors (organic) against corrosion of metals dates back several hundreds of years, and so far there are a variety of compounds that have been developed and used for corrosion inhibition. There is considerable interest in the inhibitor's efficiency in controlling corrosion and understanding the mechanism involved during the inhibition process on the metal surface (Riggs, Jr., and Nathan, 1973). Corrosion inhibitors are chemicals that are required in small quantities either intermittently or continuously by systems such as cooling water, acid, and steam systems in order to retard or slow down the undesired corrosion reactions and eventually the corrosion rate of corrosive environments.

Interaction of the inhibiting compounds with the metal on the surface of the metal is accredited to their adsorption mechanism, which may depend on several factors such as the inhibitor's chemical structure, nature and surface charge of the metal, the type of electrolyte solution, and the adsorption mode (Sherif et al., 2007). Corrosion inhibitors, as outlined by Dariva and Galio (2014), can be classified as either natural or synthetic and also through the nature of the chemical structure (as organic or inorganic), the mechanism of action (as anodic, cathodic, or the anodic-cathodic mix and by adsorption action), or as oxidants or antioxidants (Dariva and Galio, 2014) (Fig. 11.2). John and Joseph (2012) reported the effective inhibitive action of substituted triazines on mild steel in an acidic (1 M HCl) environment, which showed to be influenced by the type of functional groups (CH_2Ph, Bu^t, or Me) substituted to the triazine ring in the following decreasing efficiency order: ABTDT (CH_2Ph) > ATTDT (Bu^t) > AMTDT (Me).[1] The polarization studies indicated furthermore a mixed-type inhibitory action of the triazine molecules.

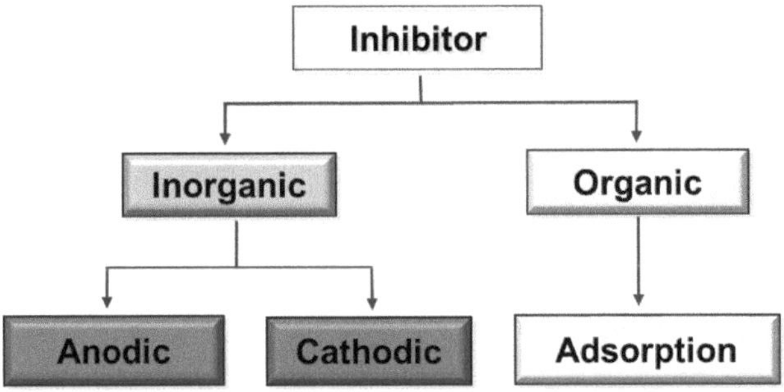

Figure 11.2 Classification of inhibitors. Reproduced from Dariva and Galio (2014) with permission.

11.3.1.3 Cathodic protection

The cathodic protection method was first applied by Sir Humphrey Davy in 1824, when he reported the successful protection of copper against corrosion in seawater (salt water) by coupling copper to iron or zinc anodes (Davy, 1824a, 1824b). The cathodic protection mechanism depends on the application of an external electric current in the polarizing cathode beyond the corrosion potential to the open circuit potential of the anode, causing both electrodes to attain the same potential thus preventing corrosion on a metal (anode). The cathode is more protected in the cathodic protection technique. This mechanism provides an effective means of reducing the

[1]AMTDT, 4-amino-6-methyl-3-thioxo-3,4-dihydro-1,2,4-triazine-5(2*H*)-one; ATTDT, 4-amino-6-(*tert*-butyl-3-thioxo-3,4-dihydro-1,2,4-triazine-5(2*H*)-one; ABTDT, 4-amino-6-benzyl-3-thioxo-3,4-dihydro-1,2,4-triazine-5(2*H*)-one; Ph, phenyl or benzyl group; Bu^t, *tert*-butyl group; Me, methyl.

corrosion rate to almost zero. Cathodic protection has been effectively applied to protect any metal material, such as to brass, copper, lead, and steel (Chung, 2004; Battocchi et al., 2006), from corrosion in corrosive environments like concrete (Pedeferri, 1996), soil, water, or any aqueous media (Touzain et al., 2005). Cathodic protection has two types of corrosion protection, which have been enormously applied: passive or sacrificial anode and active or impressed current.

11.3.1.4 Anodic protection

The anodic protection technique is relatively new and its first application came in 1954 by Edeleanu (1960). Currently, this protection technique is used on a minor scale compared to other corrosion control techniques. Anodic protection is based on passivation and protection of anode (Burleigh, 1989). The technique is accompanied by formation of a film on the metal surface when subjected to anodic polarization by externally applied anodic current equals to the corrosion rate of the system to be protected. Solutions and metals that have been found to exhibit active-passive properties and anodic protection have been shown in Table 11.1.

Table 11.1 Solutions and metals exhibiting active-passive properties and anodic protection

Solutions	Metals
Sulfuric acid (Hiroyoshi et al., 2004)	Steel (Cook et al., 2004)
Phosphoric acid (Amar et al., 2003)	Stainless steel (Zhong et al., 2006)
Nitric acid (Gray and Orme, 2007)	Nickel (Mishra and Balasubramaniam, 2004)
Nitrate solutions (Okada et al., 1976)	Nickel alloys (Morris and Scarberry, 1970)
Aqueous ammonia (Agrawal and Hindin, 1994)	Chromium (Potgieter et al., 2008)
Caustic solution (Contu et al., 2004)	Aluminum (Natishan, 2003)
Organic solutions (Ravichandran et al., 2004)	

11.4 Ceramic Nanomaterials on Corrosion Protection

Several corrosion-resistant coatings produced from ceramic materials have been found to provide good wear- and corrosion-resistant barriers for many metal materials in corrosive environments. Ceramic coatings aid in maintaining an alkaline environment at the metal–ceramic interface, where the corrosion rate is low. Ceramics mostly used for wear protection are dense or have low porosity and work efficiently in protecting the metal surface to hinder corrosion attack by porous connections. Their wide applications are brought about by their good mechanical strength and high chemical and thermal resistance properties. Most ceramics are good insulators. Traditional ceramics include materials ranging from glass to clay and earthenware, whereas modern or engineered ceramics include non-oxide-based ceramics (such as borides, carbides, and nitrides) (Al-Fadhli et al., 2006; Weimer, 2012), oxide-based ceramics (Sun et al., 1998; Shukla et al., 2002), and mixed oxides (Torre et al., 2009), and many others of a complex composition (Torre et al., 2009) can be obtained for several applications. Certain components of electronic devices are protected by silicon nitride coatings, while inorganic coatings of oxide ceramic materials such as alumina, chromia, silica, titania, yttria, and zirconia have been used in surface coatings of metal materials for improving cavitation, corrosion, erosion, and fretting wear resistance (Medvedovski, 2001; Cao et al., 2004; Rosso, 2006; Cao et al., 2008; Tian et al., 2010; Song et al., 2011; Lepule et al., 2015). The resistance of ceramics to chemical corrosion has made them ideal for industrial application (Lin et al. 2006; Sun et al., 2009). Ceramics are inorganic nonmetallic solid compounds made up of metallic and nonmetallic elements. Although there are various kinds of ceramics with different properties, generally they are chemically inert, hard, brittle, and extremely strong in compression and can withstand low tensile stress. Their low fracture toughness, brought about by their covalent and ionic bonding, which leads to a slow or limited dislocation movement of plastic deformation, has remained a drawback preventing their wider application. Efforts in developing new material design to overcome this drawback were shown by the research work of Niihara on aluminum oxide-silicon carbide (Al_2O_3/SiC) systems, which were mainly based on improving the mechanical properties, and also review studies reported by Choi and Awaji (2005) and Camargo et al. (2009) thereafter. As outlined in Camargo et al., other research studies on the system have confirmed the clear improved strength of modified fracture toughness on the resultant material consisting of a dispersed low amount of SiC particles in an Al_2O_3 matrix (Dutta, 2001; Ferroni et al., 2001). Incorporation of nanosized reinforcements aids in deflecting cracks and provides bridging features, which hinder further opening of cracks. Development of advanced ceramic materials provided more diverse hard-metal materials, beyond traditional tungsten carbide (Zum Gahr, 1998; Reyes and Neville, 2003; Katsich and Badisch, 2011). A combination of a ceramic material and a binder component forms a monolithic composite material better known as cermet. Cermets have shown good chemical stability, resistant to abrasion, and more stability in high temperatures (Zum Gahr, 1998; Torre et al., 2009; Amri et al. 2014). Modification of cermet properties can be achieved by altering the characteristic features, amount, or both for each component independently or together (Massa et al., 1998; El-Kady, 2013). Nonmetallic compounds or metalloids form a part of ceramic while the binder constitutes a metal or an alloy connected together in a 3D component. Ceramics may be produced in different sizes and shapes; nevertheless of more importance are the manufacturing techniques.

11.5 Synthetic Methods of Ceramic Nanomaterials

To synthesize novel ceramics with modified features, the chemical composition, particle size, specific particle shape, particle size distribution, homogeneity, and purity of the natural product and synthetic material must be controlled since they play a crucial role in the properties of the complete ceramic product. Processing of natural resources can provide a binary oxide ceramic, such as alumina, chromia, and titania. However, more complex oxides and nonoxide ceramics such as high-temperature superconductors, for example, $YBa_2Cu_3O_{7-x}$, require complex synthetic routes for their production. Ceramics may be produced following two approaches—mechanical (top down) and chemical (bottom up)—where each approach has types of methods classified under it, as shown in Table 11.2.

Several ways to obtain a multicomponent ceramic powder are the use of a high-temperature solid-state reaction of oxide and/or carbonate precursors (conventional route), a self-propagating high-temperature (SHS) reaction, a mechanically induced self-sustaining reaction (Setoudeh and Welham, 2011; Chen et al., 2014), mechanical alloying (Suryanarayana, 2011; Lü and Lai, 2013), carbothermic reduction (Matovic et al., 2007; Xu et al., 2013), and chemosynthesis (Khan et al., 2014).

Table 11.2 Classification of synthetic methods

Synthetic approaches of ceramics	
Top down	Physical methods (solid state)
	• Aerosol synthesis
	• High-energy ball milling
	• Laser ablation
Bottom up	Chemical methods
	• Coprecipitation
	• Hydrothermal
	• Microemulsion
	• Sol-gel

The low cost of powder production on an industrial scale, readily available oxide precursor's materials, and ease of carrying out on a laboratory (Meng et al., 2013) scale are some of the advantages of solid-state reactions. However, solid-state reactions involve high temperatures (>1000°C) for diffusion during calcination, which can result in ceramic product decomposition. Some disadvantages also associated with solid-state reactions are (Segal et al., 1997):

- Poor chemical homogeneity, particularly during doping of oxides
- Large grain sizes as a result of high-temperature firing, which could be damaging for high-strength ceramic components

- Formation of undesirable phases
- Introduction of chemical impurities during particle size reduction through milling

Ceramic powders prepared via conventional routes are somewhat not suited for the fabrication of coatings (Trapaga-Martinez and Levi, 2013). Chemical synthetic methods (soft chemistry routes) commonly used for ceramics that have been studied for modification of nanoparticles and their subsequent nanocomposite materials (Pandey and Mishra, 2011) for various coatings include coprecipitation, hydrothermal methods, microemulsion, and sol-gel. Application of the sol-gel method for the production of corrosion-protective coatings for metal materials was reported by Osborne (2001), with the study proposing consideration of using non-chromate-based conversion coatings (non-CCCs) for an aluminum substrate.

11.6 Sol-Gel Technique

Sol-gel has proved to be an efficient, environmentally friendly process involving nontoxic and noncarcinogenic pretreatment materials and coatings for corrosion protection (Hongtao and Shixi, 2003). Sol-gel-derived coatings have been reported to exhibit excellent chemical stability, controlled oxidation, and enhanced corrosion resistance at the metal–substrate interface, like steel (Aparicio et al., 2016), magnesium alloys (Wang et al., 2010), and aluminum alloys (Metroke et al., 2002; Zhang et al., 2015). The versatility of sol-gel as a surface coating process is through the intrinsic advantages of allowing coating of large areas with robotic techniques and good adhesion of the coating on the broader range of metal substrates. The ease of incorporating other components enhances the functionality of the resultant material, like adhesion protection, antireflection, antifouling, moisture resistance, corrosion protection, and UV shield. The advantages of the sol-gel process, as also outlined previously (Frenzer and Maier, 2006; Wang and Bierwagen, 2009; Pandey and Mishra, 2012), are as follows:

- It is an easily controlled process at relatively low temperatures.

- It provides coatings with controlled optical and electromagnetic properties.
- It allows production of materials in various shapes.
- It allows control of the chemical composition of the final product.
- It allows a controllable degree of porosity.
- It permits the introduction of dopants in the sol phase to be uniformly dispersed in the final product.
- It provides improved homogeneity.

The process starts with the formation of the mineral phase of the selected and hydrolyzed chemical compounds (molecular level) known as precursors, in a preferred liquid solvent (Fig. 11.3). Precursors can be of metal alkoxides, $M(OR)_4$, or inorganic salts in the presence of either mineral acids such as HCl or a base like NH_3 as the catalyst. The hydrolyzed precursor are chemically transformed from liquid phase to form sol through the Brownian motion of the colloids following an inorganic polymerization reaction.

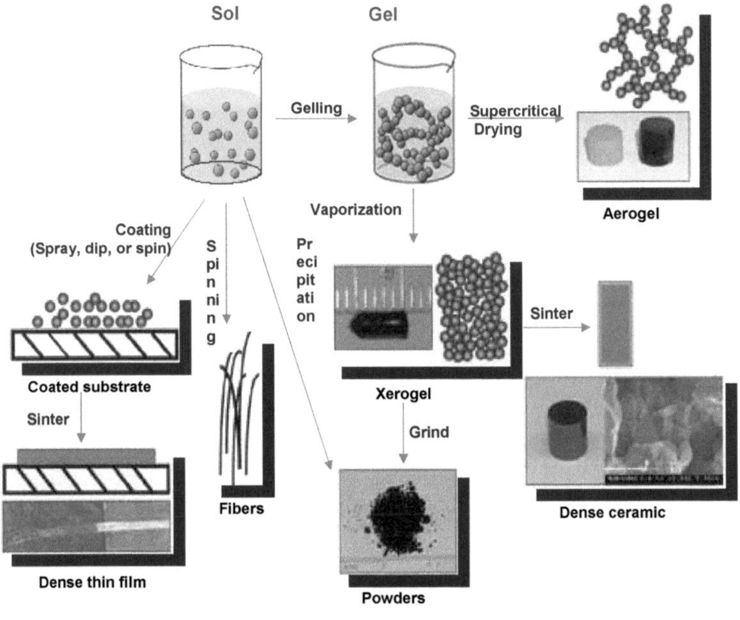

Figure 11.3 Schematic representation of the sol-gel process and the use of subsequent products in various forms.

The fundamental key steps of the sol-gel process are shown in Table 11.3, as summarized and reported by Pandey and Mishra (2012). Numerous precursors may be dissolved in various solvents, which may be aqueous, nonaqueous, and/or a combination of both. A sol can be deposited on the surface of a substrate on an industrial scale through several ways, like dip, spray, and spin coating (as shown in Fig. 11.3); capillary coating; and roll coating.

Table 11.3 Classification of synthetic methods

Step	Purpose	Important parameters
Solution chemistry	To form a gel	Type of precursor, type of solvent, pH (acid/base content), water content, precursor concentration, and temperature
Aging	To allow a gel to undergo changes in properties	Time, temperature, composition of the pore liquid (e.g., pH), and the aging environment (e.g., humidity)
Drying	To remove the solvent from a gel	Drying method (e.g., evaporation versus supercritical versus freeze drying), temperature, and pressurization rate
Calcination/ sintering	To change the physical/chemical properties of the solid, often resulting in crystallization and densification	Temperature and heating rate, time, and the gaseous environment (e.g., inert versus reactive glass)

Despite several advantages of sol-gel processing, the drawbacks of sol-gel-oxide-containing coating are that it suffers from crack formation and/or delamination of coating during high annealing or sintering temperature (>800°C), as required for obtaining dense microstructures (Han et al., 2001). Sol-gel coatings of combined inorganic and organic components offer collaborative effects contributed by properties of both components, leading to improvement of multifunctional materials. The inorganic constituent of the coating reinforces the organic constituent by improving the hardness of the coating while the presence of an organic constituent provides (i) the flexible character that made the coating less inclined to cracking, even on further heat treatment, and (ii) reduction of porosity of the coating material to make it a diffusion-less barrier.

11.6.1 Mechanism of Sol-Gel-Derived Ceramic Coatings

Ceramic materials applied as coatings not only offer scratch and wear resistance, fracture resistance, and electrical and thermal insulation but also provide moisture and diffusion barriers and corrosion resistance for protection of metals. The surface of the metal is often, if not always, coated with sol-gel films for protection against acid attack and oxidation process (Carbajal et al., 2001). The mechanism of applying corrosion-resistant coatings may generally follow the *inhibition* approach of the corrosion process (Dariva and Galio, 2014), the *barrier creation* approach between the substrate materials and the environment, the use of *sacrificial coating materials*, or recently (Battocchi et al., 2006), the *active-passive* approach. The last approach hinders the corrosion half-cell reaction and results in the completion of electrons. The inhibition approach depends on the chemisorption of the complex between the metal ion and the inhibitor at the metal interface where the efficiency of the coating is governed by the position of the terminal group. Chou et al. (2003) reported and demonstrated the possible passivation mechanism provided by sol-gel-derived silica-based hybrid coatings, which were found to be defect-free and relatively dense on a stainless steel substrate (Chou et al., 2003). On composition and microstructure, Chen et al. (2000) point out that mechanical properties of the interlayer between the coating and the substrate have a great influence on the bonding strength and the properties of coating. Furthermore, there should be no defect, such as a pore (or a microcrack) or a weak phase, produced by a reaction between coating material and substrate in the interlayer of the eligible coating (Chen et al., 2000). The earlier work by

Sanchez and Ribot (1994) reported that the hybrid network interface between organic and inorganic phases can be classified into class I and class II, in which both classes are differentiated by the types of bonding. Class I is described by the hydrogen bonds, van der Waals forces, or ionic bonds that exist between the interaction of organic and inorganic components, while in class II, organic and inorganic components of organic-inorganic hybrids are held together by covalent bonds. The stable bonds that can be obtained from sol-gel-derived ceramic coating materials are that of class II as they can be stable toward an attack of nucleophilic species, such as alcohols, hydroxylated ligands, or water, as stipulated by Figueira et al. (2015). The chemical (e.g., hydrophobicity and corrosion protection) and physical (e.g., scratch resistance and low dielectric constant) properties of the resultant hybrid materials of class II are suitable for prevention of oxidation/corrosion of metallic substrates.

11.7 Characterizations

The study of the resultant coating and/or inhibiting materials with regard to their formation, performance, internal and external surface morphological structures, chemical compositions, wettability, surface charges, etc., is made possible by the application of several analytical techniques. Such instruments include, but are not limited to, Fourier transform infrared spectroscopy (FTIR), electrochemical impedance spectroscopy (EIS), scanning electron microscopy (SEM), transmission electron microscopy (TEM), x-ray diffraction (XRD), x-ray photoelectron spectroscopy (XPS), and the contact angle as per the intended aim and objective of the produced materials. While SEM and TEM provide the surface and internal morphologies of the ceramic nanomaterials either as freestanding powders or as coated materials on the substrates, XRD provides the crystalline nature of the nanomaterials synthesized. The analysis of an SEM-/energy-dispersive x-ray spectroscopy (EDS)-coupled instrument assists in presenting the chemical composition of the element present on the surface of the synthesized material.

11.8 Conclusion

The use of polymers for the protection of metals against corrosion has been successfully brought about by the complexes they form with metal ions as a result of the functional groups they possess. There is interest in using some natural macromolecules for corrosion inhibition or protection studies. However, more work needs to be done. As per the literature, since the report on the corrosion inhibition ability of guar gum, without additional metals in protecting carbon steel against corrosion in an acidic medium, was made, several studies have been reported on the use of other natural polymers as corrosion inhibitors. Also corrosion inhibitors that are of a natural origin have shown to perform differently in various media. For example, nitrogen-based materials and their derivatives, sulfur-containing compounds, aldehydes, thioaldehydes, acetylenic compounds, and various alkaloids are used as inhibitors in an acidic medium, whereas benzoate, nitrite, chromate, and phosphate are good inhibitors in neutral media.

Acknowledgments

The authors would like to thank the Nanotechnology and Water Sustainability (NanoWS) Research Unit under the University of South Africa (UNISA) for academic and financial support toward this book chapter and the National Research Foundation (NRF) for research funding.

References

Abdallah M. (2004). Guar gum as corrosion inhibitor for carbon steel in sulfuric acid solutions. *Portugaliae Electrochim. Acta*, **22**(2): 161–175.

Agrawal AK, Hindin B. (1994). Corrosion inhibition of ammonia-water absorption chillers. US Patent, US 5342578 A.

Al-Fadhli HY, et al. (2006). The erosion-corrosion behaviour of high velocity oxy-fuel (HVOF) thermally sprayed inconel-625 coatings on different metallic surfaces. *Surf. Coat. Technol.*, **200**(20–21): 5782–5788.

Amar H, et al. (2003). A corrosion inhibition study of iron by phosphonic acids in sodium chloride solution. *J. Electroanal. Chem.*, **558**(1–2): 131–139.

Amri A, et al. (2014). Developments in the synthesis of flat plate solar selective absorber materials via sol-gel methods: a review. *Renewable Sustainable Energy Rev.*, **36**: 316–328.

Aparicio M, Jitianu A, Rodriguez G, Degnah A, Al-Marzoki K, Mosa J, Klein LC. (2016). Corrosion protection of AISI 304 stainless steel with melting gel coatings. *Electrochim. Acta*, **202**: 325–332.

Bahrami M, et al. (2015). Preparation and evaluation of corrosion behavior of GPTMS-TEOS hybrid coatings containing Zr and Ce on aluminum alloy 6061-T6. *J. Sol-Gel Sci. Technol.*, **76**(3): 552–561.

Bardal E. (2007). *Corrosion and Protection*. Springer Science & Business Media.

Battocchi D, et al. (2006). Comparison of testing solutions on the protection of Al-alloys using a Mg-rich primer. *Corros. Sci.*, **48**(8): 2226–2240.

Bradford SA. (1993). *Corrosion Control*, Van Nostrand Reinhold, New York.

Brinker CJ, et al. (1999). Evaporation-induced self-assembly: nanostructures made easy. *Adv. Mater.*, **11**(7): 579–585.

Burleigh TD. (1989). Anodic photocurrents and corrosion currents on passive and active-passive metals. *Corrosion*, **45**(6): 464–472.

Camargo HPC, Satyanarayana KG, Wypych F. (2009). Nanocomposites: synthesis, structure, properties, and new application opportunities. *Mater. Res.*, **12**(1): 1–39.

Cao X, et al. (2008). $La_2(Zr_{0.7}Ce_{0.3})_2$ O7-A new oxide ceramic material with high sintering-resistance. *Mater. Lett.*, **62**(17–18): 2667–2669.

Cao XQ, Vassen R, Stoever D. (2004). Ceramic materials for thermal barrier coatings. *J. Eur. Ceram. Soc.*, **24**(1): 1–10.

Carbajal G, et al. (2001). Corrosion-resistant coatings: a nanotechnology approach. *Anti-Corros. Methods Mater.*, **48**: 241–245.

Chen X, et al. (2014). Characterization of Ti-based solid solution cermets prepared by mechanically induced self-sustained reaction and subsequent pressureless sintering. *J. Alloys Compd.*, **583**: 523–529.

Chen Y, Ai X, Huang C. (2000). Bonding mechanism and performance of ceramic coatings by sol-gel process. *Chin. Sci. Bull.*, **45**(14): 1291–1296.

Choi S-M, Awaji H. (2005). Nanocomposites: a new material design concept. *Sci. Technol. Adv. Mater.*, **6**(1): 2–10.

Chou TP, Chandrasekaran C, Cao GZ. (2003). Sol-gel-derived hybrid coatings for corrosion protection. *J. Sol-Gel Sci. Technol.*, **26**(1–3): 321–327.

Chung DDL. (2004). Electrical applications of carbon materials. *J. Mater. Sci.*, **39**(8): 2645–2661.

Contu F, Elsener B, Böhni H. (2004). Serum effect on the electrochemical behaviour of titanium, Ti6Al4V and Ti6Al7Nb alloys in sulphuric acid and sodium hydroxide. *Corros. Sci.*, **46**(9): 2241–2254.

Cook A, et al. (2004). Corrosion protection of low carbon steel with polyaniline: passivation or inhibition? *Curr. Appl. Phys.*, **4**(2): 133–136.

Dariva CG, Galio AF. (2014). Corrosion inhibitors–principles, mechanisms and applications. In *Developments in Corrosion Protection*. InTech.

Davy H. (1824a). Additional experiments and observations on the application of electrical combinations to the preservation of the copper sheathing of ships, and to other purposes. *Philos. Trans. R. Soc. London (1776-1886)*, **114**: 242–246.

Davy H. (1824b). On the corrosion of copper sheeting by sea water, and on methods of preventing this effect; and on their application to ships of war and other ships. *Philos. Trans. R. Soc. London*, **114**(0): 151–158.

Demo J, et al. (2010). Development of a wireless miniaturized smart sensor network for aircraft corrosion monitoring. In *2010 IEEE Aerospace Conference*, pp. 1–9.

Dutta S. (2001). Fracture toughness and reliability in high-temperature structural ceramics and composites: prospects and challenges for the 21st century. *Bull. Mater. Sci.*, **24**(2): 117–120.

Eddy NO, et al. (2012). Corrosion inhibition potential of daniella oliverri gum exudate for mild steel in acidic medium. *Int. J. Electrochem. Sci.*, **7**: 7425–7439.

Edeleanu C. (1960). Corrosion control by anodic protection. *Tube Investments Res. Lab.*, (4): 86–91.

Elahee K. (2010). Heat recovery in the textile dyeing and finishing industry: lessons from developing economies. *J. Energy South Afr.*, **21**(3): 9–15.

El-Kady OAM. (2013). Effect of nano-yttria addition on the properties of WC/Co composites. *Mater. Des.*, **52**: 481–486.

Ferroni LP, et al. (2001). Determination of amorphous interfacial phases in Al_2O_3/SiC nanocomposites by computer-aided high-resolution electron microscopy. *Acta Mater.*, **49**(11): 2109–2113.

Figueira RB, Silva CJR, Pereira EV. (2015). Organic–inorganic hybrid sol–gel coatings for metal corrosion protection: a review of recent progress. *J. Coat. Technol. Res.*, **12**(1): 1–35.

France Jr WD, Greene Jr ND. (1968). Passivation of crevices during anodic protection. *Corrosion*, **24**(8): 247–251.

Frenzer G, Maier WF. (2006). Amorphous porous mixed oxides: sol-gel ways to a highly versatile class of materials and catalysts. *Annu. Rev. Mater. Res.*, **36**: 281–331.

Zum Gahr K-H. (1998). Wear by hard particles. *Tribol. Int.*, **31**(10): 587–596.

Gopi D, et al. (2014). Evaluation of the mechanical and corrosion protection performance of electrodeposited hydroxyapatite on the high energy electron beam treated titanium alloy. *J. Alloys Compd.*, **616**: 498–504.

Gray JJ, Orme CA. (2007). Electrochemical impedance spectroscopy study of the passive films of alloy 22 in low pH nitrate and chloride environments. *Electrochim. Acta*, **52**(7): 2370–2375.

Han X, et al. (2001). Synthesis and properties of Er^{3+}-doped silica glass by sol-gel processing with organic complexation. *J. Mater. Sci.*, **36**(4): 985–993.

Harvey TG, et al. (2011). The effect of inhibitor structure on the corrosion of AA2024 and AA7075. *Corros. Sci.*, **53**(6): 2184–2190.

Hiroyoshi N, et al. (2004). Synergistic effect of cupric and ferrous ions on active-passive behavior in anodic dissolution of chalcopyrite in sulfuric acid solutions. *Hydrometallurgy*, **74**(1): 103–116.

Hongtao TZLIUHYU, Shixi O. (2003). Synthesis methods of powder for microwave dielectric ceramics [J]. *Mater. Rev.*, **12**: 14.

Huang M, Zhang H, Yang J. (2012). Synthesis of organic silane microcapsules for self-healing corrosion resistant polymer coatings. *Corros. Sci.*, **65**: 561–566.

John S, Joseph A. (2012). Effective inhibition of mild steel corrosion in 1 M hydrochloric acid using substituted triazines: an experimental and theoretical study. *RSC Adv.*, **2**: 9944.

Katsich C, Badisch E. (2011). Effect of carbide degradation in a Ni-based hardfacing under abrasive and combined impact/abrasive conditions. *Surf. Coat. Technol.*, **206**(6): 1062–1068.

Khan MI, et al. (2014). Sodium silicate free geopolymer as coating material: adhesion to steel. In *Proceedings of the 1st Int. Electron. Conf. Mater., Sciforum Electronic Conference Series*, Vol. 1, b016.

Kirkpatrick A, et al. (2014). Understanding the impact of field development and changing fluid compositions on corrosion inhibitor selection and performance. In *SPE International Oilfield Corrosion Conference and Exhibition*. Society of Petroleum Engineers, SPE-169613-MS, pp. 118–125.

Koch GH, et al. (2002). *Corrosion Cost and Preventive Strategies in the United States*. Turner-Fairbank Highway Research Center.

Lee C-H, et al. (2012). Indentation deformation and microcracking in β-Si_3N_4-based nanoceramic. *J. Am. Ceram. Soc.*, **95**(4): 1421–1428.

Lepule ML, et al. (2015). Corrosion and wear behaviour of ZrO_2 modified NiTi coatings on AISI 316 stainless steel. *Surf. Coat. Technol.*, **261**: 21–27.

Lin HT, et al. (2006). Characterization of first-stage silicon nitride components after exposure to an industrial gas turbine. *J. Am. Ceram. Soc.*, **89**(1): 258–265.

Lü L, Lai MO. (2013). *Mechanical Alloying*. Springer Science & Business Media.

Maia F, et al. (2012). Silica nanocontainers for active corrosion protection. *Nanoscale*, **4**(4): 1287.

Massa TR, Van Kirk JS, Conley EV. (1998). Method of using composite cermet articles. US Patent, US 5806934 A.

Matovic B, et al. (2007). Fabrication of SiC by carbothermal-reduction reactions of diatomaceous earth. *J. Mater. Sci.*, **42**(14): 5448–5451.

Medvedovski E. (2001. Wear-resistant engineering ceramics. *Wear*, **249**(9): 821–828.

Meng F, Liang B, Wang M. (2013). Investigation of formation mechanism of Ti 3 SiC 2 by self-propagating high-temperature synthesis. *Int. J. Refract. Met. Hard Mater.*, **41**: 152–161.

Metroke TL, Kachurina O, Knobbe ET. (2002). Spectroscopic and corrosion resistance characterization of GLYMO-TEOS Ormosil coatings for aluminum alloy corrosion inhibition. *Prog. Org. Coat.*, **44**(4): 295–305.

Mishra R, Balasubramaniam R. (2004). Effect of nanocrystalline grain size on the electrochemical and corrosion behavior of nickel. *Corros. Sci.*, **46**(12): 3019–3029.

Morris PE, Scarberry RC. (1970). Anodic polarization measurements of active-passive nickel alloys by rapid-scan potentiostatic techniques. *Corrosion*, **26**(7): 269–279.

Moutarlier V, Neveu B, Gigandet MP. (2008). Evolution of corrosion protection for sol-gel coatings doped with inorganic inhibitors. *Surf. Coat. Technol.*, **202**(10): 2052–2058.

Nalwa HS. (2001). *Nanostructured Materials and Nanotechnology: Concise Edition*, Gulf Professional.

Natishan P. (2003). Introduction to methods of corrosion protection. *ASM International (The Materials Information Society), Corrosion: Fundamentals, Testing and Protection, ASM Handbook*, Vol. 13, pp. 684–686.

Niihara K. (1991). New design concept of structural ceramics. *J. Ceram. Soc. Jpn.*, **99**(1154): 974–982.

Okada H, Yukawa K, Tamura H. (1976). Transition of cracking mechanisms from active path corrosion to hydrogen embrittlement in high strength steels in a boiling nitrate solution. *Corrosion*, **32**(5): 201–203.

Osborne JH. (2001). Observations on chromate conversion coatings from a sol–gel perspective. *Prog. Org. Coat.*, **41**(4): 280–286.

Pandey S, Mishra SB. (2012). Amorphous porous mixed oxides: a new and highly versatile class of materials. In *Intelligent Nanomaterials: Processes, Properties, and Applications*. Wiley Online Library, pp. 149–181.

Pandey S, Mishra SB. (2011). Sol-gel derived organic-inorganic hybrid materials: synthesis, characterizations and applications. *J. Sol-Gel Sci. Technol.*, **59**(1): 73–94.

Pedeferri P. (1996). Cathodic protection and cathodic prevention. *Constr. Build. Mater.*, **10**(5 SPEC. ISS): 391–402.

Peter A, Obot IB, Sharma SK. (2015). Use of natural gums as green corrosion inhibitors: an overview. *Int. J. Ind. Chem.*, **6**(3): 153–164.

Pieterse N, et al. (2006). Estimating the gas permeability of commercial volatile corrosion inhibitors at elevated temperatures with thermo-gravimetry. *Corros. Sci.*, **48**(8): 1986–1995.

Pietersen D, Holtzhausen JP, Vosloo WL. (2004). An investigation into the methodology to develop an insulator pollution severity application map for South Africa. *2004 IEEE Africon. 7th Africon Conference in Africa (IEEE Cat. No.04CH37590)*, **2**: 697–703.

Popoola A, Olorunniwo OE, Ige OO. (2014). Corrosion resistance through the application of anti-corrosion coatings. *Dev. Corros. Prot.*, **2**(12): 241–270.

Popoola LT, et al. (2013). Corrosion problems during oil and gas production and its mitigation. *Int. J. Ind. Chem.*, **4**(1): 1–15.

Potgieter JH, et al. (2008). Influence of nickel additions on the corrosion behaviour of low nitrogen 22% Cr series duplex stainless steels. *Corros. Sci.*, **50**(9): 2572–2579.

Ravichandran R, Nanjundan S, Rajendran N. (2004). Effect of benzotriazole derivatives on the corrosion of brass in NaCl solutions. *Appl. Surf. Sci.*, **236**(1): 241–250.

Revie RW, Uhlig HH. (2011). *Uhlig's Corrosion Handbook*. John Wiley & Sons.

Reyes M, Neville A. (2003). Degradation mechanisms of Co-based alloy and WC metal-matrix composites for drilling tools offshore. *Wear*, **255**(7–12): 1143–1156.

Riggs Jr OL. (1973). In *Theoretical Aspects of Corrosion Inhibitors and Inhibition*. Nathan CC (ed.). NACE, Houston, TX, pp. 7–27.

Riggs O. (2012). *Anodic Protection: Theory and Practice in the Prevention of Corrosion*. Springer Science & Business Media.

Roberge PR. (2013). *Handbook of Corrosion Engineering*, 2nd ed. McGraw-Hill, New York, NY.

Roberge PR. (2000). *Handbook of Corrosion Engineering*. McGraw-Hill, New York, NY.

Rosso M. (2006). Ceramic and metal matrix composites: routes and properties. *J. Mater. Process. Technol.*, **175**(1–3): 364–375.

Saji VS, Thomas J. (2007). Nanomaterials for corrosion control. *Curr. Sci.*, **92**(1): 51–55.

Sanchez C, Ribot F. (1994). Design of hybrid organic-inorganic materials synthesized via sol-gel chemistry. *Appl. Organomet. Chem.*, **18**: 1007–1047.

Segal D, et al. (1997). Chemical synthesis of ceramic materials. *J. Mater. Chem.*, **7**(8): 1297–1305.

Setoudeh N, Welham NJ. (2011). Ball milling induced reduction of $SrSO_4$ by Al. *Int. J. Miner. Process.*, **98**(3): 214–218.

Shaw A, et al. (2013). Challenges and solutions in PGM furnace operation: high matte temperature and copper cooler corrosion. *J. South. Afr. Inst. Min. Metall.*, **113**(3): 193–216.

Sherif ESM, Erasmus RM, Comins JD. (2007). Corrosion of copper in aerated synthetic sea water solutions and its inhibition by 3-amino-1,2,4-triazole. *J. Colloid Interface Sci.*, **309**(2): 470–477.

Shukla S, et al. (2002). Effect of nanocrystallite morphology on the metastable tetragonal phase stabilization in zirconia. *Nano Lett.*, **2**(9): 989–993.

Song X, et al. (2011). High-temperature thermal properties of yttria fully stabilized zirconia ceramics. *J. Rare Earths*, **29**(2): 155–159.

Sun C, et al. (2009). Progress in corrosion resistant materials for supercritical water reactors. *Corros. Sci.*, **51**(11): 2508–2523.

Sun Y, et al. (1998). Unlubricated friction and wear behaviour of zirconia ceramics. *Wear*, **215**(1–2): 232–236.

Suryanarayana C. (2011). Synthesis of nanocomposites by mechanical alloying. *J. Alloys Compd.*, **509**: S229–S234.

Thompson NG, Yunovich M, Dunmire D. (2007). Cost of corrosion and corrosion mainenance strategies. *Corros. Rev.*, **25**(3–4): 247–262.

Thostenson ET, Li C, Chou TW. (2005). Nanocomposites in context. *Compos. Sci. Technol.*, **65**(3–4): 491–516.

Tian W, Wang Y, Yang Y. (2010). Three body abrasive wear characteristics of plasma sprayed conventional and nanostructured Al_2O_3-13% TiO_2 coatings. *Tribol. Int.*, **43**(5): 876–881.

Torre GC, et al. (2009). Study of ceramic and hybrid coatings produced by the sol-gel method for corrosion protection. *Open Corros. J.*, **2**(1): 197–203.

Touzain S, Thu Q, Bonnet G. (2005). Evaluation of thick organic coatings degradation in seawater using cathodic protection and thermally accelerated tests. *Prog. Org. Coat.*, **52**(4): 311–319.

Trapaga-Martinez LG, Levi CG. (2013). *Multiphase-Multifunctional Ceramic Coatings*. DTIC Document, Mexico.

Umoren SA, et al. (2006). Gum arabic as a potential corrosion inhibitor for aluminium in alkaline medium and its adsorption characteristics. *Anti-Corros. Methods Mater.*, **53**(5): 277–282.

Voevodin NN, et al. (2003). Nanostructured coatings approach for corrosion protection. *Prog. Org. Coat.*, **47**(3–4): 416–423.

Wang D, Bierwagen GP. (2009). Sol-gel coatings on metals for corrosion protection. *Prog. Org. Coat.*, **64**(4): 327–338.

Wang H, Akid R, Gobara M. (2010). Scratch-resistant anticorrosion sol-gel coating for the protection of AZ31 magnesium alloy via a low temperature sol-gel route. *Corros. Sci.*, **52**(8): 2565–2570.

Wang Y, et al. (2000). Abrasive wear characteristics of plasma sprayed nanostructured alumina/titania coatings. *Wear*, **237**(2): 176–185.

Weimer AW. (2012). *Carbide, Nitride and Boride Materials Synthesis and Processing*. Springer Science & Business Media.

West JM. (1986) *Basic Corrosion and Oxidation*, 2nd ed., Ellis Horwood, pp. 312–337.

Xu Y, et al. (2013). Synchronistic preparation of fibre-like SiC and cubic-ZrO_2/SiC composite from zircon via carbothermal reduction process. *Mater. Res. Bull.*, **48**(1): 7–11.

Zäch M, et al. (2006). Nanoscience and nanotechnology for advanced energy systems. *Curr. Opin. Solid State Mater. Sci.*, **10**(3–4): 132–143.

Zhang L, Zhang J, Liu R. (2016). Organic–inorganic hybrid sol–gel coatings for corrosion protection of aluminum alloys. *Surf. Innovations*, **4**(2): 51–69.

Zheludkevich ML, et al. (2007). Anticorrosion coatings with self-healing effect based on nanocontainers impregnated with corrosion inhibitor. *Chem. Mater.*, **19**(3): 402–411.

Zhong L, et al. (2006). Application of polyaniline to galvanic anodic protection on stainless steel in H_2SO_4 solutions. *Corros. Sci.*, **48**(12): 3960–3968.

Chapter 12

TiO_2 Nanomaterials for Photocatalytic Applications

Nityananda Agasti

Department of Chemistry, Deen Dayal Upadhyaya College, University of Delhi, New Delhi 110078, Delhi, India

nnagasti@gmail.com

TiO_2, a semiconductor having a variety of applications, has drawn wide attention as an important class of ceramic material. In this chapter various methods of preparation and applications of TiO_2 nanomaterials of different sizes and morphologies have been discussed. The effects of variation of reaction parameters like pH of the medium, temperature, and solvent on the size and shape of nanostructured TiO_2 have been cited with examples. Keeping their shape-dependent applications in mind, preparation of TiO_2 nanomaterials of different shapes, such as nanorods, nanotubes, nanobelts, and nanowires, has been included. Facet-specific TiO_2 nanocrystals are also cited, with their methods of preparations and applications. Photocatalytic applications of TiO_2 nanomaterials with detailed mechanism and examples have been covered. Besides pure TiO_2, preparation and applications of doped TiO_2

Smart Ceramics: Preparation, Properties, and Applications
Edited by Ajay Kumar Mishra

ISBN 978-981-4774-30-7 (Hardcover), 978-1-315-16359-8 (eBook)
www.panstanford.com

nanomaterials have been covered. Preparation and application of TiO_2 nanomaterials doped with noble metals have been discussed, with mechanistic details of their improved photocatalytic activities in comparison to undoped TiO_2. Various applications of TiO_2 nanomaterials having industrial importance, like hydrogen production, oxidation of alcohols, reduction of CO_2, and degradation of organic and inorganic pollutants are a few key elements of this chapter.

12.1 Introduction

TiO_2, a ceramic material, has attracted extensive studies due to its excellent stability and interesting physical and chemical properties. Its catalytic activity was realized in 1972 by the discovery of the water-splitting phenomenon on a TiO_2 electrode (Fujishima and Honda, 1972). TiO_2 offers high efficiency in the field of catalysis and energy applications in comparison to its other oxide counterparts. Being a semiconductor TiO_2 exhibits a bandgap (E_g), which is the energy gap between the valence band (VB) and the conduction band (CB). This bandgap determines the light absorption property and redox behavior of TiO_2. When photon with energy equal to or greater than E_g is absorbed by TiO_2, then electrons from the VB get excited to the CB, leaving holes in the VB. These photogenerated electron-hole pairs' migration leads to conductivity of the TiO_2 semiconductor. Thus bandgap plays an important role in the physical and chemical properties of TiO_2. By varying the E_g properties of TiO_2 can be altered. Variation in the bandgap of TiO_2 can be achieved by adding suitable dopants to TiO_2 and also by preparing nanostructured TiO_2 materials modified with metal nanoparticles (NPs) and metal oxide NPs. TiO_2 exhibits mainly four different phases: three naturally existing phases, i.e., anatase (tetragonal), rutile (tetragonal), brookite (orthorhombic), and a fourth one, TiO_2-B (monoclinic), a less common structure derived from titanates, also called TiO_2 bronze. All these phases contain TiO_6 octahedra but differ in the distortion of the octahedron units and share edges and corners in different manners. Figure 12.1 shows all four phases of TiO_2.

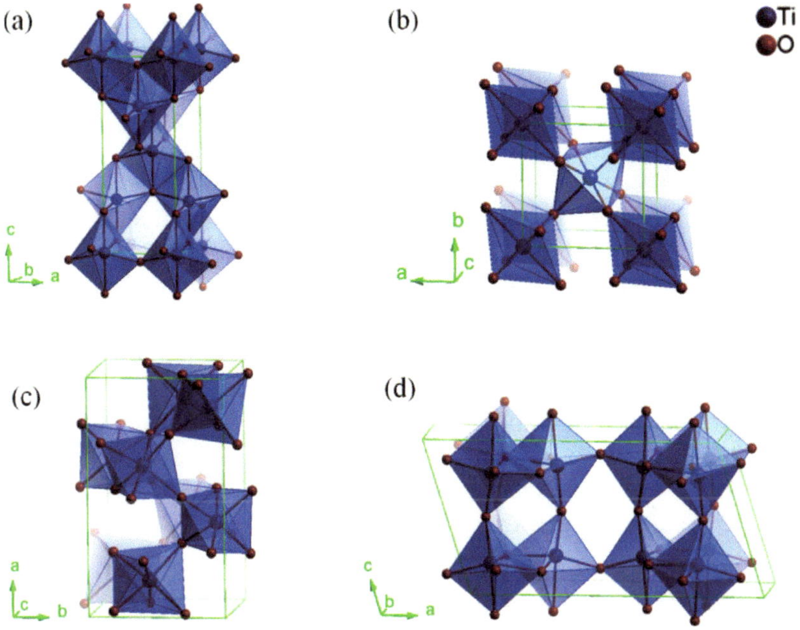

Figure 12.1 Crystalline structures of TiO_2 in different phases: (a) anatase, (b) rutile, (c) brookite, and (d) TiO_2-B. Reprinted with permission from Ma Y, et al. (2014). Titanium dioxide-based nanomaterials for photocatalytic fuel generations. *Chem. Rev.*, **114**: 9987. Copyright (2014) American Chemical Society.

In anatase, the octahedra arranged in zigzag chains along {221} share four edges; in rutile, the octahedra share only two edges and connect in linear chains parallel to {001} (Gopal et al., 1997), while in brookite both corners and edges are connected (Carp et al., 2004). TiO_2-B is mainly derived from the layered titanates (Feist and Davies, 1992). These different phases of TiO_2 exhibit different bandgaps. Anatase (3.0 eV) has a smaller bandgap than rutile (3.2 eV). Out of these, rutile is generally considered to be stable in the bulk phase, while anatase and brookite are stable at the nanoscale. Most studied crystal forms of TiO_2 are anatase and rutile.

This chapter covers various methods of preparation of nanostructured TiO_2 with different morphologies and its photocatalytic applications.

12.2 Preparation of TiO_2 Nanomaterials

Preparation of nanostructured TiO_2 materials from precursor salts by using physical methods or solution-based methods has gained significant attention from researchers. On the basis of the methods of preparation, TiO_2 nanomaterials of desired properties with various morphologies can be obtained. Besides methods of preparation, experimental parameters such as temperature and pH also affect not only crystalline forms of TiO_2 but also its shape and morphologies. To obtain nanomaterials of improved properties, TiO_2 can be doped by addition of suitable metallic or nonmetallic dopants. Many methods have been reported for preparation of TiO_2 nanocrystals of different morphologies, like nanotubes, nanorods, nanowires, nanobelts, nanosheets, and nanofibers. This chapter covers mostly used methods, like hydrothermal, solvothermal, sol-gel, template-assisted, and chemical vapor deposition (CVD).

12.2.1 The Hydrothermal Method

It's a simple and cost-effective approach to synthesize TiO_2 nanomaterials. As the name suggests, this method involves heating of a titanium precursor in an aqueous medium. On the basis of the medium of reaction, the hydrothermal method can be categorized as acid hydrothermal or alkali hydrothermal. In the acid hydrothermal method an inorganic acid is used with a titanium salt and in the alkali hydrothermal method NaOH is used with a titanium salt. Both methods produce TiO_2 nanomaterials of different morphologies and crystallinities. In the hydrothermal method the reaction is conducted in a polytetrafluoroethylene (Teflon)-lined autoclave that can sustain high temperatures and a corrosive alkaline medium. By heating the sealed vessel, the inner temperature can be elevated to above the boiling point of water, producing high pressure. One-dimensional TiO_2 nanostructures can be obtained under such harsh reaction conditions. TiO_2 NPs of different morphologies can be prepared under different conditions. For the first time in 1998 Kasuga et al. (1998) prepared nanotubular TiO_2 by the hydrothermal method, where amorphous TiO_2 was treated at an elevated temperature in a highly concentrated alkaline medium to produce TiO_2 nanotubes.

Manorama and coworkers (Nag et al., 2007) prepared rutile titania nanocrystals using the acid hydrothermal method. In a typical reaction process 2 mL of 0.5 M $TiCl_4$ (50% HCl) was added to 26 mL of distilled water having a pH less than 1. This highly acidic aqueous titanium(IV) chloride solution was stirred and transferred to the Teflon-lined stainless steel autoclave and held at a constant temperature for a definite time interval. They have conducted a series of reactions at temperatures ranging from 40°C to 150°C and time duration of 4 to 48 h under constant autogenerated pressure to prepare TiO_2 nanocrystals.

In another report TiO_2 nanomaterials were prepared by the alkaline hydrothermal method (Bavykin et al., 2005). A mixture of solid titanium dioxide and NaOH solution was heated at 140°C to produce white, powdery TiO_2. Zhang et al. (2015) reported preparation of hollow TiO_2 NPs by the hydrothermal method. An aqueous solution of titanium(IV) oxalate and ammonia was heated at 115°C in an autoclave. The product obtained was calcined at 450°C to yield nanostructured TiO_2. The concentration of ammonia was varied to control morphologies and crystalline forms of TiO_2. Figure 12.2 represents the formation of TiO_2 by the alkaline hydrothermal method.

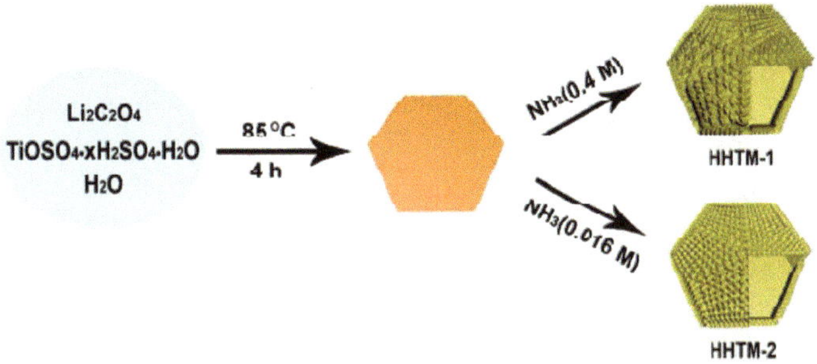

Figure 12.2 Schematic illustration of the preparation of TiO_2 from a Ti(IV) oxalate complex in different concentrations of ammonia solution. Reproduced from Zhang Y, et al. (2015). Ti(IV) oxalate complex-derived hierarchicalhollow TiO_2 materials with dye degradation properties in water. *Dalton Trans.*, **45**: 265, with permission from The Royal Society of Chemistry.

In hydrothermal methods, besides acidic and alkaline conditions, other reaction parameters, like the precursor concentration, temperature, and time, have an important effect on the structures and morphologies of TiO_2 nanomaterials. Morgan et al. (2008) reported the preparation of TiO_2 nanomaterials of different morphologies by the alkaline hydrothermal method. They investigated the effect of the concentration of NaOH and the temperature of the reaction on the morphology of TiO_2. TiO_2 nanostructured materials have been prepared from alkaline hydrothermal treatment of Degussa P25 (commercially available) TiO_2 for 20 h at a temperature range of 100°C–200°C and 5–10 $mol \cdot dm^{-3}$ NaOH. It was observed that at low concentrations of NaOH (5 $mol \cdot dm^{-3}$) NPs are formed at a low temperature but as the temperature increases nanotubes are formed. Nanotube formation was found to be more dependent on the temperature than concentration of NaOH. At an elevated temperature Morgan et al. observed phase transition of TiO_2 from nanotubes to nanoribbons. When the temperature is increased above 180°C nanoribbons are formed. But unlike nanotubes, nanoribbon formation was found to be more dependent on the concentration of NaOH than temperature. They examined over the NaOH concentration range (5 and 7.5 $mol \cdot dm^{-3}$) requiring a greater temperature to produce nanoribbons (220°C and 200°C, respectively). But complete conversion to nanoribbons was observed only in the 10 $mol \cdot dm^{-3}$ NaOH concentration at 180°C, whereas lower NaOH concentrations exhibited coexistence of both phases. This indicates that high concentrations of free Na^+ promotes the formation of nanoribbons at lower temperatures.

Using the alkaline hydrothermal method, Bavykin et al. (2004) prepared TiO_2 nanomaterials of different morphologies influenced by different reaction conditions, like the temperature of synthesis and the molar ratio of TiO_2 to NaOH. They suggested a mechanism of formation of TiO_2 nanotubes that includes the wrapping of multilayered titanate nanosheets. The driving force for the curving of nanosheets is believed to be mechanical stress due to the imbalance in the width of layers in the multilayered nanosheets.

12.2.2 The Solvothermal Method

In the hydrothermal method an aqueous solution of a titanium precursor is used. However, when the same reactions are conducted in organic solvents like ethanol, ethylene glycol (EG), and hexane, the method of preparation is called the solvothermal method.

Mesoporous TiO_2 hollow spheres were prepared by Hu et al. (2016). TiO_2 was synthesized with tetrabutyl titanate (TBT) as precursor and dimethyl formamide as solvent. TBT and anhydrous oxalic acid dissolved in *N,N*-dimethylformamide were heated in an autoclave to 170°C for 10 h. The as-synthesized samples were washed with ethanol and dried at 60°C, and the products were refluxed with an ethylenediamine aqueous solution (pH 11–12) for 48 h at 80°C to 90°C. The obtained samples were calcined at 600°C for 3 h in air to remove the organic species and produce white mesoporous TiO_2 hollow spheres.

Like pure TiO_2, doped TiO_2 nanomaterials can also be prepared by the solvothermal method. Wang W. et al. (2016) have reported the synthesis of hierarchical porous nitrogen doped TiO_2 microspheres by solvothermal method. As per their synthesis procedure, TBT was added to polyethylene pyrrole (PVP, K-30) and acetic acid solution with continuous stirring for 5 min. The obtained white suspension was heated at 150°C for 24 h in a Teflon-lined stainless steel autoclave. Calcination of dried products at 400°C for 2 h in air produces the porous TiO_2 microspheres. Heating TiO_2 microspheres with urea at 400°C in air for 3 h produces N-TiO_2. Different TiO_2-to-urea ratios were used to incorporate varied amounts of nitrogen, resulting in the formation of different types of N-TiO_2 nanomaterials.

Almeida et al. (2016) investigated Eu-modified TiO_2 (Eu-TiO_2) as a pressure-dependent luminescent material that can have potential use as an optical sensor for pressure devices. They synthesized Eu-modified TiO_2 nanomaterial by the solvothermal method. Salts of both Ti and Eu were added to the solvent ethanol, followed by heat treatment in a Teflon-lined autoclave. Finally, annealing the sample yields Eu-TiO_2.

12.2.3 The Sol-Gel Method

It is a wet-chemical process widely used for the preparation of TiO_2 nanomaterials. Controlling the particle size and morphology through experimental conditions is considered to be the advantage of this method. TiO_2 NPs are synthesized from the hydrolysis of a titanium alkoxide precursor such as titanium isopropoxide, titanium butoxide, or tetra-n-butyl titanate. In this method hydrolysis of the precursor solution is done to form the complex sol, followed by heat treatment and aging to obtain the gel and, finally, annealing at an elevated temperature to obtain the required oxide. Sugimoto et al. (2003) reported the synthesis of TiO_2 by the sol-gel method. First titanium isopropoxide is mixed with tetraethanol amine in a 2:1 ratio to produce a sol of an organometallic complex and then refluxed and aged under H_2O at 100°C for 24 h to obtain a Ti-containing gel, which is subsequently aged for 72 h at 140°C to produce anatase TiO_2 NPs. Pottier et al. (2001) prepared brookite TiO_2 NPs by thermolysis of $TiCl_4$ in an aqueous HCl solution. They investigated the influence of the physicochemical conditions, like the acidity of the medium, the nature of the anions, and the ageing of suspensions, on the formation of brookite with varied sizes and morphologies. In their report TiO_2 particles were formed by adding $TiCl_4$ dropwise to hydrochloric or perchloric acid and the solution was heated and aged at 100°C in an oven to produce brookite TiO_2.

By combining sol-gel and hydrothermal method three phases of TiO_2 can be prepared. Reyes-Coronado et al. (2008) reported the syntheses of phase-pure anatase, rutile, and brookite TiO_2 NPs. Amorphous TiO_2 NPs were first prepared by the sol-gel technique using titanium alkoxide. Further hydrothermal treatment of TiO_2 NPs at three different acid concentrations and temperatures yields phase-pure products of anatase, rutile, and brookite phases separately. They observed that the hydrothermal conditions for the synthesis of anatase phase are heat treatment at 200°C for 6 h under 1.5 M acetic acid solution, for rutile phase are heat treatment at 200°C for 8 h under 4 M HCl solution, whereas the same for obtaining brookite phase are heat treatment at 175°C for 7 h under 3 M HCl solution. Zhang et al. (2016) reported the preparation of mesoporous TiO_2 by acid-assisted sol-gel synthesis. The process starts with the slow hydrolysis of titanium alkoxides precursor in EG to form antiaggregated titanium glycolate, which is stabilized by the electrostatic repulsive force from their positive surface charge. Then, as-prepared titanium glycolate microspheres underwent heat treatment to remove EG and evaporate the organic groups, yielding the monodisperse mesoporous titania microspheres with an ordered structure.

12.2.4 Template-Assisted Synthesis

Besides the nanodimension, the nanoarchitecture of TiO_2 is important for the enhancement of properties and thus applications. The method utilizes materials of known morphology as templates to construct new materials with a similar morphology. By adjusting the morphology of the template material, it is possible to prepare numerous new materials with regular and controlled morphologies on the nano- and microscale. In most cases, the template material is sacrificial and needs to be destroyed after synthesis. The preparation of TiO_2 nanotubes by template-assisted synthesis usually involves reaction of titanium compounds in the presence of templating agents, followed by the polymerization of TiO_2 in the self-assembled template molecules or deposition of TiO_2 onto the surface of the template. The next stages are selective removal of the templating agent and separation of the TiO_2 material.

To prepare TiO_2 nanomaterials of desired shapes and morphologies, various templates are used. Xu et al. (2013) reported template-directed preparation of NPs, nanotubes, and nanofibers of TiO_2. Electrospun fibers were used as the template. Electrospun fibers and thermoplastic plaster elastomer (TPPE) were immersed into an alcoholic solution of TBT in a basic pH medium. Then calcination of TPPE coated with a titanium precursor at 550°C yields nanostructured TiO_2.

Sander et al. (2004) reported template-assisted synthesis of TiO_2 nanotubes. They used an aluminum oxide nanoporous membrane as the template for the preparation of TiO_2 nanotubes. An alumina template is prepared by anodization of an Al film that is evaporated onto the surface of a silicon substrate. After template formation TiO_2 is deposited on the substrate using atomic layer deposition (ALD) of $TiCl_4$ and H_2O at 105°C. Once TiO_2 is deposited, the alumina template can be partially or completely removed by chemical etching, resulting in the deposition of TiO_2 nanotubes on the silicon surface.

Preparation of mesoporous TiO_2-B microspheres using template-assisted synthesis is reported by Liu et al. (2011). Nanosized silica colloids are used as hard templates to create mesopores. As reported, the titania-silica composite precursor was prepared by a template-assisted ultrasonic spray pyrolysis method by adding colloidal silica into a titanium(IV) bis(ammonium lactato) dihydroxide solution. The titania-silica powders were then refluxed in a sodium hydroxide solution at 100°C for 3 days in order to etch silica and insert sodium ions into titania. The sample was washed by hydrochloric acid to remove silicate residue from the pores and exchange sodium ions with protons. The hydrated sample ($H_2Ti_2O_5 \cdot H_2O$) was then heat-treated at 400°C and 500°C in a flowing argon gas to form $H_2Ti_3O_7$ and TiO_2-B, respectively.

12.2.5 Chemical Vapor Deposition

This technique produces high-quality TiO_2 nanomaterials although it requires expensive equipment and the cost is too high for mass production. One of the earlier reported vacuum-based deposition techniques is metal-organic chemical vapor deposition (MOCVD) method. In this technique an organometallic precursor is used for the deposition of TiO_2 on a substrate in vacuum. Okuyama et al. (1989) used thermal decomposition of titanium tetraisopropoxide (TTIP) vapor in a controlled cylindrical furnace to produce ultrafine TiO_2 NPs in the size range of 10–60 nm. The CVD reactor is a quartz tube with a porous quartz disk at one end. The precursor TTIP was introduced into the CVD reactor using nitrogen as the carrier gas. Li et al. (2002) prepared transition-metal-and-lanthanide-doped TiO_2 NPs by the MOCVD method using TTIP as precursor. For doping of TiO_2, low-melting or volatile precursors of metals, like neodymium(III) acetylacetonate, palladium(II) acetylacetonate, platinum(IV) acetylacetonate, and iron(III) acetylacetonate, were used. In the process of MOCVD of TiO_2, the liquid titanium precursor TTIP was introduced into a reactor chamber having a temperature in the range of 1000°C, with the help of argon as the carrier gas, and TiO_2 film was deposited on a substrate. Dopant precursors, typically in the powder form, were placed inside the reactor chamber, which was vaporized and incorporated into TiO_2 film. By controlling the flow rate of the precursor and the temperature, the thickness of TiO_2 deposition can be controlled.

Wang M. et al. (2016) reported preparation of high-quality crystalline anatase TiO_2 thin films on $[LaAlO_3]_{0.3}[SrAl_{0.5}Ta_{0.5}O_3]_{0.7}$ (LSAT) (001) substrates by MOCVD in the substrate temperature range of 500°C to 650°C. Tetrakis-dimethylamino titanium

(TDMAT) is used as the organometallic precursor of titanium. A clean and dried substrate LSAT was introduced inside the growth chamber. The titanium precursor was carried inside the MOCVD chamber by using ultra-high-purity N_2 as the carrier gas. High-purity O_2 (5N) was injected using a separate delivery line into the chamber as an oxidant. The oxidation reaction that takes place for the deposition of TiO_2 film is as follows:

$$Ti(N(CH_3)_2)_4 + 15O_2 \rightarrow TiO_2 + 8CO_2\uparrow + 2N_2\uparrow + 12H_2O\uparrow$$

Chimupala et al. (2014) prepared mixed-phase TiO_2 bronze (B)/anatase titania thin films by low-pressure chemical vapor deposition (LPCVD) onto a range of different substrates, including a silicon wafer, fused quartz, highly ordered pyrolytic graphite (HOPG), and pressed graphite flake. The CVD precursor was TTIP (97%) kept in a bubbling chamber set in an oil bath at 90°C–100°C. TTIP vapor was introduced by N_2 as the carrier gas (flowing at the rate of 1 l/min.) into the reactor with 400°C–600°C inside temperature. TiO_2 gets deposited onto the target substrate, a soda-lime microscope slide, by allowing the precursor flow for specific duration and a coated substrate can be obtained after cooling the reaction chamber to room temperature. A schematic illustration of LPCVD can be shown as in Fig. 12.3.

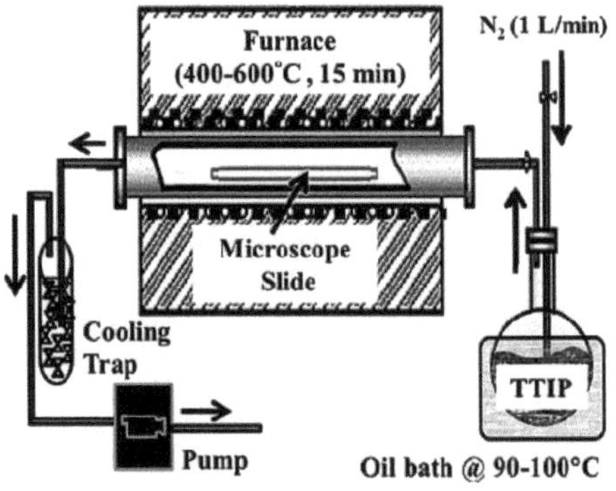

Figure 12.3 Schematic representation showing the LPCVD system for preparing TiO_2. Reproduced from Chimupala Y, et al. (2014). Universal synthesis method for mixed phase TiO_2(B)/anatase TiO_2 thin films on substrates via a modified low pressure chemical vapour deposition (LPCVD) route. *J. Mater. Chem. A*, **4**: 5685, with permission from The Royal Society of Chemistry.

12.3 Preparation of TiO_2 Nanostructures

12.3.1 Preparation of TiO_2 Nanostructures of Various Shapes/Morphologies

One-dimensional TiO_2 nanostructures may be grouped generally into four typical morphologies: nanotubes, nanorods, nanowire/nanofibers, and nanobelt/nanoribbons. Here preparations of TiO_2 with these four morphologies are covered. It is observed that TiO_2 nanotubes are primarily formed through the alkaline hydrothermal treatment and nanoribbons are formed under harsher alkaline hydrothermal treatment, like a higher concentration of NaOH and a high temperature.

TiO_2 nanotubes have been prepared by Bavykin et al. (2004) using alkali hydrothermal synthesis. The possible mechanism of nanotube formation is suggested as wrapping of intermediate multilayered titanate nanosheets. The driving force for wrapping is considered to be the mechanical stress arising during crystallization/dissolution. Variation of experimental conditions like concentration of precursor and temperature led to the formation of TiO_2 of different morphologies. A mixture of titanium dioxide (anatase) and NaOH solution was heated in a Teflon-lined autoclave for 22 h to yield white, powdery TiO_2. Multiwalled TiO_2 nanotubes were obtained when 0.75 g of anatase was heated in 300 mL of 10 M NaOH at 120°C, whereas 0.25 g of anatase in 300 mL of 10 M NaOH at 140°C resulted in TiO_2 nanotubes of smaller diameters and 1 g of anatase in 300 mL of 10 M NaOH at 190°C results in TiO_2 nanofibers. It is also observed that at a fixed concentration of anatase, increase in temperature increases the average diameter of the nanotubes.

Kolen'ko et al. (2006) reported hydrothermal synthesis of TiO_2 nanorods from amorphous $TiO_2 \cdot nH_2O$ gel and a concentrated NaOH solution. By varying heating temperature and atmosphere of precursors, different nanorods were obtained. $TiO_2 \cdot nH_2O$ gel was mixed with a NaOH aqueous solution and placed in a Teflon-lined autoclave at 200°C for 20 h. The products were isolated by centrifugation and then washed several times with distilled water. After heating at 200°C sodium trititanate ($Na_2Ti_3O_7$) nanorods are formed. Dispersing this ($Na_2Ti_3O_7$) in a HCl aqueous solution gives $H_2Ti_3O_7$ nanorods. Again thermal treatment of ($Na_2Ti_3O_7$) samples at 500°C under an atmosphere of hydrogen and air gives nanorods of sodium hexatitanate ($Na_2Ti_6O_{13}$) and metastable TiO_2-B phases, respectively.

TiO_2 nanobelts were prepared using the hydrothermal method by Sarkar and Chattopadhyay (2014). They controlled the size of nanobelts by varying the reaction temperatures and reaction times. Anatase TiO_2 dispersed in a 10 M NaOH aqueous solution was mixed in ethanol and heated at 200°C in a Teflon-lined stainless steel autoclave and then cooled to room temperature naturally to yield sodium titanate nanobelts. Sodium titanate nanobelts on treatment with HCl get transformed to hydrogen titanate nanobelts. They observed that the samples heated at 200°C for 5, 12, and 72 h exhibit bundles of nanobelts with short lengths and large diameters. However, the sample heated for 8 h results in nanobelts of long lengths, a smooth surface, and small diameters. They also prepared branched nanobelts of different phases of TiO_2 using chemical bath deposition just by varying annealing temperatures. The prepared TiO_2 nanobelts were heated with stirring with a titanium butoxide, $Ti(OC_4H_9)_4$, solution at 60°C for 6 h. By varying the amount of $Ti(OC_4H_9)_4$, branched nanobelts of different surface areas and densities are formed. Further annealing of TiO_2 nanobelts at different temperatures results in the formation of branched TiO_2 nanobelts of both single and multiple phases. At an annealing temperature of 400°C TiO_2-B, at 500°C mixed phases of TiO_2-B/anatase/rutile, and at 700°C mixed phases of anatase/rutile are formed.

Titanate nanowires were synthesized through the hydrothermal synthesis route by Horváth et al. (2007). TiO_2 nanowires were prepared from anatase TiO_2 and a 10 M aqueous NaOH solution heated at 130°C for 1 to 72 h in a rotating Teflon-lined autoclave. It was observed that during the first 18 h of reaction nanotubes were formed with an increase in surface area but after 18 h till 72 h of reaction the surface area decreased with the formation of nanowires.

12.3.2 Preparation of TiO_2 Nanostructures of Different Crystal Facets

Besides morphology, TiO_2 nanomaterials of different dominant crystal facets (having maximum surface area and energy) can be prepared by varying experimental conditions. Properties of TiO_2, in particular photocatalytic activity, are influenced by the dominant crystal facets. Therefore, preparation of facet-controlled TiO_2 is important. Zhang et al. (2014) prepared different anatase TiO_2 with [101], [001], and [111] crystal facets by the hydrothermal route, using TBT [$Ti(OBu)_4$], NH_4F, HF (40%), and deionized water with different mole ratios as precursors. A fixed concentration of titanium precursor with varied concentrations of other reactants yields TiO_2 of different facets. $Ti(OBu)_4$ mixed with NH_4F, HF, ammonia (28%), and deionized water and heated in a Teflon-lined autoclave at 180°C for 24 h yields TiO_2. Zhang et al. (2014) prepared three samples with different NH_4F, HF, ammonia, and deionized water content mixed with 25 mL of $Ti(OBu)_4$, T_1 (2.74 mL ammonia and 2.36 mL deionized water), T_2 (0.75 g NH_4F, 0.78 mL HF, and 4.22 mL deionized water), and T_3 (1.56 mL HF and 3.44 mL deionized water). All three powders were calcined at 600°C before characterization. An x-ray diffraction study shows all three are anatase TiO_2. The Brunauer–Emmett–Teller (BET) specific surface area measurement shows T_1 having a (101) facet with a 95% surface area, T_2 with a 70% surface area and a (111) facet, and T_3 having a 76% surface area with a (001) facet. All these are found to exhibit different extents of photocatalytic activities.

Another report is made for the preparation of TiO_2 with different dominant facets by the hydrothermal method using $TiOSO_4$ (titanium oxysulfate) as the precursor of titanium and HF as a morphology controlling agent (Cheng et al., 2013). Unlike the above method, here the concentration of the titanium precursor is varied keeping that of HF constant. They dissolved 64 and 70 mg of $TiOSO_4 \cdot xH_2O$ powder in 40 mL aqueous solutions of HF and heated these at 180°C for 12 h in a Teflon-lined autoclave to produce $TiO_2(101)$ and $TiO_2(001)$ samples. X-ray diffraction patterns of both $TiO_2(101)$ and $TiO_2(001)$ show the formation of anatase TiO_2. High-resolution transmission electron microscopy (HRTEM) and selected-area electron diffraction (SAED) show TiO_2(101) having a 70% [101] facet and TiO_2(001) with a 60% (001) facet. Field-emission scanning electron microscopy (FESEM) images of both TiO_2 of different dominant facets are shown in Fig. 12.4.

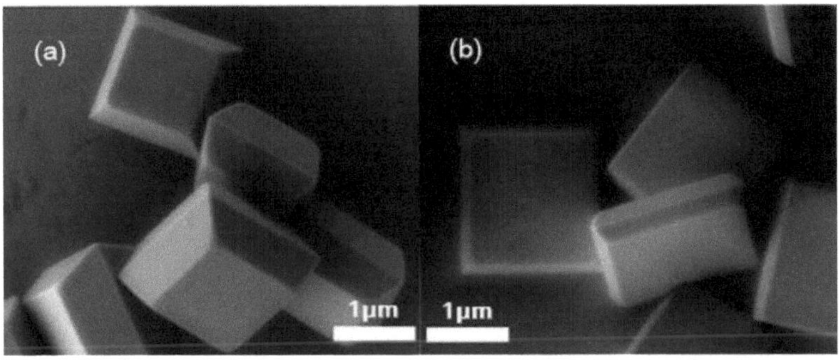

Figure 12.4 FESEM images of anatase (a) TiO_2(101) and (b) TiO_2(001) samples. Reprinted with permission from Cheng K, et al. (2013). Sonochemical deposition of Au nanoparticles on different facets dominated anatase TiO_2 single crystals and resulting photocatalytic performance. *J. Phys. Chem. C*, **117**: 14600. Copyright (2013) American Chemical Society.

12.4 Photocatalytic Application of TiO_2 Nanomaterials

12.4.1 Mechanism of Photocatalysis in TiO_2

TiO_2 is a semiconductor having a filled VB and an empty CB. Absorption of a photon ($h\nu$) equal to or greater than the bandgap energy (E_g = 3.2 eV) promotes an electron from the VB to the CB, resulting in free photoelectrons in the CB and photoholes in the VB (Eq. 12.1). These electron-hole pairs subsequently dissipate the energy in the form of heat or react with the donor or acceptor adsorbed on the surface. This light-/photoinduced promotion of an electron makes TiO_2 a potential photocatalyst.

$$TiO_2 + h\nu = h^+ \text{ (hole)} + e^- \text{ (electron)} \tag{12.1}$$

To enhance the photocatalytic activity of TiO_2 the E_g of TiO_2 can be reduced by doping with suitable metals and nonmetals, loading noble metal NPs on the surface of TiO_2, or by doping with another semiconductor. Thind (2012) reported enhanced photocatalytic activity of nitrogen (N)- and tungsten (W)-doped TiO_2. Photocatalytic degradation of organic pollutant rhodamine B was significantly enhanced by doping. Enhanced photocatalytic activity is attributed to reducing the bandgap and increasing the surface area of TiO_2 as a result of N and W doping. Another report shows carbon doping in TiO_2 increases its photocatalytic activity (Dong et al., 2011), where sucrose is used as a source of carbon. Post-thermal treatment of C-doped TiO_2 significantly enhances photocatalytic activity by inhibiting the recombination of electron-hole pairs generated due to photoexcitation. Christoforidis and Fernández-García (2016) reported Cu- and V-doped TiO_2 for photocatalytic applications. Photo-oxidation of toluene was drastically improved by Cu^{2+} and V^{4+} doping in anatase TiO_2. These dopants enhance photocatalytic activity by increasing the charge separation efficiency, acting as efficient electron scavengers, and increasing the surface availability of the key charge carriers, the holes. Apart from C and N, nonmetals like S and F were also introduced into the TiO_2 lattice by various methods, such as hydrothermal, sputtering, and thermal treatment methods.

Apart from doping, surface modification of TiO_2 by noble metal NPs also enhances photocatalytic activity of TiO_2. Noble metal NPs adsorbed on TiO_2 have been proved to be efficient photocatalysts. Due to their photostability and characteristic surface plasmon resonance bands noble metals on the surface of TiO_2 act as photosensitizers. The Fermi levels of Ag, Au, and Pt are lower than the CB of TiO_2. So these metal NPs effectively suppress the recombination of photogenerated electron-hole pairs in TiO_2 and thus enhance photocatalytic efficiency of TiO_2. Silva et al. (2011) have provided a mechanism to explain the enhancement of photocatalytic activities by Au NPs deposited on TiO_2. As shown in the Fig. 12.5, Au NPs exhibit a dual role, as (i) light harvesters injecting electrons into the semiconductor CB and (ii) catalytic sites for H_2 gas generation. In Fig. 12.5a, upon visible light illumination, the excitation of the Au surface plasmon band generates electron-hole pairs and the electrons get readily transferred to the CB of TiO_2, while the holes remain at the highest occupied molecular orbital (HOMO) of Au NPs for the oxidation of ethylene diamine tetra-acetic acid (EDTA), enhancing the efficiency of the charge separation process. In Fig. 12.5b, under UV light illumination, the electrons from TiO_2 VB are excited to the CB of TiO_2, which subsequently migrate to the surface of Au NPs to participate in the redox reactions, while the photogenerated holes oxidize the oxidizing agent on the surface of TiO_2.

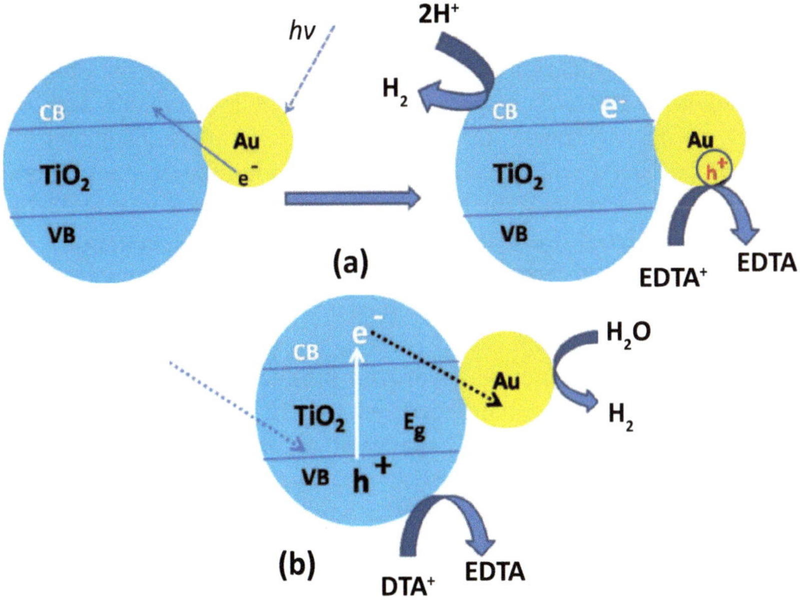

Figure 12.5 Mechanism of photocatalytic activity of Au/TiO_2 (a) upon excitation of the gold surface plasmon band and (b) under UV light excitation (Silva et al., 2011).

Au, Ag, Pd, and Pt coated on TiO_2 nanomaterials have been prepared by many methods, like chemical reduction, plasma sputtering, electrodeposition, and hydrothermal methods. Li et al. (2015) reported Au-, Pt-, and Pd-modified TiO_2 NPs by chemically reducing the corresponding metal precursors using an aqueous solution of $NaBH_4$ as the reducing agent. They have found that Pt-modified TiO_2 (Pt@TiO_2) exhibits the highest photocatalytic dehydrogenation of cyclohexane. Ghaffari et al. (2012) prepared Au NPs deposited on TiO_2 nanorods by the UV irradiation reduction method. Au NPs prevented photogenerated electrons from approaching the surface of TiO_2 nanorods by forming a barrier, which improves the conversion efficiency of a dye-sensitized solar cell.

12.4.2 Photocatalytic Applications

12.4.2.1 Photocatalytic water splitting and hydrogen production

To meet the challenges of fossil fuel depletion, hydrogen energy is considered as an attractive alternative energy source. Photocatalyzed water splitting has become a promising way of generating hydrogen. In comparison to other semiconductors TiO_2 has been widely considered as a photocatalyst due to low cost and nontoxicity. In this process splitting of water to H_2 is being catalyzed by TiO_2. Photoinduced electrons generated in TiO_2 get excited from the VB to the CB. The holes generated at the VB react with water to produce H^+, which gets reduced to H_2 by electrons at the CB.

Splitting of water for hydrogen production by using TiO_2 as a photocatalyst has been reported by Krishnappa et al. (2015). TiO_2 nanomaterials have been used as efficient photocatalysts, yielding up to 1304 $\mu mol \cdot g^{-1}$ of hydrogen. They prepared TiO_2 from hydrolysis of $TiCl_4$ in an ionic liquid, 1-methyl-3-[2-(2-methoxy(ethoxy)ethyl]imidazolium methane sulfonate (M(MEE) I$\cdot CH_3SO_3$). During photocatalyzed water splitting with TiO_2, they used ethanol as the sacrificial agent to prevent the recombination of photogenerated excited electrons in the CB and positive holes in the VB in the TiO_2, leading to larger quantum efficiency and hydrogen production. The excited electrons from the CB reduce the protons from water molecules, producing H_2, and ethanol is oxidized by positive holes.

Though TiO_2 is used as a photocatalyst, it has limitations because of the recombination of photogenerated electron-hole pairs. To improve photocatalytic efficiency, recombination of electron-hole pairs has to be prevented. Addition of metals and metal oxides to TiO_2 is an effective strategy to prevent electron-hole pair recombination, thus enhancing photocatalytic efficiency of TiO_2.

Introducing magnetic NPs of metal oxide into TiO_2 not only enhances photocatalytic activity but also facilitates the separation of the photocatalyst from the reaction and thus increases the chances of reuse. Kim et al. (2014) reported a magnetically separable core-shell-structured $NiFe_2O_4$@TiO_2 nanomaterial as an efficient photocatalyst for the production of hydrogen by water splitting. Enhancement of photocatalytic activity is due to effective charge transfer from TiO_2 to $NiFe_2O_4$, which suppresses electron-hole pair recombination and facilitates the splitting of water for hydrogen production. They observed that hydrogen production from water splitting by using $NiFe_2O_4$@TiO_2 is 10 times greater than that produced using pure TiO_2 as the photocatalyst.

Pu et al. (2013) reported Au NP–decorated TiO_2 nanowires for photocatalytic water splitting. TiO_2 nanowires were synthesized on a fluorine-doped tin oxide glass surface. Then calcination of the TiO_2 substrate dipped in a $HAuCl_4$ solution led to the deposition of Au NPs on the TiO_2 surface. A TiO_2 electrode decorated with Au NPs exhibited enhanced photoelectrochemical water oxidation. They observed the enhanced photoactivity of Au NP–decorated TiO_2 in the UV region is partially because of increased optical absorption of TiO_2 due to excitation of CB electrons of Au NPs by incident photons, which is known as surface plasmon resonance of Au NPs.

Similarly TiO_2 modified with another noble metal NP also exhibits enhanced photocatalytic activity due to their surface plasmon resonance phenomenon. Pt-modified TiO_2 also acts as an efficient photocatalyst for water splitting. Banerjee et al. (2015) used Pt-doped TiO_2 for splitting of water under solar irradiation. An ionic-liquid-mediated approach is used to prepare TiO_2 nanocrystals followed by Pt doping by a chemical impregnation method. They observed Pt doping significantly enhances the hydrogen production by water splitting. This is due to a decrease in the bandgap of TiO_2 and increase in the absorption of TiO_2 due to surface plasmon resonance of Pt NPs.

12.4.2.2 Photocatalytic oxidation of alcohols

The selective oxidation of alcohols into carbonyl compounds by using dioxygen (O_2) as the oxidant in the presence of light can be catalyzed by TiO_2. This photocatalytic oxidation process involves selective cleavage of the C–O bond of alcohol to form a new C–O bond in the product carbonyl compound, where the O atom comes from dioxygen. Zhang et al. (2009) studied photocatalytic oxidation of alcohols by TiO_2 using isotope labelled dioxygen in the organic solvent benzotrifluoride (BTF). An isotope study confirms the oxygen atom in the oxidation product comes from the dioxygen, as shown in the reaction here:

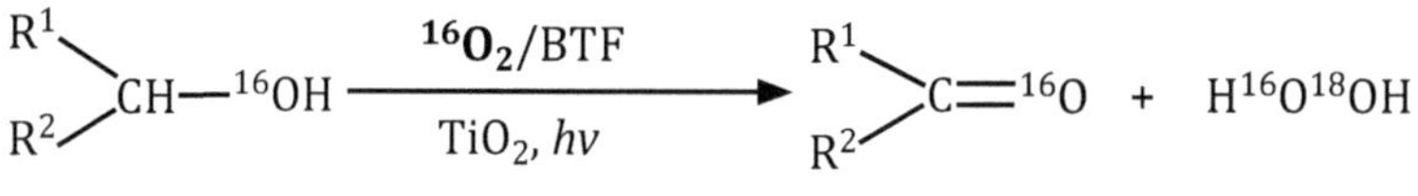

Zhang et al. (2009) proposed a mechanism, as shown in Fig. 12.6. An alcohol molecule adsorbs onto the surface of TiO_2 to form the structure I via a deprotonation process. When UV light falls on TiO_2, h^+/e- pairs are formed. The photogenerated hole reacts with the adsorbed alcohol to form a carbon radical, and the photogenerated electron gets absorbed by Ti^{4+} to form Ti^{3+}(II). Both the carbon radical and the Ti^{3+} combine with the dioxygen to form the oxygen bridge structure III via two possible routes: (i) the electron/Ti^{3+} (in the CB) first reduces O_2 to the superoxide, which then attacks the carbon radical to form intermediate III, or (ii) the carbon radical first combines with the dioxygen to form an organic superoxide radical, which additionally reacts with Ti^{3+} to form structure III. After cleavage of C–O and O–O bonds, carbonyl is formed as an oxidized product and the TiO_2-bound peroxide bridge structure IV may then combine with the protons to form H_2O_2.

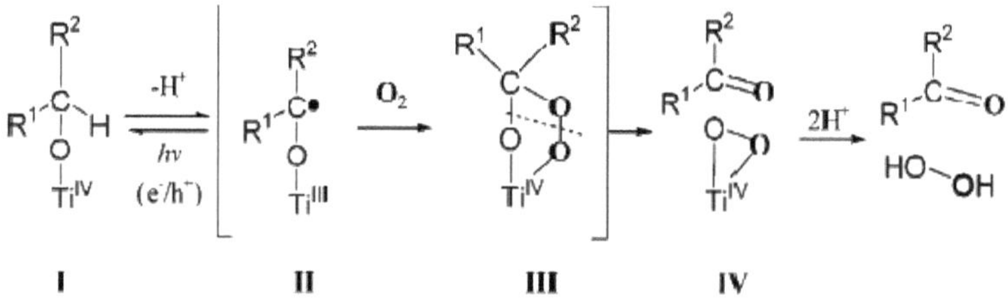

Figure 12.6 The proposed oxygen transfer process in the TiO_2 photocatalytic oxidation of alcohols in a benzotrifluoride solvent in the presence of O_2. Reproduced from Zhang M, Wang Q, Chen C, Zang L, Ma W, Zhao J. (2009). Oxygen atom transfer in the photocatalytic oxidation of alcohols by TiO_2: oxygen isotope studies. *Angew. Chem. Int. Ed.*, **48**: 6081, with permission from Wiley.

Jiang et al. (2015) reported Au and Au-Pd alloy NPs deposited on TiO_2 heterostructured (comprising both TiO_2-B and anatase) nanobelts as effective plasmonic photocatalysts for benzyl alcohol oxidation. Bimetallic Au-Pd-alloy-modified TiO_2 exhibited significant improvement of catalytic activity in comparison to monometallic Au-modified TiO_2. This is mainly due to plasmon-mediated charge distribution within alloy NPs in bimetallic nanostructures. Au and Au-Pd alloy NPs were deposited on TiO_2 nanobelts by the deposition precipitation method. TiO_2 nanobelts were dispersed in an aqueous solution of $HAuCl_4$ followed by the addition of a precipitating agent such as urea and calcination led to the formation of Au NPs deposited on TiO_2 nanobelts. Similarly by using a mixture of $PdCl_2$ and $HAuCl_4$, TiO_2 nanobelts decorated with Au-Pd-alloy NPs were prepared. Oxidation of benzyl alcohol was performed by the photoelectrochemical process using three electrodes, where Au/TiO_2 and Au-Pd/TiO_2 were used as the photoanode, Pt as the counterelectrode, and saturated calomel as the reference electrode. Efficiency of the catalyst was assessed from the photocurrent response produced by aerobic oxidation of benzyl alcohol to benzaldehyde after irradiation of visible light. Oxygen molecules adsorbed on TiO_2 can capture the electrons injected from Au NPs and form $O–O^-$ species, which can promote oxidation of benzyl alcohol.

A TiO_2-Au core-shell nanomaterial has been used as the catalyst for photo-electro-oxidation of methanol by Pandikumar et al. (2010). Au NPs were deposited on the surface of TiO_2 by using an *N*-[3-(trimethoxysilyl)propyl]ethylenediamine (EDAS)-derived silicate sol-gel matrix. EDAS/(TiO_2-Au) NPs have been used to prepare electrodes for photoelectrochemical oxidation reaction. The functionalized aminosilicate acts as a stabilizer and a support for the core-shell TiO_2-Au nanomaterials. They observed that on dipping the indium tin oxide (ITO)/EDAS/TiO_2 electrode in a mixture of Na_2SO_4 and CH_3OH, there was no significant current observed in a dark condition but a large increase in the current under illumination. The methoxy radicals ($\cdot CH_2OH$, CH_2O, $CHO\cdot$, HCOOH, and $HCOO^-$) formed during methanol oxidation can further inject electrons and contribute to the increased photocurrent generation. This is because of the electron-hole separation at the TiO_2 and scavenging of holes by methanol. Performance of the EDAS/(TiO_2-Au)-NP-modified electrode showed a 12-fold enhancement in the catalytic activity toward photo-electro-oxidation of methanol in comparison to TiO_2 dispersed in an EDAS silicate sol-gel matrix. This enhancement in catalytic activity is due to interfacial charge transfer between Au and TiO_2 resulting in the suppression of electron-hole pair recombination.

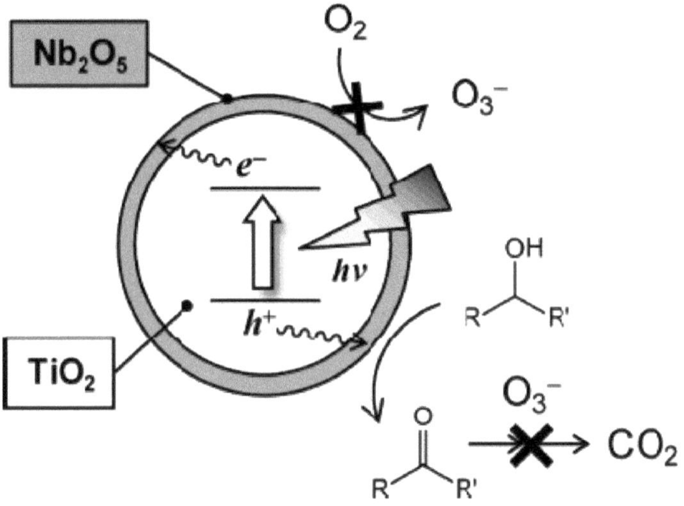

Figure 12.7 Strategy for the selective photocatalytic oxidation of alcohols over Nb_2O_5/TiO_2. Reprinted with permission from Furukawa S, et al. (2012). Photocatalytic oxidation of alcohols over TiO_2 covered with Nb_2O_5. *ACS Catal.*, **2**: 175. Copyright (2012) American Chemical Society.

Selective oxidation of alcohols can be achieved by surface modification of TiO_2. Furukawa et al. (2012) observed photocatalytic oxidation of alcohols over TiO_2 covered with Nb_2O_5. They have proposed the mechanism of photocatalysis, as shown in the Fig. 12.7. Addition of Nb_2O_5 to TiO_2 does not change the bandgap, but Nb_2O_5 coated on TiO_2 inhibits the production of O^{3-}. During photocatalytic oxidation of alcohols by TiO_2, undesired oxidation to carbon dioxide (CO_2) is observed due to the presence of highly oxidizing radical species: positive holes trapped by the surface lattice oxygen (O^-), ozonide ion (O^{3-}), or hydroxyl radical (HO·) on the surface of TiO_2. O^{3-} is formed upon the combination of O^- and O_2, and HO· is released via the oxidation of a surface hydroxyl group by a hole. Therefore, this is an approach to cover the surface of TiO_2 by Nb_2O_5 in order to inhibit the production of highly oxidizing O^{3-} on the surface and thus facilitates selective oxidation of alcohols.

12.4.2.3 Photocatalytic reduction of CO_2

Due to its strong reduction ability, TiO_2 is used as an efficient photocatalyst to reduce CO_2 into energy fuels. Photoinduced reduction of CO_2 into organic compounds such as formaldehyde, methanol, and methane in the presence of TiO_2 nanomaterials carries significant importance. Inoue et al. (1979) proposed the following mechanism of photoreduction of CO_2 by semiconductor catalysts suspended in water.

$$\text{Catalyst} + h\nu \rightarrow \text{Catalyst*}\ (e^-_{CB} + h^+_{VB})$$

For an oxidation reaction:

$$H_2O + 2h^+_{VB} \rightarrow 1/2O_2 + 2H^+$$

For reduction reactions:

$$CO_2\,(aq) + 2H^+ + 2e^-_{CB} \rightarrow HCOOH$$

$$HCOOH + 2H^+ + 2e^-_{CB} \rightarrow HCHO + H_2O$$

$$HCHO + 2H^+ + 2e^-_{CB} \rightarrow CH_3OH$$

$$CH_3OH + 2H^+ + 2e^-_{CB} \rightarrow CH_4 + H_2O$$

Photogenerated holes in the VB (h^+_{VB}) of semiconductor oxidize water to O_2 and release H^+, and photogenerated electrons in the CB (e^-_{CB}) reduce CO_2 by a sequence of reactions to produce HCOOH, HCHO, CH_3OH, CH_4, etc.

TiO_2 and modified TiO_2 with metals and metal oxides exhibit photocatalytic reduction of CO_2. Nanocomposites comprising mixed-phase TiO_2 have shown improved photocatalytic reduction of CO_2 in comparison to single-phase TiO_2 (Li et al., 2008). It has been reported that anatase-rutile nanocomposites exhibit higher photocatalytic reduction of CO_2 to CH_3OH and CH_4. They have proposed the effective charge separation between anatase and rutile as the key factor for improved catalytic property. Ag NPs deposited on anatase-brookite mixed-phase TiO_2 exhibited superior performance for photocatalytic reduction of CO_2 to CH_4 and H_2. The enhanced activity of Ag(0)/TiO_2 (anatase-brookite) is attributed to their large surface area, small and well-dispersed Ag(0) NPs, and an enhanced interfacial charge transfer between the anatase and brookite nanocrystals (Liu et al., 2013).

Au NPs deposited on TiO_2 (Au/TiO_2) exhibited remarkable enhancement of photocatalytic reduction of CO_2. When the wavelength of the incident photon matches the surface plasmon resonance wavelength of Au NPs it exhibits enhanced photocatalytic activity. The intense local fields created by the surface plasmon resonance of Au NPs excite the electron-hole pair in TiO_2 at a much higher rate than the incident light. This may enhance the sub-bandgap absorption in TiO_2, leading to enhanced photocatalytic performance (Hou et al., 2011).

Xie et al. (2013) found TiO_2 loaded with Pt and MgO on its surface exhibits improved photocatalytic reduction of CO_2 to CH_4 in the presence of water. It is proposed that MgO enhances the density of CO_2 molecules on the catalyst surface by adsorption, followed by reduction of CO_2 by enriched electron density on Pt. MgO is proposed to have a dual role in photoreduction of CO_2: (i) capture the holes and facilitate the separation of photogenerated electrons and holes, and (ii) hinder the direct contact of CO or CH_4 with TiO_2 surfaces, reducing the reoxidation possibilities for CO and CH_4.

12.4.2.4 Photocatalytic degradation of pollutants

Removal of organic and inorganic pollutants from air or water has been an area of significant interest. Semiconductor photocatalysts like TiO_2 are suitable for degradation of pollutants because of their redox behavior, chemical and biological inertness, nontoxicity, and comparatively low cost. When TiO_2 absorbs light equal to or greater than its bandgap energy electrons jump from the VB to the CB, leaving electrons in the CB and holes in the VB. Oxygen molecules adsorbed on the surface of TiO_2 are reduced by these electrons to generate $O{-}O^-$ and also holes can react with water, generating $OH^{\cdot}$; as a result pollutants get decomposed or degraded to small molecules like CO_2, NH_3, and H_2O.

A combination of magnetic NPs with TiO_2 results in magnetically separable and recyclable photocatalysts with improved catalytic performance. Li et al. (2014) observed a magnetic composite of TiO_2 for photocatalytic degradation of rhodamine B in water. The composite Fe_3O_4@SiO_2@TiO_2@Pt comprises a SiO_2-coated Fe_3O_4 core inside a TiO_2 shell along with a Pt layer deposited on the surface of TiO_2, as illustrated in the Fig. 12.8. The layer of SiO_2 in between magnetic Fe_3O_4 and TiO_2 acts as a barrier to prevent photoinduced electrons from being trapped by Fe_3O_4. Besides exhibiting high photocatalytic reduction of rhodamine B, this catalyst can be easily separated from the reaction solution by applying an external magnetic field. When UV light falls on Fe_3O_4@SiO_2@TiO_2@Pt composites, electrons in the VB of TiO_2 get excited to the CB, leaving holes in the VB to oxidize the organic reagents. Pt having a lower Fermi level than the CB of TiO_2, electrons from the CB of TiO_2 jumps to the Fermi level of Pt (Fig. 12.8). As a result the recombination of electron-hole pairs is suppressed, leading to improved catalytic performance. Li et al. (2014) have also observed less photocatalytic activity of the nanocomposite without Pt on the surface of TiO_2. Rhodamine B decomposition can be monitored by measuring UV-Vis absorption spectra of a solution containing the catalyst and rhodamine B, after irradiating the solution by a source of light.

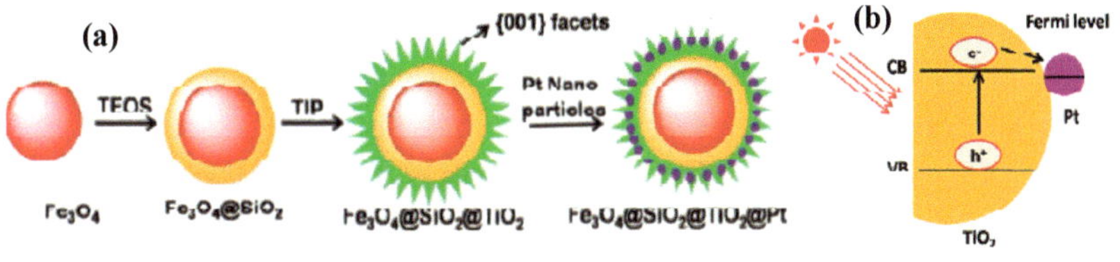

Figure 12.8 Schematic illustration of (a) the synthesis of Fe_3O_4@ SiO_2@ TiO_2@Pt microspheres and (b) the energy structure between TiO_2 and Pt of the Fe_3O_4@SiO_2@TiO_2@Pt composites. Reprinted with permission from Li X, et al. (2014). Fe_3O_4@SiO_2@TiO_2@Pt hierarchical core–shell microspheres: controlled synthesis, enhanced degradation system, and rapid magnetic separation to recycle. *Cryst. Growth Des.*, **14**: 5506. Copyright (2014) American Chemical Society.

Methylene blue (MB) is a nonbiodegradable dye widely used in textile industries and can be degraded by using TiO_2 nanomaterials as photocatalyst. Cu_2O-loaded TiO_2 nanomaterials have been found to be effective photocatalysts for the degradation of pollutant MB (Han et al., 2015). By irradiating visible light on the reactor containing the MB solution, with Cu_2O-TiO_2 as the photocatalyst, degradation of MB can be estimated quantitatively using UV-Vis spectroscopy. It was observed that the rate of degradation of MB by TiO_2 is less ln comparison to Cu_2O-TiO_2. This is due to a reduction in the bandgap of TiO_2 in the presence of Cu_2O on its surface.

Haw et al. (2016) reported enhanced photocatalytic degradation of MB by $CoFe_2O_4$-doped TiO_2. $CoFe_2O_4$-TiO_2 reduces the rate of recombination of electron-hole pairs. In TiO_2 electrons produced due to photoexcitation are transferred to the CB, leaving holes in the VB. These electrons get transferred to the CB of $CoFe_2O_4$, which is adsorbed on the surface of TiO_2. As a result recombination of electron-hole pairs is hindered and mobility of electron gets increased. Electron acceptors on the surface of $CoFe_2O_4$ accept these electrons and initiate catalytic activity.

In addition to organic pollutants TiO_2 nanomaterials can effectively adsorb and reduce toxic metal ions like Hg(II), As(II), Cd(II), and Cr(VI) into less toxic metallic or ionic states.

Not only adsorption of metal oxides on TiO_2 surface but also facet-specific TiO_2 nanostructures exhibit improved photocatalytic activities. The synthesis of the reactive (001) facet to expose anatase TiO_2 single crystals has also been an alternative route to enhance photocatalytic efficiency. Yang et al. (2014) reported enhanced photocatalytic activities of TiO_2 hollow spheres for the degradation of toxic metal ions Cr(VI) in solution. They found anatase TiO_2 hollow spheres with highly reactive (011) facets exhibit enhanced photocatalytic reduction of Cr(VI) in comparison to normal TiO_2. The (011) facets are more effective in the adsorption of Cr(VI) than the commonly exposed (101) facets in normal TiO_2 structures. In their experiment, UV light was irradiated to a suspension containing TiO_2 and an aqueous solution of Cr(VI) at a pH range of 2.5 to 4. After specific duration of irradiation, the sample was taken for spectrophotometric determination of Cr(VI) concentration. At a low pH, enhanced catalytic activity is due to two reasons: (i) the potential difference between the CB of TiO_2 and Cr(VI)/Cr(III) increases with a decrease in the pH and thus the thermodynamic driving force for the reduction of Cr(VI) is enhanced at a low pH, and (ii) in an acidic condition, the positively charged TiO_2 surface favors the adsorption of negatively charged Cr(VI) groups (CrO_4^{2-} or $Cr_2O_7^{2-}$) on TiO_2 due to electrostatic attraction, which enhances the Cr(VI) removal. In the high pH region, due to electrostatic repulsion between the negatively charged group (CrO_4^{2-} or $Cr_2O_7^{2-}$) and TiO_2, adsorption decreases and thus catalytic activity gets reduced.

References

Almeida NAF, et al. (2016). Pressure dependent luminescence in titanium dioxide particles modified with europium ions. *Sens. Actuators B*, **234**: 137.

Banerjee B, et al. (2015). Green synthesis of Pt-doped TiO_2 nanocrystals with exposed (001) facets and mesoscopic void space for photo-splitting of water under solar irradiation. *Nanoscale*, **7**: 10504.

Bavykin DV, et al. (2004). The effect of hydrothermal conditions on the mesoporous structure of TiO_2 nanotubes. *J. Mater. Chem.*, **14**: 3370.

Bavykin DV, et al. (2005). Reversible storage of molecular hydrogen by sorption into multilayered TiO_2 nanotubes. *J. Phys. Chem. B*, **109**: 19422.

Carp O, et al. (2004). Photoinduced reactivity of titanium dioxide. *Prog. Solid State Chem.*, **32**: 33.

Cheng K, et al. (2013). Sonochemical deposition of Au nanoparticles on different facets dominated anatase TiO_2 single crystals and resulting photocatalytic performance. *J. Phys. Chem. C*, **117**: 14600.

Chimupala Y, et al. (2014). Universal synthesis method for mixed phase TiO_2(B)/anatase TiO_2 thin films on substrates via a modified low pressure chemical vapour deposition (LPCVD) route. *J. Mater. Chem. A*, **4**: 5685.

Christoforidis KC, Fernández-García M. (2016). Photoactivity and charge trapping sites in copper and vanadium doped anatase TiO_2 nano-materials. *Catal. Sci. Technol.*, **6**: 1094.

Dong F, et al. (2011). Enhancement of the visible light photocatalytic activity of C-doped TiO_2 nanomaterials prepared by a green synthetic approach. *J. Phys. Chem. C*, **115**: 13285.

Feist T, Davies, PK. (1992). The soft chemical synthesis of TiO_2 (B) from layered titanates. *J. Solid State Chem.*, **101**: 275.

Fujishima A, Honda K. (1972). Electrochemical photolysis of water at a semiconductor electrode. *Nature*, **238**: 37.

Furukawa S, et al. (2012). Photocatalytic oxidation of alcohols over TiO_2 covered with Nb_2O_5. *ACS Catal.*, **2**: 175.

Ghaffari M, et al. (2012). Effect of Au nano-particles on TiO_2 nanorod electrode in dye-sensitized solar cells. *Electrochim. Acta*, **76**: 446.

Gopal M, et al. (1997). Room temperature synthesis of crystalline metal oxides. *J. Mater. Sci.*, **32**: 6001.

Han T, et al. (2015). The study on preparation and photocatalytic activities of Cu_2O/TiO_2 nanoparticles. *J. Environ. Chem. Eng.*, **3**: 2453.

Haw C, et al. (2016). The design of new magnetic-photocatalyst nanocomposites ($CoFe_2O_4$–TiO_2) as smart nanomaterials for recyclable-photocatalysis applications. *New J. Chem.*, **40**: 1124.

Horváth E, et al. (2007). Hydrothermal conversion of self-assembled titanate nanotubes into nanowires in a revolving autoclave. *Chem. Mater.*, **19**: 927.

Hou W, et al. (2011). Photocatalytic conversion of CO_2 to hydrocarbon fuels via plasmon-enhanced absorption and metallic interband transitions. *ACS Catal.*, **1**: 929.

Hu W, et al. (2016). Facile strategy for controllable synthesis of stable mesoporous black TiO_2 hollow spheres with efficient solar-driven photocatalytic hydrogen evolution. *J. Mater. Chem. A*, **4**: 7495.

Inoue T, et al. (1979). Photoelectrocatalytic reduction of carbon dioxide in aqueous suspension of semiconductor powders. *Nature*, **277**: 637.

Jiang T, et al. (2015). Gold and gold–palladium alloy nanoparticles on heterostructured TiO_2 nanobelts as plasmonic photocatalysts for benzyl alcohol oxidation. *Nanoscale*, **7**: 209.

Kasuga T, et al. (1998). Formation of titanium oxide nanotube. *Langmuir*, **14**: 3160.

Kim HS, et al. (2014). Synthesis of magnetically separable core@shell structured $NiFe_2O_4$@TiO_2 nanomaterial and its use for photocatalytic hydrogen production by methanol/water splitting. *Chem. Eng. J*, **243**: 272.

Kolen'ko YV, et al. (2006). Hydrothermal synthesis and characterization of nanorods of various titanates and titanium dioxide. *J. Phys. Chem. B*, **110**: 4030.

Krishnappa M, et al. (2015). Mesoporous foam TiO_2 nanomaterials for effective hydrogen production. *Chem. Eur. J.*, **21**: 17624.

Li W, et al. (2002). Metallorganic chemical vapor deposition and characterization of TiO_2 nanoparticles. *Mater. Sci. Eng. B*, **96**(3): 247.

Li G, et al. (2008). Synthesizing mixed-phase TiO_2 nanocomposites using a hydrothermal method for photo-oxidation and photoreduction applications. *J. Catal.*, **253**: 105.

Li X, et al. (2014). Fe_3O_4@SiO_2@TiO_2@Pt hierarchical core–shell microspheres: controlled synthesis, enhanced degradation system, and rapid magnetic separation to recycle. *Cryst. Growth Des.*, **14**: 5506.

Li L, et al. (2015). Simple and efficient system for combined solar energy harvesting and reversible hydrogen storage. *J. Am. Chem. Soc.*, **137**: 7576.

Liu H, et al. (2011). Mesoporous TiO_2–B microspheres with superior rate performance for lithium ion batteries. *Adv. Mater.*, **23**: 3450.

Liu L, et al. (2013). Silver-incorporated bicrystalline (anatase/brookite) TiO_2 microspheres for CO_2 photoreduction with water in the presence of methanol. *Appl. Catal. A*, **467**: 474.

Ma Y, et al. (2014). Titanium dioxide-based nanomaterials for photo catalytic fuel generations. *Chem. Rev.*, **114**: 9987.

Morgan DL, et al. (2008). Determination of a morphological phase diagram of titania/titanate nanostructures from alkaline hydrothermal treatment of degussa P25. *Chem. Mater.*, **20**: 3800.

Nag M, et al. (2007). Low-temperature hydrothermal synthesis of phase-pure rutile titania nanocrystals: time temperature tuning of morphology and photocatalytic activity. *Mater. Res. Bull.*, **42**: 1691.

Okuyama K, et al. (1989). Experimental control of ultrafine TiO_2 particle generation from thermal decomposition of titanium isopropoxide vapour. *Chem. Eng. Sci.*, **44**(6): 1369.

Pandikumar A, et al. (2010). Functionalized silicate sol-gel-supported TiO_2-Au core-shell nanomaterials and their photoelectrocatalytic activity. *ACS Appl. Mater. Interfaces*, **2**(7): 1912.

Pottier A, et al. (2001). Synthesis of brookite TiO_2 nanoparticles by thermolysis of $TiCl_4$ in strongly acidic aqueous media. *J. Mater. Chem.*, **11**: 1116.

Pu YC, et al. (2013). Au nanostructure-decorated TiO_2 nanowires exhibiting photoactivity across entire UV-visible region for photoelectrochemical water splitting. *Nano Lett.*, **13**: 3817.

Reyes-Coronado D, et al. (2008). Phase-pure TiO_2 nanoparticles: anatase, brookite and rutile. *Nanotechnology*, **19**: 145605.

Sander MS, et al. (2004). Template assisted fabrication of dense, alligned arrays of titania nanotubes with well controlled dimensions on substrates. *Adv. Mater.*, **16**: 2052.

Sarkar D, Chattopadhyay KK. (2014). Branch density-controlled synthesis of hierarchical TiO_2 nanobelt and tunable three-step electron transfer for enhanced photocatalytic property. *ACS Appl. Mater. Interfaces*, **6**: 10044.

Silva CG, et al. (2011). Influence of excitation wavelength (UV or visible light) on the photocatalytic activity of titania containing gold nanoparticles for the generation of hydrogen or oxygen from water. *J. Am. Chem. Soc.*, **133**: 595.

Sugimoto T, et al. (2003). Synthesis of uniform anatase TiO_2 nanoparticles by gel–sol method formation process and size control. *J. Colloid Interface Sci.*, **259**(1): 43.

Thind SS. (2012). Significant enhancement in the photocatalytic activity of N, W co-doped TiO_2 nanomaterials for promising environmental applications. *Nanotechnology*, **23**: 475706.

Wang M, et al. (2016). Preparation and characterization of single crystalline anatase TiO_2 films on LSAT (001) by MOCVD. *RSC Adv.*, **6**: 4867.

Wang W, et al. (2016). Nitrogen-doped TiO_2 microspheres with hierarchical micro/nanostructures and rich dualphase junctions for enhanced photocatalytic activity. *RSC Adv.*, **6**: 40923.

Xie S, et al. (2013). Photocatalytic reduction of CO_2 with H_2O: significant enhancement of the activity of Pt–TiO_2 in CH_4 formation by addition of MgO. *Chem. Commun.*, **49**: 2451.

Xu GR, et al. (2013). Template directed preparation of TiO_2 nanomaterials with tunable morphologies and their photocatalytic activity research. *Appl. Surf. Sci.*, **279**: 103.

Yang Y, et al. (2014). Enhanced photocatalytic activity of hierarchical structure TiO_2 hollow spheres with reactive (001) facets for the removal of toxic heavy metal Cr(VI). *RSC Adv.*, **4**: 34577.

Zhang M, et al. (2009). Oxygen atom transfer in the photocatalytic oxidation of alcohols by TiO_2: oxygen isotope studies. *Angew. Chem. Int. Ed.*, **48**: 6081.

Zhang J, et al. (2014). Regulating photocatalytic selectivity of anatase TiO_2 with {101}, {001}, and {111} facets. *J. Am. Ceram. Soc.*, **97**(12): 4005.

Zhang Y, et al. (2015). Ti(IV) oxalate complex-derived hierarchical hollow TiO_2 materials with dye degradation properties in water. *Dalton Trans.*, **45**: 265.

Zhang Y, et al. (2016). Prolonged electron lifetime in ordered TiO_2 mesophyll cell-like microspheres for efficient photocatalytic water reduction and oxidation. *Small*, **12**(17): 2291.

Chapter 13

Sol-Gel-Fabricated Bioceramics for Clinical Application

Neetu Talreja and Dinesh Kumar

Department of Chemistry, Banasthali University, Banasthali 304022, Rajasthan, India

dsbchoudhary2002@gmail.com

Bioceramics have played a very important role in the development of new materials for clinical applications. The development of porous ceramics is a pivotal target for many relevant industrial applications, particularly in the field of bone surgery. The development of bone substitutes for load-bearing bone parts represents one of the most challenging applications, especially due to the difficulty of expressing high bioactivity and bone-like mechanical properties simultaneously. The development of bioceramic materials using the sol-gel process allows the preparation of bioceramic glasses with different chemical compositions within the nanometric range, with modified morphologies and pore texture. The different categories of bioceramics and their applications in the field of biomedical science,

Smart Ceramics: Preparation, Properties, and Applications
Edited by Ajay Kumar Mishra

ISBN 978-981-4774-30-7 (Hardcover), 978-1-315-16359-8 (eBook)
www.panstanford.com

especially in tissue engineering, dental application, and drug and antigen delivery using modified ceramics, have been discussed in this chapter. Detailed control of the process for the manufacturing of bioceramics has also been discussed.

13.1 What Are Bioceramics?

A bioceramic is a specially designed, nonmetallic inorganic ceramic material for the reconstruction or repair of parts of the human body. Bioceramics are widely used in different biological applications, including medicine, because of their extraordinary biocompatibility and insignificant cytotoxicity. The wider applications of bioceramics in clinical uses are, for example, reconstruction of middle ear ossicles, powder for bone filling, injectable bone cement, porous scaffolds, hydroxyapatite (HA)-coated screws, knees, hips, ligaments, spinal fusion, teeth, periodontal, maxillofacial reconstruction, jaw bone stabilizing, and tendons. Several types of bioceramic materials, such as bioinert (alumina, zirconia), bioresorbable (tricalcium phosphate), and bioactive (HA, bioactive glasses, glass ceramics) bioceramics, are used in the various applications. Bioceramics are widely used in the reconstruction or repair of the musculoskeletal system. The fabrication of several forms (microsphere, thin-layer-coated implants, porous matrix, polished surface) of bioceramics depends on the end applications. The surface of the solids is altered by decreasing the particle size up to nanometers. The nanorange bioceramics enhance the properties that have potential in newer end applications. The ceramic nanostructure has a high surface-to-volume ratio compared to other forms or micron-size materials (Regi, 2010; Soon et al., 2016). The various methods for constructing such ceramics are electrospinning, self-assembly, template-assisted synthesis, sol-gel, etc. (Dapporto et al., 2016).

13.2 Classification of Bioceramics

Bioceramic-based materials have been used in various clinical applications. Cytocompatibility, mainly the cellular response on the surface of bioceramic implants, is a major concern. Cellular responses are different with different types of bioceramic materials. Bioceramic materials are categorized mainly into four types on the basis of cell adhesion on implants: (i) inert bioceramics, (ii) porous bioceramics, (iii) bioactive ceramics, and (iv) resorbable bioceramics. Figure 13.1 shows the classification of bioceramics.

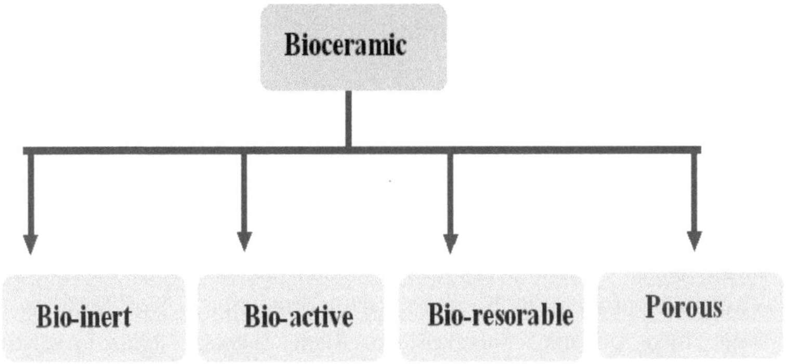

Figure 13.1 Classification of bioceramics.

13.2.1 Inert Bioceramics

Inert-bioceramic-based materials are used as implants that depend on the mechanical properties of the materials, which in turn are a function of grain size and purity. The inert implant is a nonadherent fibrous capsule and inert in the body or against tissues. Several inert bioceramics, such as alumina (Al_2O_3), zirconia (ZrO_2), titania (TiO_2), and silicon nitride (Si_3N_4), have low reactivity in comparison to other metals and polymeric materials. Inert bioceramics are noncytotoxic, nonallergic, and noncarcinogenic, which makes them potentially able materials for biomedical applications (Dorozhkin, 2016). Several factors, such as the host tissue, the implant's condition, the motion condition, the mechanical load, and the mechanical stability of the material affect the thickness of the produced fibers. Alumina has exceptional biocompatibility, corrosion resistance, high strength, and high wear resistance and is widely used in orthopedic surgery, dental implants, and load-bearing prostheses. The advantages of uses of alumina in clinical applications are mainly acceptance of cells and tissues of human beings, low friction coefficient, and minimal scar formation, which makes cementless fixation of prostheses possible. Alumina is mainly used in various orthopedic implants, for example, middle ear bone, knee prostheses, bone screws, cornea, jaw bone, segmental bone, screws, blades, and dental implants, because of its excellent properties.

13.2.2 Porous Bioceramics

Porous-bioceramic-based materials play a vital role in tissue engineering applications due to their inertness, combined with the mechanical stability of the complex interface developed when bone starts growing into the bioceramic. This adhesion is known as "biological fixation." Tissue engineering of the bone is a process of proliferation of the cells into a porous bioceramic. The main process of bone tissue engineering is proliferation of cells into porous materials to form tissues. Therefore, the physicochemical and biological properties of porous bioceramic materials play a significant role in regeneration and tissue reconstruction. Porous-bioceramic-based materials have low mechanical stability. They are used for low-load- or non-load-bearing applications, mainly growth of bone using interconnected pores of the biomaterials. Therefore, the implants provide a scaffold for the formation of bones. Some of the porous bioceramics have ideal microstructures with controlled pore sizes (140–160 and 200–1000 μm). Synthesis of porous bioceramics mainly involves the incorporation of soluble salts into the matrix, followed by the removal of these salts, which forms the desired porous structure of the materials. The materials can be converted into porous simply be selecting the foaming agent that evolves gases during heating (Jiang et al., 2013).

13.2.2.1 Disadvantages of porosity

Porosity can significantly weaken the mechanical strength of a material. On increasing the porosity, the strength of the material decreases rapidly. The implant material is subjected to a condition of corrosion compared to other materials, as the surface area is now more exposed. Most of the pores of the implants are not covered by the bones; these pores can enhance the possibility of bacterial infections.

13.2.3 Bioactive Ceramics

The bioactive ceramic materials form bonds between tissues or cells and implant materials after initiation of biological responses at the surface of materials. Glass and sintered HA are the first synthetic bioactive ceramic materials and have properties in between inert and resorbable bioceramic. Bioactive ceramics are widely used for grafting or filling of bone defects and periodontal irregularities due to their fragile nature. Environmental compatibility is created by the bioactive material for bone growth by providing a mineralizing interface for a normal bonding junction between body and material. Bioactive ceramics show chemical reactivity with the natural living environment within the body as they are composed of HA, glass ceramics, and functional ceramics, which significantly contribute in bond formation, simultaneously healing any wound. The type of adhesion is called "bioactive fixation." The fixation of the implants made of the inert ceramics has to rely on their correct shape accounting for the remodeling ability of the adjacent bony tissue under the influence of the stress and strain field created by the insertion of such implants. All bioactive bioceramic materials actively contribute to bond formation; therefore, they are known as bioactive ceramics. This is the reason none of these ceramics are actually stable in the body environment and the formation of a bond between the tissue and the implant surface might be stimulated. The interlocking of the two protects the implant from dislocation resulting from imposed stresses (Lee et al., 2014).

13.2.4 Resorbable Bioceramics

Resorbable ceramic materials are used for the development of new tissue or cells as these materials are temporary space fillers or scaffolds. These materials are composed of either inert ceramics or may be those that are easily processed through normal metabolic pathways and gradually leach out bone-growth-stimulating ions responsible for natural tissue reconstruction and simultaneous resorption. Resorbable implants perfectly fit in with the living bone. The transient nature of their properties is the feature of resorbable bioceramics. Resorbable bioceramics are extensively used in different implants, such as tendons, periodontal, maxillofacial defects, and plates of bone. The regeneration of tissues depends on the dissolution trends of the bioceramic materials and the dissolution is replaced with tissue. The advantage of restorable ceramics is that they have high mechanical strength with a small porous texture compared to large porous materials. On dissolution, the porosity increases, allowing high tissue growth to occur. Consequently, mechanical strength is preserved with minimum stress concentration. Rapidly degrading or slowly degrading ceramic materials both have their individual advantages and applications. Tissue replacement often needs fast-degrading materials, for example periodontal defects and open spaces in bone resulting from surgical procedures. Gradual degrading is essential when tissue needs to be augmented (Chang et al., 2015).

13.3 Sol-Gel Process for Bioceramics

The sol-gel method for the preparation of ceramics is a very attractive process as it is a simple, economical, and reliable method. The process involves the preparation of a sol using a monomer or an organometallic oxide as a precursor of the starting material, followed by the addition of a viscosity changer and binders. A small amount of acid at some constant temperature is provided to the reaction mixture to avoid significant particle aggregation, which is required for the stabilization of the sol. The acid is used to maintain a stable dispersion by producing similar electric charges as the particles have. This developed charge causes the particles to repel each other. Later, the thickened sol is collected and dried under a certain temperature to form a gel. The final process involves calcination and sintering before the composite ceramic is obtained (Owens et al., 2016).

13.4 Steps Involved in the Sol-Gel Process

13.4.1 Agitation

This part involves the formation of a colloidal solution through mechanical agitation of colloidal particles with water as a solvent at a pH that prevents precipitation. The hydrolysis and polycondensation reaction that takes place between the metal alkoxide precursor $M(OR)_4$ and water results in the formation of a small-particle colloidal suspension, which finally converts into a 3D scaffold of the analogous inorganic oxide.

13.4.2 Casting

A low-viscosity sol can be easily cast into a mold and can be easily converted into any shape. An appropriate mold can provide the desired shape and also prevent the adhesion of the gel to the container used for casting.

13.4.3 Gelation

On process propagation, colloidal particles and condensed silica start forming a 3D network. Gel properties are decided by the size of particles and the gelation process. The gelation process involves increase in the viscosity, which results in the shape of the casting vessel. During the process as the gelation occurs, fibers are twisted together with controlled change in viscosity. The formation of gelation is due to the interactions among the components, which can be electrical in nature. They interact together to form aggregates with colloidal particle, which results in gelation.

13.4.4 Syneresis

Aging is also known as syneresis. This process consists of the preservation of the cast body for a specific duration, which can be from hours to days. Throughout the aging process, precipitation of the gel network starts due to polycondensation, subsequently enhacing the thickness of interparticle necks and decreasing porosity of the materials. Therefore, the strength of the gel is enhanced and the produced product is resistive to cracking while drying.

13.4.5 Drying

The removal of excessive solvent from the complex 3D network is known as drying. When there is small porous texture large capillary stresses are produced, which cause the gels to crack instantly. The stress can only be stopped by decreasing the liquid amount during the process. This can only be controlled by indirect interaction of solid and liquid by using elimination of very small pores by adding surfactants. The hypercritical evaporation can stop the interaction of solid and liquid.

13.4.6 Densification

The densification of gel is done by heating the produced gel at elevated temperatures. The porous texture is also eliminated from the gel during the process and marks the density roughly around fused silica or quartz. The temperature should be contingent on the dimension of the pore, the pore connectivity, and its surface area during densification.

13.5 Biocompatibility of Ceramics

The human body has its own defense mechanisms designed to repel intruders. A "foreigner" may be pushed back out through the skin, dissolved, or, failing these, surrounded by a capsule of scar tissue, the thickness of which is inversely related to the severity of the body's reaction. This reaction depends upon the sterility of the intruder and the nature of the products of any chemical or physical breakdown. A severe local reaction can abolish physiological function, which leads to tissue damage. Thrombus formation in blood vessels, calcification and gross fibrous encapsulation in heart valves, and gross fibrous encapsulation of silica or asbestos particles in the lung lead to malignancy. Immune responses are also possible, especially from skin, bone, or organ replacements from another person. After a few hours or days, the foreign element stimulates an immunological response, which enhances the production of toxins and loss of adherence (Kharaziha and Fathi, 2010; Ghosh et al., 2016).

13.6 Biomedical Application

Recently, ceramics are attracting growing interest for applicability in biomedicine, such as drug and protein delivery, tissue engineering, implantable medical devices, noninvasive imaging, bone grafting, and biosensing. Herein, the unique properties of the ceramics, mainly high porosity and surface area and high mechanical strength, together with the versatility of sol-gel chemistry, play important roles.

13.6.1 Biosensors

Biosensing is a process where a biological analyte interacts with a physical transducer, which transforms the signal for measurement. Biosensors are mainly used for response time, analytical range, sensitivity, selectivity, detection limit, and reproducibility of the materials. The sensing mechanism is based on the biorecognized elements and their binding ability to the analytes by using a specific process of detection. Substantial variations in the analytical recital can be consequent from changes in these parameters. Typically, reduced approachability leads to lesser sensitivity, which can modify the analytical range. The lesser sensitivity can be controlled by using polymeric composite. Additional approaches comprise the capability to produce matrices that are either electrochemically active or optically transparent. The higher sorption ability tunes the porosity and polarity of the matrix. The higher biocompatibility of the materials modulates the composition of the materials and reduces leaching of biomolecules. Recently, most of the studies aim to synthesize hybrid-sol-gel and monolith materials for electrochemical and optical sensing on inorganic molecules. Several materials, mainly mesoporous silica nanoparticles, zeolites and mesoporous silicates, and sol-gel and template-assisted mesoporous materials, were used for biosensing, electrochemical detection, and fluorescence-based sensors, respectively. Mesoporous silicates are widely used in the sensing of various analytes (vapor phase and liquid) due to their versatility. Despite these versatile behaviors, the utility of the sensor, response time, reproducibility, and recovery time are still concerns. These factors have significant roles in medical diagnostic treatment using a sol-gel-based sensing material or device. The sol-gel-based materials have various properties, and compatibility is one of the main properties of concern in the development of biosensors.

13.6.2 Tissue Engineering or Scaffolds

Tissue engineering is an ideal approach facilitating the development of mimic structures for the bone matrix. The technique has potential to combat the severe issue of the availability of tissue and organ for transplantation. Body cells are guided for their cell proliferation and differentiation into the specific tissue or organ by using a temporary structure having a well-defined architecture known as a biomaterial scaffold. A scaffold is usually incorporated with growth factors and other biomolecules, along with the cells, to support and control the cellular functions during regeneration of tissues or organs. Scaffolds are usually made from biodegradable polymeric materials or ceramics for tissue engineering, which may be of synthetic or natural origin. Recently, scaffolds invented from inorganic ceramic materials such as CaP-based bioceramics and silica-based glass ceramics have higher mechanical strength than other polymeric and biometallic scaffolds (Gomes et al., 2010). Both these materials are still under investigation (Khan et al., 2014; Dorozhkin, 2015). The use of biodegradable synthetic polymer scaffolds for regeneration of load-bearing bones is challenging due to several disadvantages, such as their low mechanical strength and absence of biofunctionality (Zhang et al., 2007; Baino et al., 2015).

Several attempts have been made to fabricate nanocomposites of biodegradable polymers with biocompatible inorganic ceramics and of metallic scaffolds, but their corrosion behavior and other factors currently remain a major concern. However, the biocompatibility of such materials is not a major challenge as the surface coating of the material facilitates the biocompatibility of the implant materials (Mousa et al., 2015).

The major focus of this chapter is ceramic scaffolds, which have attracted a huge amount of attention over the past few years. The chapter covers a wide range of biocompatible ceramic materials and their composites. These materials have unique properties and are also known as silica-based sol-gel bioactive glasses. Bone grafting is one of the major areas of profound impact of bioactive materials. Bone grafts are used for new bone formation in the healing of skeletal fractures, healing of bones between a damaged joint, replacement and regeneration of lost or misplaced bone in the case of infection or disease, improvement of the bone healing response, and also regeneration of the tissue at the place of surgically implanted devices. "Osteogenic," "osteoconductive," and "osteoinductive" potential are the terms used for tissue regeneration capacity of these grafts (Bock et al., 2015; Prati and Gandolfi, 2015).

Biomimetic scaffolds as biomimetics stimulate the formation of natural tissues on the surface of the scaffolds. It is a design that requires understanding the bone biology and clinical physiology and involves an artificial building block for the development of bone; bone is the composition of osteogenic cells, extracellular matrix (ECM) proteins, growth factors, mineral in the form of hydroxycarbonate apatite (HCA), and a complex vascular system and can be seen as an open cell. The osteoprogenitor cells of mesenchymal origin are responsible for bone formation and represent about 10% of the total volume. Bone progenitor cells are preosteoblasts stimulated by specific growth factors. The bone progenitor cells located in the periosteum, endosteum, and Haversian canals originate from mesenchymal stem cells (MSCs). In general, the scaffold-based tissue engineering process provides provisional support for cell proliferation to synthesize new tissue of desired size and shape (Heimann, 2013; Ulum et al., 2014; Gopi et al., 2015). Figure 13.2 shows the schematics of tissue engineering.

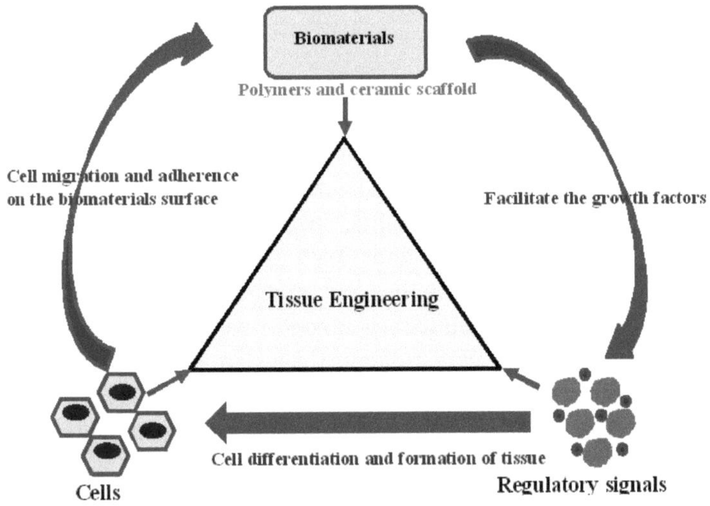

Figure 13.2 Three essential components of tissue engineering.

13.6.3 Dental Applications

Infections by microorganisms in the mouth are a leading cause of loss of teeth and supporting gum tissue. The demineralization and dissolution of teeth form dental cavities (caries), with metabolic activity in plaque (mucus film, which entraps microorganisms, especially bacteria, on the teeth surface), which leads to the loss of teeth. Several materials are used for the development of teeth and segments of teeth. Nowadays, dental composites have been accepted as a healing biomaterial. Compared to dental amalgams, the composites have better aesthetics and fewer safety concerns and have demonstrated good clinical results. Dental composites consist of resin matrices and different inorganic fillers. Calcium phosphates are widely used as a substitute for bones in clinical applications due to their biocompatibility, low density, chemical stability, and compositional resemblance to the mineral phase of bone. HA is the prime component of bone cements because it can be characterized as osteoconductive, biocompatible, noninflammatory, nontoxic, nonimmunogenic, and bioactive (Kaygili et al., 2012; Rena et al., 2012), and hence its ability to form chemical bonds with living tissues. However, the poor mechanical strength of HA limits its clinical applications under load-bearing conditions (Zhu et al., 2014; Cucuruz et al., 2015).

13.6.4 Orthopedics

Orthopedic implant devices are the leading application for bioceramic-based biomaterials. Arthritis affects the structure of freely movable (synovial) joints, such as the hip, knee, shoulder, ankle, and elbow. The joint pain includes pain in weight-bearing joints, such as the hip and knee, and can be considerable, and the effects on ambulatory function can be quite distressing. Nowadays, with the advent of anesthesia, antisepsis, and antibiotic drugs, the replacement of these joints and the restoration of mobility and pain relief have become commonplace for thousands of patients. Bioceramic-based biomaterials, using a variety of metals and ceramics, are used for orthopedic implant devices (Furko et al., 2016).

13.6.5 Cardiovascular Applications

Cardiovascular diseases are one of the leading causes of morbidity and mortality globally. The damage to valves and blockage of arteries require immediate treatment or repair. Various treatment methods are available, such as bypass surgery of arteries and heart transplantation. Recently, biomaterials have been used for the treatment of cardiovascular therapies. Bioceramics are widely used in blood-contacting devices such as cardiac valves and cardiac-assisted devices. Carbon-based materials are also used for prosthetic valves. Almost all bioceramic-based materials are used for blood-contacting devices without lysis of erythrocytes, except titanium and aluminum nitride.

13.6.6 Wound Dressing

A significant part of wound care is to control microorganism infections of the wound. Several efforts have been made to combat the incidence of bacterial infections using wound care techniques or dressing materials. Skin plays an important role in preventing the entry of undesirable substances and micro-organisms into the bloodstream. Thereby, proper dressing is required for wound infections. However, the absence of such dressing in the case of damaged skin can lead to a prolonged healing process and even death. Therefore, there is a necessity to develop newer materials or modify existing materials that protect wounds from various pathogens and improve the healing process. In this context, bioceramics are remarkable materials that are used in the form of fracture fixation devices, plates for bones, screws, nails, rods, wires, etc., for the treatment of fractures. Recently, nonmetallic materials such as carbon-carbon composite bone plates have been used for the treatment of fractures.

13.6.7 Ceramics for the Drug Delivery System

The development of newer drug molecules is time consuming and economically expensive. Therefore, existing drug molecules have been tried using several methods, such as drugs delivered at a specific rate (controlled or slow) and targeted delivery. The drug delivery system is one of the most demanding and interesting research areas nowadays. The customary theory for the administration of drugs is based on the systemic process, in which molecules of the drug are distributed in the body through the absorption of molecules in the bloodstream. The major drawback of the systematic process is poor penetration into tissue or cells and cytotoxicity associated with the body organs, mainly kidneys and liver. The drug delivery system is used in different areas such as tissue engineering and bone infections. Figure 13.3 shows the drug delivery applications of bioceramic-based materials.

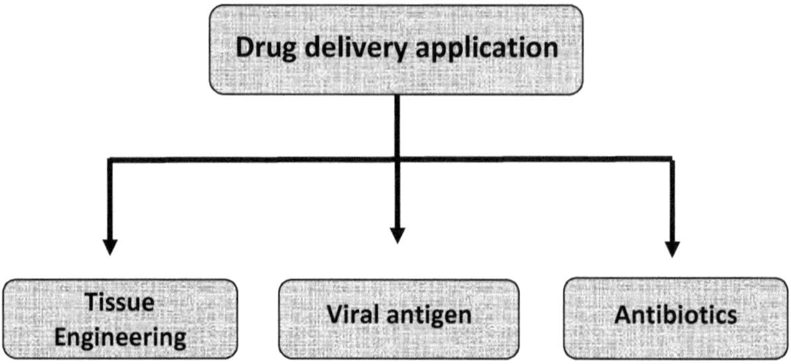

Figure 13.3 Drug delivery applications of bioceramic-based materials.

Bone tissue engineering is based on biodegradable 3D scaffolds containing high pores. Moreover, the release of growth scaffolds that produce biological responses leads to the migration of cells and triggers various the healing process. On the other hand, infections by microorganisms are one of the major devastating threats with replacement and bone filling.

To overcome these issues is the drug delivery system, which releases antimicrobial drugs at the site. Several bioceramic materials, such as mesoporous silica with an HA composite, nanoporous HA with polyvinyl alcohol, bone ash–reinforced gelatin/alginate/hyaluronic acid, and calcium polyphosphate, are used for drug delivery applications. Various mesoporous bioceramics are used in the drug delivery system, such as vancomycin, gentamycin, naproxen, aspirin, diflunisal, amoxicillin, captopril, and alendronate. Various bioceramic-based drug delivery systems have been developed that release drugs by diffusion and degradation of the scaffold. Figure 13.4 shows the release of drugs in a drug delivery system.

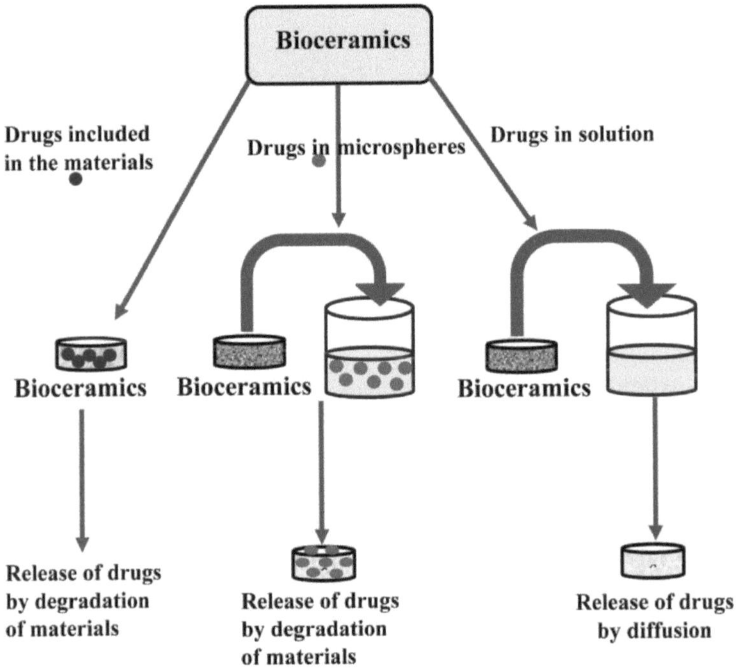

Figure 13.4 Release of drugs in a drug delivery system.

13.6.8 Bioceramics as a Carrier for Viral Antigen

Bioceramics are widely used in various biomedical applications, including s carriers for viral antigens or proteins. Silica-based bioceramic materials have been used as a carrier of drugs as well as for regeneration of bone. The mesoporous-silica-based bioceramics manage the loading and release of biologically active compounds. Thereby, pore size, surface area, and pore volume are important parameters for the adsorption and kinetics of biologically active molecules. Serum albumin delivery in protein delivery is most highlighted because these proteins are the main components in the human body. The surface modification or functionalized bioceramic materials are used as carriers for viral antigens.

In this approach the surface of the ceramic is functionalized with a complex multicomponent structure on which the drug is immobilized. The functionalized nanoceramics release the antigen at the specific site.

13.7 Conclusions and Future Perspectives

Sol-gel synthesis is an efficient method that enhances the homogeneity of chemicals, subsequently producing bioceramics. The development of bioceramic-based materials has provided a newer platform because of their unique structures. The biocompatibility of a metallic substrate is enhanced by using surface coatings with bioceramics materials, which also improves various healing processes. The sol-gel process is versatile, works at extended composition ranges, provides better homogeneity, is energy efficient, and does not require special or expensive equipment. Another main advantage of lower-temperature synthesis of sol-gel is that it prevents an incompatible thermal expansion coefficient, which is found in traditional coatings and leads to cracks and poor bonding. The tunable chemistry of the sol-gel process permits pore size control and morphology and surface modification for clinical applications. Drug adsorption and release in the drug delivery system depend on the surface area and pore diameter of the materials. The sol-gel process is simple, stable, economically viable, and accessible and facilitates various industrial and clinical applications. These unique properties become a suitable candidate for various clinical applications. In general, bioceramic-based materials created by using the sol-gel process are an efficient tool for clinical applications and safer use.

References

Baino F, Gand N, Brovarone CV. (2015). Bioceramics and scaffolds: a winning combination for tissue engineering. *Front. Bioeng. Biotechnol.*, **3**: 1–17.

Bock RM, McEntire BJ, Bal BS, Rahaman MN, Boffelli M, Pezzotti G. (2015). Surface modulation of silicon nitride ceramics for orthopaedic applications. *Acta Biomater.*, **26**: 318–330.

Chang MP, Tsung YC, Hsu HC, Tuan WH, Lai PL. (2015). Addition of a small amount of glass to improve the degradation behavior of calcium sulfate bioceramic. *Ceram. Int.*, **41**: 155–1162.

Cucuruz AT, Andronescu E, Ficai A, Ilie A, Iordache F. (2015). Synthesis and characterization of new composite materials based on poly(methacrylic acid) and hydroxyapatite with applications in dentistry. *Int. J. Pharm.*, **16**: 30061–30068.

Dapporto M, Sprio S, Fabbi C, Figalloc E, Tampieria A. (2016). A novel route for the synthesis of macroporous bioceramics for bone regeneration. *J. Eur. Ceram. Soc.*, **36**: 2383–2388.

Dorozhkin SV. (2015). Calcium orthophosphate bioceramics. *Ceram. Int.*, **41**: 13913–13966.

Dorozhkin SV. (2016). Multiphasic calcium orthophosphate (CaPO4) bioceramics and their biomedical applications. *Ceram. Int.*, **42**: 6529–6554.

Furko M, Jiang Y, Wilkins T, Balázsi C. (2016). Development and characterization of silver and zinc doped bioceramic layer on metallic implant materials for orthopedic application. *Ceram. Int.*, **42**: 4924–4931.

Ghosh R, Sarkar R, Paul S, Pal SK. (2016). Biocompatibility and drilling performance of beta tricalcium phosphate:yttrium phosphate bioceramic composite. *Ceram. Int.*, **42**: 8263–8273.

Gomes S, Renaudin G, Mesbah A, Jallot E, Bonhomme C, Babonneau F, Nedelec JM. (2010). Thorough analysis of silicon substitution in biphasic calcium phosphate bioceramics: a multi-technique study. *Acta Biomater.*, **6**: 3264–3274.

Gopi D, Shinyjoy E, Kavitha L. (2015). Influence of ionic substitution in improving the biological property of carbon nanotubes reinforced hydroxyapatite composite coating on titanium for orthopedic applications. *Ceram. Int.*, **41**: 5454–5463.

Heimann RB. (2013). Structure, properties, and biomedical performance of osteoconductive bioceramic coatings. *Surf. Coat. Technol.*, **233**: 27–38.

Jiang B, Hu X, Huang Z. (2013). Porous bio-ceramic coating on zirconia formed through freeze-drying. *Mater. Lett.*, **109**: 66–69.

Kaygili O, Tatar C, Yakuphanoglu F. (2012). Structural and dielectrical properties of Mg_3–$Ca_3(PO_4)_2$ bioceramics obtained from hydroxyapatite by sol–gel method. *Ceram. Int.*, **38**: 5713–5722.

Khan A Fa, Saleema M, Afzal A, Ali A, Khan A, Khan AR. (2014). Bioactive behavior of silicon substituted calcium phosphate based bioceramics for bone regeneration. *Mater. Sci. Eng. C*, **35**: 245–252.

Kharaziha M, Fathi MH. (2010) Improvement of mechanical properties and biocompatibility of forsterite bioceramic addressed to bone tissue engineering materials. *J. Mech. Behav. Biomed. Mater.*, **3**: 530– 537.

Lee WH, Loo CY, Rohanizadeh R. (2014). A review of chemical surface modification of bioceramics: effects on protein adsorption and cellular response. *Colloids Surf. B*, **122**: 823–834.

Mousa HM, Husseinc KH, Woo HM, Parka CH, Kim CS. (2015). One-step anodization deposition of anticorrosive bioceramic compounds on AZ31B magnesium alloy for biomedical application. *Ceram. Int.*, **41**: 10861–10870.

Owens GJ, Singh RK, Foroutan F, Alqaysi M, Han CM, Mahapatra C, Kim HW, Knowles JC. (2016). Sol–gel based materials for biomedical applications. *Prog. Mater. Sci.*, **77**: 1–79.

Prati C, Gandolfi MG. (2015). Calcium silicate bioactive cements: biological perspectives and clinical applications. *Dent. Mater.*, **31**: 351–370.

Regi MV. (2010). Evolution of bioceramics within the field of biomaterials. *C.R. Chimie*, **13**: 174–185.

Rena LM, Todo M, Arahira T, Yoshikawa H, Myoui A. (2012). A comparative biomechanical study of bone ingrowth in two porous hydroxyapatite bioceramics. *Appl. Surf. Sci.*, **262**: 81– 88.

Soon G, Pingguan-Murphy B, Lai KW, Akbar SA. (2016). Review of zirconia-based bioceramic: surface modification and cellular response. *Ceram. Int.*, **42**: 12543–12555.

Ulum MF, Arafat A, Noviana D, Yusopa AH, Nasutiona AK, Kadir MRA, Hermawan H. (2014). In vitro and in vivo degradation evaluation of novel iron-bioceramic composites for bone implant applications. *Mater. Sci. Eng. C*, **36**: 336–344.

Zhang F, Chang J, Lu J, Lin K, Ning C. (2007). Bioinspired structure of bioceramics for bone regeneration in load-bearing sites. *Acta Biomater.*, **3**: 896–904.

Zhu L, Yang J, Zhang J, Lei D, Xiao L, Cheng X, Lin Y, Peng B. (2014). In vitro and in vivo evaluation of a nanoparticles bioceramic paste for dental pulp repair. *Acta Biomater.*, **10**: 5156–5168.

Chapter 14

Sol-Gel-Based Bioceramics: From Materials to Medicine

Bharti Arora,[a] Ji Hoon Park,[b] Eun Ha Choi,[b] and Pankaj Attri[b]

[a]*Department of Applied Science, NorthCap University, Sector 23(A), Gurgaon, Haryana, India*
[b]*Plasma Bioscience Research Center/Department of Electrical and Biological Physics, Kwangwoon University, Seoul, Korea*

chem.pankaj@gmail.com

The progress of nanotechnology has opened the path for the development of new biomaterials. Recent developments have demonstrated that sol-gel bioceramics play an important role in the biomedical field due to their superior biological and mechanical properties. A bioactive hydroxyapatite (HA) is a common bioceramic that attaches to living tissues in vivo at body temperature. Sol-gel processes are now being used to produce bioactive coatings, powders, and substrates that can facilitate control over biological behavior of proteins and cells with broad clinical applications. It has been found that HA particles can down-regulate the expression of some cancerogenic genes in

Smart Ceramics: Preparation, Properties, and Applications
Edited by Ajay Kumar Mishra

ISBN 978-981-4774-30-7 (Hardcover), 978-1-315-16359-8 (eBook)
www.panstanford.com

tumor cells. The nano-HA particles from the biodegradation of HA ceramics could penetrate the membranes of tumor cells. A rapid and economic sol-gel method has been designed to synthesize sodium silicate (Na_2SiO_3) into silica gel in large quantities. Sol-gel-based bioceramics have the potential to present unexpected excellent biological functions. Thus, the understanding and control of the interactions between bioceramics and biological entities may play one of the leading roles in the development of nanomedicine.

14.1 Introduction

Medical science offers the most exciting and challenging applications for scientists working in areas of materials science. Dental filling materials, breast implants, heart valves, pacemakers, fracture fixation plates, nails and screws in orthopedics, orthodontic wires, and total joint replacement prostheses are a few examples from daily life. In the last few years, bone-related diseases have become more prevalent that need to be treated through implants. Since the implantable items need to be accepted by the biological systems, they are produced from a special class of materials, commonly known as biomaterials (Dorozhkin, 2011). The complexity of the interactions between a biomaterial and the living tissue occurring on varied scales (nano, micro, and macro) has garnered the attention of the scientific community. Biomaterials may be categorized as metals, ceramics, polymer, or composites (Fig. 14.1). Biomaterials offer the additional advantage of enduring in a biological environment without getting damaged and without any detrimental effect on the surroundings. There has been significant advancement in the synthesis of materials in the past 50 years, especially the innovative ceramic materials for applications in skeletal repair and reconstruction. Such class of materials for medical implants is often referred to as bioceramics, and the usage for medical applications has been proven in terms of the scientific literature available.

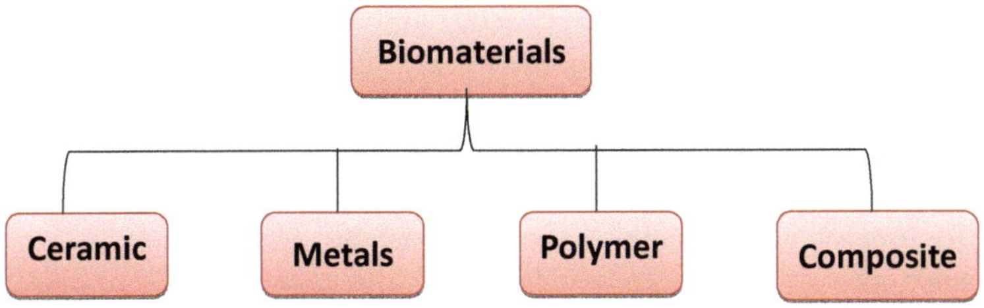

Figure 14.1 Classification of biomaterials.

14.2 Bioceramics

Since the success rate of implants is decreasing day by day, the use of bioceramics, especially in clinical applications, such as the repair, reconstruction, and replacement of diseased or damaged parts of the musculoskeletal system, has rapidly increased in the last few years (Hench and Wilson, 1996).

There could be different mechanisms by which the tissues may interact with the implants on a nanoscale inside the human body. On the basis of this nanoscale interaction of tissues, bioceramics may be categorized (Fig. 14.2) as bioinert, bioactive, and biodegradable (Suvarna et al., 2009). Bioceramics either bioinert or bioactive are available in different forms in the market and are widely used in a large number of different applications in the field of medicine. These mechanisms are thought to be governed by the response to the implant surface, which in turn is dependent on their crystallinity, composition, and the thickness of coatings (Cao and Hench and Wilson, 1996). Ceramics with different ratios have been tested for clinical usages, but few of them are being used presently. Calcium phosphates have been used primarily as metal alloy coatings or as porous materials for bone repair; ZrO_2 and Al_2O_3 have been used for total joint replacement; and glass ceramics and bioactive glasses have been used as powders to repair bone or for replacement of teeth, vertebrae, or ear bones (Cao and Hench, 1996).

Various methodologies are being used for the preparation of bioceramics, such as the hydrothermal technique, precipitation technique, electrodeposition technique, biomimetic deposition technique, and sol-gel approach. Recently, much attention of the scientists has been drawn to the fundamental research and applications of the sol-gel technique due to its well-known characteristic advantages in generating glass, ceramics, and glass ceramics (Cheng et al., 2001, 2003; Weng et al., 2002, 2003). An overview of sol-gel chemistry will be discussed in the next section.

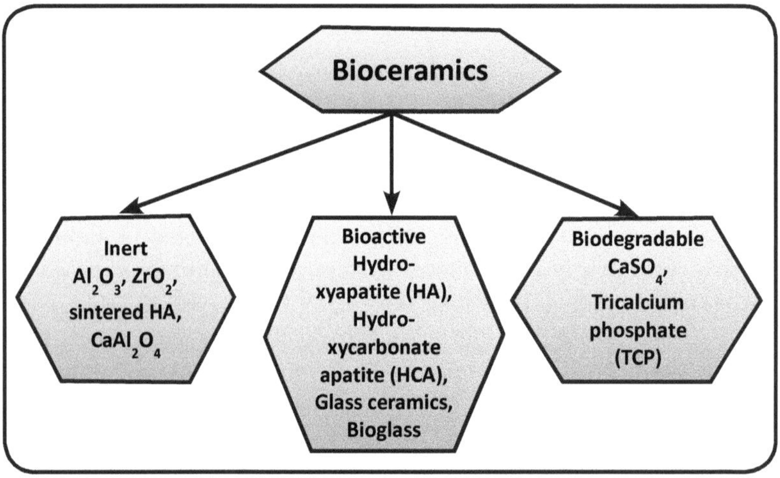

Figure 14.2 Classification of bioceramics according to their bioactivity.

14.3 Sol-Gel Methodology and Applications

The term "sol-gel" was first conceived in 1864 by Graham while working on silica sols (Graham, 1864). The sol-gel methodology implies the formation of a ceramic or ceramic composite material via a low-temperature production route. There has been enormous development in scientific knowledge of sol-gel chemistry and its applications in the last decades (Ebelmen, 1844). In 1846, a metal alkoxide was first obtained from $SiCl_4$ and ethanol that gelled on exposure to the atmosphere (Ebelmen, 1846; Wright and Sommersjik, 2000). A rapid and economic sol-gel method has been designed to synthesize sodium silicate (Na_2SiO_3) into silica gel in large quantities (Patrick, 1919). Kistler (1931) reported that hydrolytic polycondensation of silicic acid ($Si(OH)_4$) could result in the formation of an *aerogel* by drying the gel under supercritical conditions. Both these scientific developments in the field of sol-gel chemistry have paved the way for various biomedical applications.

Classical silica glasses, as well as certain nonsilicate oxide glass-like materials (ZrO_2^-, etc.), can be prepared using the sol-gel process based on various alkoxides. Sol-gel methods also enable powderless processing of ceramics and thin films or fibers directly from a solution. A precursor for these can be a metal/metalloid element surrounded by various reactive ligands. Tetramethoxysilane (TMOS) and tetraethoxysilane (TEOS) being readily soluble in water are preferred as metal alkoxides (Stachs et al., 1999).

The synthesis of gels from mixtures of liquid reagents (sols) in sol-gel chemistry at ambient temperatures mainly occurs in three steps: (i) preparation of inorganic networks, (ii) production of a sol, and (iii) solidification of the sol in a continuous liquid phase to form a network (gel). The most popular ones are sol-gel materials that are silica based. First, a solvent is added to the silicate precursor (TEOS or TMOS) and stirred with a catalyst for a few hours so that the resultant gels get stabilized. This, in turn, increases the level of their mechanical strength and transparency. In this process, the Si–O–R linkages get hydrolyzed in the presence of acids or bases. All these processes occur simultaneously. Gelation is followed by *aging*. Here, the solvent (water or ethanol) is removed from the resultant gel material, the latter being referred to as *syneresis*. Dry gels, or xerogels, are formed when drying occurs under atmospheric conditions (<100°C). Xerogels (alcohol- or water-based gels) are generally cracked, shrunken or sometimes powdered, transparent, and porous in nature. The pore size is found to be dependent on duration, the temperature at which hydrolysis is carried out, and the presence of a catalyst. The main reason for cracking is proposed as the decrease in the gel volume due to drying. Addition of dimethylformamide or dimethylsulfoxide to hydrolyzate helps in enhancing mechanical strength and transparency as they provide stability to the resultant gels (Podbielsk and Ulatowska-Jar˙Za, 2005). The above-discussed process for sol-gel chemistry is shown as a flowchart in Fig. 14.3. However, it is possible to dry the gel in supercritical conditions, which is referred to as an aerogel.

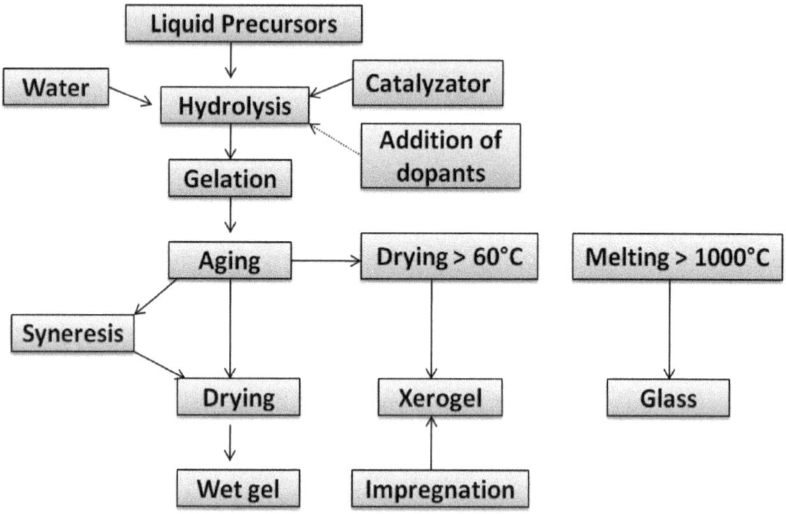

Figure 14.3 The sol-gel methodology.

The biggest advantages of the sol-gel process over conventional methods are as follows:

- Mixing of two liquids results in homogeneity at the molecular level (Kaur et al., 2013).
- A much higher purity can be ensured by the exact stoichiometry of all the components in the homogeneous mixture (Kaur et al., 2013).
- An easy production methodology at ambient temperatures helps in achieving radiation-inert and chemical porous materials (Conroy et al., 2000).

Sol-gel-processed materials have been exploited for a varied range of applications (Fig. 14.4), for example, intelligent coatings, microelectronics, batteries, solar cells, and catalysis, as well as for clinical applications. The biomedical application of sol-gel-based bioceramics will be discussed in the next section.

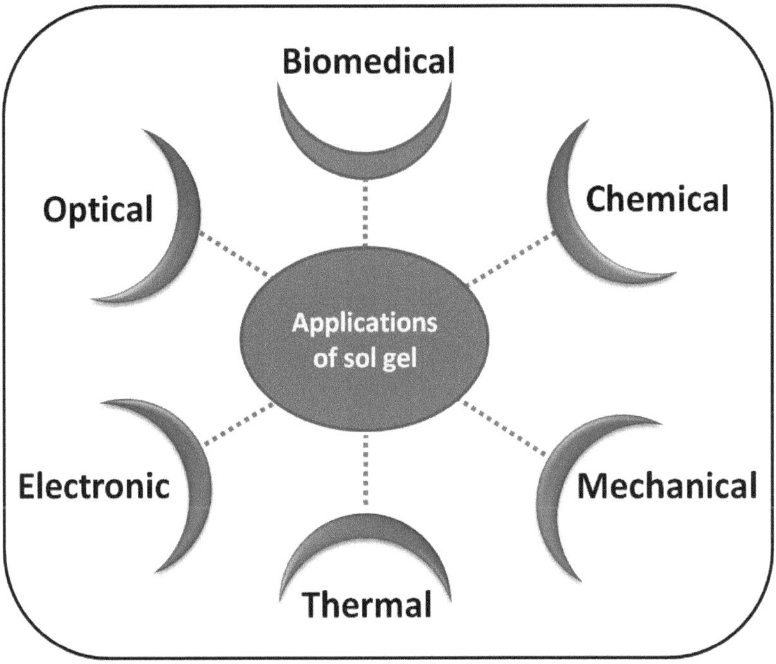

Figure 14.4 Different applications of the sol-gel methodology.

14.4 Biomedical Applications of Sol-Gel Bioceramics

Bioactive materials are bioceramics being developed to achieve specific biological responses in implant materials (Hubbell, 1999). This further governs a strong chemical linkage between the implant material and the tissue. Inert and porous ceramics are attached to the bone on growth of the tissue or biological fixation. Surface-reactive and dense bioceramics attach by forming a chemical bond with the bone or bioactive fixation. Inert ceramics attach to the bone by above-mentioned mechanisms and gradually replace the bone. Such bioceramics include tricalcium phosphate, Al_2O_3, hydroxyapatite (HAp), and HAp-coated porous implants (Ratner et al., 2004). A few examples of bioceramics used for clinical applications are listed in Table 14.1.

Table 14.1 Examples of bioceramics and their applications

Types of ceramics	Applications
Zirconia (bioinert)	Replacement for knees, hips, teeth, and tendons and ligaments; as bone fillers after tumor surgery; and repair in the case of periodontal disease
Alumina (bioinert)	Dental implant for tooth fixation, artificial total joint replacement, acetabular and femoral components, vertebrae spacers and extensors, and orthodontic anchors
Calcium phosphates (biodegradable)	Dental implants; skin treatments; jawbone reconstruction; orthopedics; facial surgery; and ear, nose, and throat repair
Bioglass ceramics (bioactive)	Tooth replacement implants, dental implants, middle ear implants, heart valves, artificial total joint replacement, bone plates, screws, wires, and spinal fusion

Source: Patel and Gohil (2012).

However, a detailed discussion of applications of sol-gel-based bioceramics in the field of medicine is as mentioned next.

14.4.1 Bioactive Sol-Gel Coatings and Implants

Various materials may be surface-coated with sol-gel hydrolyzate (liquid), resulting in a change of their attributes and improved biocompatibility (Paterson and Ben-Nissan, 1996; Sakka and Kazuka, 1998; Que et al., 2002; Liu et al., 2002, 2003). HA is widely used for coatings because it shows bioactivity, biocompatibility, and inertness and stimulates the formation of bone (Hench, 1982). Silica-based sol-gel coatings have been used in laser therapy for the production of fiber optic applicators (Podbielska et al., 2004). To minimize the invasive destruction of pathologic tissues in liver and brain, a new treatment modality has been devised, known as interstitial laser thermotherapy. Here, a fiber optic laser applicator has been used with laser light: The applicator is injected into the pathologic lesion, and curing laser light is guided through the fiber (Muller and Rogan, 1995; Rohde et al., 2001).

Generally, a biologically active HA layer is formed on the implant surface so that the implant fixes to the surface of the tissue through some chemical linkages. This kind of fixation is termed as "bioactive fixation" (Hench, 2000). One of the remarkable characteristics of bioactive fixation is that the strength of the bond between implant the bone (i.e., bioactive bond) is almost greater than bone after 3–6 months. It has been proposed that the aggregation of a dense layer of hydroxycarbonate apatite (HCA) nanoscale crystals binds to collagen fibrils, which provides strength to the bioactive bonds. It was found that implant materials bond to the bone to a certain extent, depending on the compositions of bioactive ceramics. Additionally, this affects the growth rate of the bone in the presence of bioceramics.

The sol-gel technique offers flexibility in terms of geometry of the substrate and can be used to create a coating, as reported by several authors. HA coatings have been known to increase the bone bonding, thereby increasing biocompatibility and reducing the lethal impact of bioimplants on living tissues. Kim et al. (2004) prepared sol-gel-based HA and fluor-hydroxyapatite (FHA) films (~5 mm thickness) on a titanium substrate, which improved the biocompatibility of Ti-based biomedical implants. Kim et al. (2005) reported that 30 mol% of HA composites with titania (TiO_2) when coated on a titanium (Ti) substrate by sol-gel chemistry resulted in enhanced adhesion strength of the coating layers with increasing concentration of TiO_2. This concludes that sol-gel-based HAp-TiO_2 composite coatings possess excellent characteristic properties that can be exploited for hard tissue applications. Exceptional bioactivity was exhibited by Ti/Zr alloy coated with HAp/TiO_2 in a simulated body fluid (SBF) using the sol-gel methodology. Titanium-zirconium alloy was coated with HAp/titania sol-gel coatings for applications in field of biomedicine, as reported by Wen et al. (2007).

Sato et al. (2005) prepared titanium substrates coated with TiO_2, HA, and poly (DL lactic/glycolic acid) hydrothermally and investigated through scanning electron microscopy (SEM), energy-dispersive x-ray spectroscopy (EDS), atomic force microscopy (AFM), and x-ray diffraction (XRD) techniques. It was confirmed that these coatings exhibited good resistance to cracking with a high degree of crystallinity, which can be exploited to strengthen the bonding of titanium implants to juxtaposed bone as osteoblast adhesion is a prerequisite for subsequent formation of bone (Sato et al., 2005).

14.4.2 Bioactive Sol-Gel Glasses

Since bioglasses ensure a strong bonding with human bone tissues, they are used for the reconstruction of diseased bone tissues and tissue repair (Liu and Miao, 2004). A bioactive glass comprising SiO_2-CaO-P_2O_5 has been prepared using the sol-gel methodology. The various steps involved in the processing of gels are shown in Fig. 14.3. Sol-gel glasses exhibit a higher percentage of HCA formation and bioactivity index (Li et al., 1992; Pereira et al., 1994). This is accredited to an enhanced release of soluble silica that agglomerates HCA crystals in the pores of the gel-glass (Greenspan et al., 1997; Radin et al., 1997). Culture experiments of rat marrow stromal cells have been carried out by Radin et al. (1997) to explain the bone results in vivo. The results depicted an increase in cellular differentiation on the sol-gel glasses (Radin et al., 1997), which is in correlation with bioactive glass-ceramic substrates (Ohgushi et al., 1996).

The sol-gel technique was used to produce a 3D hierarchical structure comprising micropores and macropores (Li and Miao, 2004). The 58S bioglass powder produced by the sol-gel method, having the composition SiO_2-CaO-P_2O_5, was pushed into the sintered discs at temperatures of 500°C, 800°C, 1000°C, and 1200°C. A densitometer was used to test the apparent density of the sintered bioglass, a Vickers hardness tester was used for calculating Vickers hardness, dilatometry and differential scanning calorimetry were used to examine thermal properties, and SEM and XRD were used to monitor crystallization of the bioglass. The bioglass surface after immersion in body fluid solution was investigated using SEM, and Fourier transform infrared spectroscopy (FTIR) was then utilized to confirm the apatite layer formation. It was observed that bioactive amorphous bioglass formed a dense apatite layer, whereas bioglass when subjected to crystallization formed an incomplete/porous layer. Remarkable mechanical properties and apatite-forming ability shown by macroporous bioglasses make these biomaterials suitable for tissue repair and engineering applications (Li and Miao, 2004). The bioresorption of 58S and 77S gel glasses was in correlation with animal data, as evidenced by in vitro studies (Wheeler and Stokes, 1997).

14.4.3 Encapsulation within Sol-Gel Matrices

The use of sol-gel-based biomaterials for encapsulation in antibodies, enzymes, or any other proteins has proved to be a landmark in the history of sciences. Insights into human cells in titania sol-gels have been possible through the development of an immune process. The direct investigation of the cell behavior inside human pluripotential mesenchymal stem cells and chondrosarcoma cells from bone marrow was done with the help of a fluorescent agent combined with an autoimmune reaction. Several other authors, on the basis of in vitro results, have confirmed enhanced biological levels of cell adhesion and proliferation in the case of TiO_2 surfaces (Manso et al., 2002).

Sol-gel-based bioceramics are effective as bone substitutes due to requirement of low temperatures. The production of bioactive ceramics by the sol-gel process for bone production has been discussed (Xynos et al., 2001). A silica matrix modified with calcium and phosphate ions was prepared using the sol-gel technique. Precursors calcium nitrate, triethyl phosphate, and tetraethylorthosilicate were subjected to hydrolysis followed by condensation to form a matrix. The next step involved drying and heating the samples at 110°C or 900°C and then injected in SBF. After 3 days, infrared spectroscopy, XRD, and SEM analyses confirmed that calcium phosphate silicate and HA were present in the sample treated at 110°C before and after contact with SBF, respectively. The HA phase was found to be prominent in the sample treated at 900°C before as well as after contact with SBF, whereas the crystalline phase was only observed in the latter stage (Ribeiro et al., 2014).

The literature is rich in reports where antibodies and enzymes have been encapsulated using a sol-gel matrix, resulting in enhanced biological activity of the molecules inside the matrix (Avnir et al., 1997; Livage, 1997). Roux and Livage (1997) described that sol-gel matrices can also be used to execute antigen–antibody reactions. A silica gel was chosen for encapsulation of antigens such that the antibodies can diffuse through the gel pores. This offered specific antigen–antibody interaction within the gel. Some others demonstrated the interactions of cytochrome c with both thin-film and bulk silica gel matrices. Dave et al.

(1997) reported that denaturation of proteins is a reversible process with restricted aggregation.

Fennouh et al. (1999) demonstrated the entrapment of bacteria *Escherichia coli* cells within silica sol-gel matrices. It was observed that the bacteria were residing comfortably within the cellular organization. He latter remarked that encapsulation enhanced the enzymatic activity of concentrated *E. coli* suspensions. This was attributed to the fact that clustering of the bacteria was restricted by sol-gel encapsulation (Fennouh et al., 1999). It has been proposed that pore size and interconnections in the ceramic phase, growth of cells, and the circulation of nutrients regulate the cell-material interactions (Dalby et al., 2000; Karageorgio and Kaplan, 2005; Meyer et al., 2005). Very recently, osteogenic compounds, for example, bone morphongenic protein (BMP) growth factors, were entrapped in a porous ceramic scaffold to achieve a desirable biological response and were also tested by coating on a material surface (Kuboki et al., 1998; Liu et al., 2005).

Additionally, silica xerogels prepared at room temperature were investigated as controlled release carriers for model proteins. Since the size (21 kD) of the trypsin inhibitor is comparable to bone growth factors, it was used as a model protein. It was revealed that the liberation of the trypsin inhibitor was time and dose dependent in vitro and the excellent characteristics exhibited by xerogel materials make them useful for bone repair applications (Santos et al., 1999). The insulin secretory response of the silica-gel-encapsulated pancreatic islets in blood sugar levels of nonobese diabetic mice in vivo demonstrates the feasibility of this novel concept (Pope, 1997).

One of the major problems associated with implants is bacterial infection, which acts as a driving force for the premature expulsion of implant material, leading to painful and lethal situations for patients. This is a serious issue that results in the premature removal of the implant. However, the innovative methods of tissue regeneration may be preferred over conventional tissue replacement by designing a new generation of bioceramics that activate natural biological repair mechanisms and stimulate proliferation.

14.5 Conclusion

Different materials have been used for clinical applications for bone implant, restorative tooth coating, diagnosis, drug delivery, biosensor, and scaffolding. Biocompatible and bioactive materials through sol-gel methodology have gained importance due to the production of materials with controlled stoichiometry and particle size.

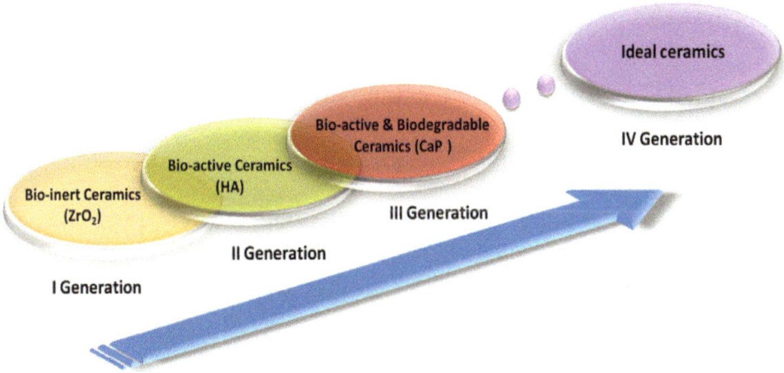

Figure 14.5 Different generations of bioceramics.

Investigations of the biological events at the interface between biomaterials and human body have led to research in bioceramics for the development of bioinert and bioactive devices and the emergence of the fourth generation of ideal bioceramics (Fig. 14.5). Several research groups are working on different aspects of sol-gel-based bioceramics, making these clinically viable strategies for various applications.

References

Avnir D, Klein Lisa C, Levy D, Schubert U, Wojcik AB. (1997). Organosilica sol-gel materials. In *The Chemistry of Organosilicon Compounds Part 2*. Apeloig Y, Rappopoti Z (eds.). Wiley & Sons, Chichester.

Cao W, Hench LL. (1996). Bioactive materials. *Ceram. Int.*, **22**: 493–607.

Cheng K, Shen G, Weng W, Han G, Ferreira JMF, Yang J. (2001). Synthesis of hydroxyapatite/fluoroapatite solid solution by a sol-gel method. *Mater. Lett.*, **51**: 37–41.

Cheng K, Weng W, Han G, Du P, Shen G, Yang J, Ferreira, JMF. (2003). Sol-gel derived fluoridated hydroxyapatite films. *Mater. Res. Bull.*, **38**: 89–97.

Conroy JF, Power ME, Norris PM. (2000). Applications for sol-gel-derived materials in medicine and biology. *J. Lab. Autom.*, **5**(1): 52–57.

Dalby MJ, Silvio LD, Davies GW, Bonfield W. (2000). Surface topography and HA filler volume effect on primary human osteoblasts *in vitro*. *J. Mater. Sci. Mater. Med.*, **11**: 805–810.

Dave BC, Miller JM, Dunn B, Valentine JS, Zink JI. (1997). Encapsulation of proteins in bulk and thin film sol-gel matrices. *J. Sol-Gel Sci. Technol.*, **8**: 629–634.

Dorozhkin SV. (2011). Medical application of calcium orthophosphate bioceramics. *BIO*, **1**: 1–51.

Ebelmen JJ. (1844). Sur les éthers siliciques. *C.R. Acad. Sci.*, **19**: 398–400.

Ebelmen M. (1846). Research on combinations of boric and silicic acids with ethers. *Ann. Chim. Phys.*, **16**: 166.

Fennouh S, Guyon S, Jourdat C, Livage J, Roux C. (1999). Encapsulation of bacteria in silica gels. *C.R. Acad. Sci. Paris*, **2**: 625–630.

Graham TJ. (1864). On the properties of silicic acid and other analogous substances. *Chem. Soc.*, **17**: 318–327.

Greenspan DC, Zhong JP, LaTorre GP, Booher M. (1997). The *in-vitro* degradability of sol-gel bioglass. In *Transactions of the 23rd Annual Meeting of the Society for Biomaterials*. Society for Biomaterials, New Orleans, LA.

Hench LL. (1982). CRC Handbook of Bonfield W, Tanner KE. (1997). *Bioactive Ceramics*, Vol. 1. Yamamuro T, Hench LL, Wilson J. (eds.). CRC Press, Boca Raton.

Hench LL. (2000). The challenge of orthopaedic materials. *Curr. Orthop.*, **14**: 7–14.

Hench LL, Wilson J. (1996). *Clinical Performance of Skeletal Prostheses*. Chapman and Hall, London.

Hubbell JA. (1999). Bioactive biomaterials. *Curr. Opin. Biotech.*, **10**: 123–129.

Jie LQ, Zhu S, Wang R. (2004). A new route to prepare macroporous bioactive sol-gel glasses with high mechanical strength. *Mater. Lett.*, **58**: 2747–2750.

Karageorgio V, Kaplan D. (2005). Porosity of 3D biomaterial scaffolds and osteogenesis. *Biomaterials*, **26**: 5474–5491.

Kaur S, Bala N, Khosla C. (2013). Preparation and deposition of hydroxyapatite on biomaterials by sol-gel technique: a review. *Chitkara Chem. Rev.*, **1**(2): 59–69.

Kim H-W, Kim HE, Knowlesb JC. (2004). Fluor-hydroxyapatite sol–gel coating on titanium substrate for hard tissue implants. *Biomaterials*, **25**(17): 3351–3358.

Kim HW, Kim HE, Salih V, Knowlesb JC. (2005). Hydroxyapatite and titania sol–gel composite coatings on titanium for hard tissue implants; mechanical and *in vitro* biological performance. *J. Biomed. Mater. Res. Part B*, **72B**: 1–8.

Kistler SS. (1931). Coherent expanded aerogels and jellies. *Nature*, **127**: 741.

Kuboki Y, Takita H, Kobayashi D, Tsugura E, Inoue M, Muruta M, Nagai N, Dohi Y. (1998). BMP-induced osteogenesis on the surface of hydroxyapatite with geometrically feasible and nonfeasible structures: topology of osteogenesis. *J. Biomed. Mater. Res.*, **39**: 190–199.

Li R, Clark AE, Hench LL. (1992). Effect of structure and surface area on bioactive powders made by sol-gel process. In *Chemical Processing of Advanced Materials*. Hench LL, West JK. (eds.). John Wiley and Sons, New York, pp. 627–633.

Liu DM, Yang Q, Troczynski T. (2002). Sol-gel hydroxyapatite coatings on stainless steel substrates. *Biomaterials*, **23**: 691–698.

Liu J, Miao X. (2004). Sol-gel derived bioglass as a coating material for porous alumina scaffolds. *Ceram. Int.*, **30**: 1781–1785.

Liu J-X, Yang D-Z, Shi F, Cai Y-J. (2003). Sol-gel deposited TiO_2 film on NiTi surgical alloy for biocompatybility improvement. *Thin Solid Films*, **429**: 225–230.

Liu Y, Degroot K, Hunziker EB. (2005). BMP-2 liberated from biomimetic implant coatings induces and sustains direct ossification in an ectopic rat model. *Bone*, **36**: 745–757.

Livage J. (1997). Sol-gel processes. *Curr. Opin. Solid State Mater. Sci.*, **2**: 132–136.

Manso M, Ogueta S, Perez-Rigueiro J, Garcia JP, Martinez-Duart JM. (2002). Testing biomaterials by in-situ evaluation of cell response. *Biomol. Eng.*, **19**: 239–242.

Meyer U, Buchter A, Wiesmann HP, Joos U, Jones DB. (2005). Basic reactions of osteoblasts on structured material surfaces. *Euro Cells Mater.*, **9**: 39–49.

Muller G, Rogan A. (1995). *Laser-Induced Interstitial Thermotherapy*. SPIE Optical Engineering Press, Bellingham, WA.

Ohgushi H, Dohi Y, Yoshikawa T, Tamai S, Tabata S, Okunaga K, Shibuya T. (1996). Osteogenic differentiation of culture marrow stromal stem cells on the surface of bioactive glass ceramics. *J. Biomed. Mater. Res.*, **32**: 341–334.

Patel NR, Gohil PP. (2012). A review on biomaterials: scope, applications and human anatomy significance. *Int. J. Emerging Technol. Adv. Eng.*, **2**(4): 91–101.

Paterson MJ, Ben-Nissan B. (1996). Multilayer sol-gel zirconia coatings on 316 stainless steel. *Surf. Coat. Technol.*, **86–87**: 153–158.

Patrick WA. (1919). Silica gel and process of making same. Google patents.

Pereira MM, Clark AE, Hench LL. (1994). Homogeneity of bioactive sol-gel derived glasses in the system CaO-P_2O_5-SiO_2. *J. Mater. Synth. Process.*, **2**: 189–196.

Podbielska H, Ulatowska-Jarza A. (2005). Sol-gel technology for biomedical engineering. *Bull. Pol. Acad. Sci. Tech. Sci.*, **53**(3): 261–271.

Podbielska H, Ulatowska-Jarza A, Holowacz I. (2004). Sol-gel applicators for medical light therapy. *Physica Med. XX*, **1**: 43–45.

Pope EJA. (1997). Bioartificial organs I: silica gel encapsulated pancreatic islets for the treatment of diabetes mellitus. *J. Sol-Gel Sci. Technol.*, **8**: 635–639.

Que W, Sun Z, Zhou Y, Lam YL, Cheng SD, Chan YC, Kam CH. (2000). Preparation of hard optical coatings based on an organic/inorganic composite by sol-gel method. *Mater. Lett.*, **42**: 326–330.

Radin S, Ducheyne P, Reicin KE, Lee MH. (1997). The effect of composition and *in vitro* immersion on the formation of a bioactive surface on sol-gel derived glass. In *Transactions of the 23rd Annual Meeting of the Society for Biomaterials*. Society for Biomaterials, New Orleans, LA.

Ratner BD, Hoffman AS, Schoen FJ, Lemons JE. (2004). *Biomaterials Science: An Introduction to Materials in Medicine*, 2nd ed. Elsevier Academic Press, San Diego, CA.

Ribeiro TJ, De Lima OJ, Faria EH, Rocha LA, Calefi PS, Ciuffi KJ, Nassar EJ. (2014). Sol-gel as methodology to obtain bioactive materials. *Ann. Braz. Acad. Sci.*, **86**(1): 27–36.

Rohde E, Mesecke von Rheinbaben I, Podbielska H, Hopf M, Mueller G. (2001). Interstitial laser-induced thermotheraphy (LITT):

comparison of in-vitro irradiation effects of Nd:YAG (1064 nm) and diode (940 nm) laser. *Med. Laser Appl.*, **16**: 81–90.

Roux C, Livage J. (1997). Antibody-antigen reactions in porous sol-gel matrices. *J. Sol-Gel Sci. Technol.*, **8**: 663–666.

Sakka S, Kozuka H. (1998). Sol-gel preparation of coating films containing noble metal colloids. *J. Sol-Gel Sci. Technol.*, **13**: 701–705.

Santos EM, Radin S, Ducheyne P. (1999). Sol-gel derived carrier for the controlled release of proteins. *Biomaterials*, **20**: 1695–1700.

Sato M, Slamovich EB, Webster TJ. (2005). Enhanced osteoblast adhesion on hydrothermally treated hydroxyapatite/titania/poly(lactide-co-glycolide) sol-gel titanium coatings. *Biomaterials*, **26**: 1349–1357.

Stachs O, Gerber Th, Petkov V. (1999). The structure formation of zirconium oxide gels in alcoholic solutions. *J. Sol-Gel Sci. Technol.*, **15**: 23–30.

Suvarna V, Shetye S, Chordia A. (2009). Bio-ceramics. *J. Pharm. Res.*, **2**(10): 1620–1626.

Wen CE, Xu W, Hu WY, Hodgson PD. (2007). Hydroxyapatite/titania sol-gel coatings on titanium-zirconium alloy for biomedical applications. *J. Acta Biomater.*, **3**: 403–410.

Weng W, Han G, Du P, Shen G, Yang, J. (2002). The effect of citric acid addition on sol-gel preparation of apatite films. *Mater. Chem. Phys.*, **77**: 578–582.

Weng W, Zhang S, Cheng K, Qu H, Du P, Shen G, Yuan J, Han G. (2003). Sol–gel preparation of bioactive apatite films. *Surf. Coat. Technol.*, **167**: 292–296.

Wheeler DL, Stokes KE. (1997). In vivo evaluation of sol-gel biogiass. Part I: histological findings. In *Transactions of the 23rd Annual Meeting of the Society for Biomaterials.* Society for Biomaterials, New Orleans, LA.

Wright JD, Sommersjik NAJM. (2000). *Sol-Gel Materials: Chemistry and Application.* CRC Press, Taylor & Francis Group, USA.

Xynos ID, Edgar AJ, Buttery LDK, Hench LL, Polak JM. (2001). Gene-expression profiling of human osteoblasts following treatment with the ionic products of Bioglass 45S5 dissolution. *J. Biomed. Mater. Res.*, **55**: 151–157.

Index